AGRICULTURAL STRATEGIES

Cotsen Advanced Seminar Series

The published results of the Cotsen Advanced Seminars, where scholars explore cross-disciplinary themes in conferences periodically sponsored by the Cotsen Institute.

CAS 2 *Agricultural Strategies*. Joyce Marcus and Charles Stanish (editors)

CAS 1 *Theory and Practice in Mediterranean Archaeology: Old World and New World Perspectives*. John K. Papadopoulos and Richard M. Levanthal (editors)

Agricultural Strategies

edited by

JOYCE MARCUS
&
CHARLES STANISH

Cotsen Institute of Archaeology
University of California, Los Angeles
2006

THE COTSEN INSTITUTE OF ARCHAEOLOGY at UCLA is a research unit at the University of California, Los Angeles that promotes the comprehensive and interdisciplinary study of the human past. Established in 1973, the Cotsen Institute is a unique resource that provides an opportunity for faculty, staff, graduate students, research associates, volunteers and the general public to gather together in their explorations of ancient human societies.

Former President and CEO of Neutrogena Corporation Lloyd E. Cotsen has been associated with UCLA for more than 30 years as a volunteer and donor and maintains a special interest in archaeology. Lloyd E. Cotsen has been an advisor and supporter of the Institute since 1980. In 1999, The UCLA Institute of Archaeology changed its name to the Cotsen Institute of Archaeology at UCLA to honor the longtime support of Lloyd E. Cotsen.

Cotsen Institute Publications specializes in producing high-quality data monographs in several different series, including Monumenta Archaeologica, Monographs, and Perspectives in California Archaeology, as well as innovative ideas in the Cotsen Advanced Seminar Series and the Ideas, Debates and Perspectives Series. Through the generosity of Lloyd E. Cotsen, our publications are subsidized, producing superb volumes at an affordable price.

This book is set in 10-point Janson Text, with titles in 25-point Chaparrel.
Edited by Joe Abbott
Designed by William Morosi

Library of Congress Cataloging-in-Publication Data

Agricultural strategies / edited by Joyce Marcus & Charles Stanish.
p. cm. -- (Cotsen advanced seminar)
Includes bibliographical references and index.
ISBN: 1-931745-24-2 (cloth) 1-931745-22-6 (paper)
1. Irrigation. 2. Agricultural intensification. 3. Agricultural innovations. I. Marcus, Joyce. II. Stanish, Charles, 1956- III. Series.

S613.A37 2006
630.9--dc22

2005031833

CONTENTS

FIGURES

TABLES

CONTRIBUTORS

Robert McC. Adams
Department of Anthropology
University of California, San Diego

Clark L. Erickson
Department of Anthropology
University of Pennsylvania

Gary M. Feinman
Department of Anthropology
The Field Museum

Thomas F. Glick
Department of History
Boston University

Patrick V. Kirch
Department of Anthropology
University of California, Berkeley

Lisa J. Lucero
Department of Anthropology
New Mexico State University

Joyce Marcus
Museum of Anthropology
University of Michigan

Heather M.-L. Miller
Department of Anthropology
University of Toronto at Mississauga

Kathleen Morrison
Department of Anthropology
University of Chicago

Vernon L. Scarborough
Department of Anthropology
University of Cincinnati

Charles Stanish
Department of Anthropology
University of California, Los Angeles

Tina L. Thurston
Department of Anthropology
University of New York at Buffalo

Tony Wilkinson
School of Arts, Culture
and Environment
University of Edinburgh

Patrick Ryan Williams
Department of Anthropology
The Field Museum

INTRODUCTION

JOYCE MARCUS AND CHARLES STANISH

With the modern world's concerns about water shortages, "water management" has become an active topic of discussion and debate among politicians, development specialists, and a wide range of scholars (e.g., Gleick 1998; Kahrl 1982; Postel 1997; Raskin et al. 1995; Uphoff 1986; Wunderlich and Prins 1987). It is not an exaggeration to say that the effective control and distribution of fresh water will be one of the primary challenges facing the global community in the twenty-first century, as the world reaches a projected population of 10 billion.[1] Providing an anthropological, archaeological, and historical background to this debate is the central goal of this book.

Anthropologists have renewed their interest in water management and have begun to refine their understanding of agricultural strategies and economies as *long-term processes*. There is a sense of urgency about conducting fieldwork since environmental conditions and indigenous practices are being altered all over the world. For their part archaeologists are working to understand how a wide range of subsistence systems managed to endure over millennia, undergoing constant modification to meet changing socioeconomic, climatic, and political conditions. The goal of one group of archaeologists is to model long-term relationships among hydraulic technologies, economic strategies, sociopolitical groups, and multiple levels of decision making. Such modeling will require fine-grained regional chronologies to date each sequence of developments and the analysis of a growing body of quantitative and qualitative data from many parts of the world.

All these recent developments suggested to us that the time had arrived for the Cotsen Institute of Archaeology to host a symposium on water management and agricultural strategies. We chose the phrase "water management" because it seemed broader and more inclusive than "hydraulic civilization," a term perhaps too tied to arid environments and to large-scale irrigation systems, particularly in Asia. "Water management" includes strategies both for times when there is too little water and times when there is too much water. Water management covers topics like the draining off of excess water, the diversion of rain runoff, the control of floodwater, the control of salinization, and water storage. It also can refer to soil erosion management when intensive rains threaten to remove topsoil.

In many parts of the ancient world irrigation seems to have been preceded by (and perhaps inspired by) drainage. In Mexico's Oaxaca Valley, for example, the first canals were small, and their purpose seems to have been to divert rain runoff away from the house, in some cases into a large cistern for water storage (Flannery and Marcus 2005; Marcus, this volume). These simple early ditches were excavated by individual households to remove *excess* water, but they provided the model for later, larger canals used in conjunction with brush-and-boulder dams to divert water from piedmont streams to the alluvial lands below. Likewise, in the Titicaca Basin modest raised field and irrigation technology existed at least a millennium prior to the massive constructions that characterized the Tiwanaku state fields (Stanish, this volume).

The Cotsen symposium's goals were (1) to take a new look at a wide range of agricultural strategies; (2) to develop more comprehensive models to explain how those strategies operated over long spans of time; (3) to document how the associated societies brought about, and adjusted to, changes in their economic, social, and political agendas; (4) to establish linkages among important variables and to see if cause and effect could be determined; and (5) to integrate "man-man" and "man-land" relations.

As the reader will see, the authors in this volume (and in other studies such as Adams and Anderson 1988; Brookfield 1986; Connelly 1994; Guyer and Lambin 1993; Hakansson 1998; Harrower 2003; Hauser-Schaublin 2003; Hunt 2000; Janusek and Kolata 2004; Lansing 1987; Mabry 1996, 2000; Mitchell and Guillet 1993; Scarborough 2003; Sheridan 2002; Trawick 2001a, 2001b; Treacy 1994; Zuidema 1986) are analyzing a wider array of topics than ever before—strategies of intensification versus disintensification, short-term versus long-term planning, centralized versus decentralized decision making, local versus regional political agendas, multicropping cycles, diverse technologies, crops grown for household consumption versus crops destined for elite or state use, agricultural rituals, low-risk versus high-risk decisions, and the coordination of labor. The authors are also using recent ethnographic and archaeological data to generate more sophisticated models about the long-term relationships among those variables.

THE HISTORY OF HYDRAULIC STUDIES

A long line of distinguished scholars has focused on hydraulic societies. Many of the key works were stimulated, in one way or another, by Karl Wittfogel, a pioneer who valued both comparative and diachronic approaches. Wittfogel's work was in turn influenced by Max Weber, Friedrich Engels, and Karl Marx, among others who wrote about the "Asiatic mode of production." At a time when most scholars were using Western models, Wittfogel drew our attention to non-Western case studies, and such a departure continues to make us aware that we must try to understand the strategies selected by each culture. Not all of these cultures' strategies were necessarily efficient, adaptive, and low-risk; in fact, some societies had other goals—they might have been looking for high return in the short run or to establish political control of products and people.

Wittfogel's early works—including his 1929 article "Geopolitics, Geographical Materialism, and Marxism" and his 1938 "Die theorie der orientalischen gesellschaft" (The Theory of Asian Society)—influenced scholars such as Julian Steward long before 1957, when Wittfogel's *Oriental Despotism: A Comparative Study of Total Power* was translated from German into English. For example, Steward says:

> The concept of multilinear evolution was provisionally formulated in "Culture and Process" prepared for the 1952 symposium of the Wenner-Gren Foundation (Steward 1953). The germ of this concept, however, was stated nearly twenty years earlier in "The Social and Economic Basis of Primitive Bands" (Steward 1936). This approach received tremendous stimulus when Karl Wittfogel formulated the cross-culturally recurrent characteristics of the type that John Stuart Mill had called the "Orient State." After devoting many years to detailed analysis of Chinese culture history and thoroughly aware of the importance of water control in several key world areas, Wittfogel (1938) described this type of state as an Hydraulic Society or Hydraulic State.
>
> In 1949, I undertook to extend Wittfogel's formulation by exploring the possibility that the hydraulic or irrigation societies began a parallel evolution with the first use of domesticated plants and that the development of local communities, technology, and even intellectual, aesthetic and religious achievements as well as the economic and political patterns ran similar courses. (Steward 1955:1–2)

Even from the vantage point of the twenty-first century, Wittfogel and his work continue to be of importance, not only because they stimulated generations of scholars to undertake the fieldwork necessary to date hydraulic systems but also because Wittfogel linked large-scale hydraulic systems to the development of hierarchy and managerial control. Many scholars still agree with Wittfogel that there *is* a link between hierarchy and hydraulics. Some archaeologists, however, argue that the greatest investments in large-scale hydraulic systems *postdated* the rise of centralized control (Adams 1974). Other scholars suggest that "large scale" is not the key variable. For example, William Mitchell (1976:37) says, "The problems of analysis can be avoided if the hypothesis is reformulated to change the independent variable from the *size* of the irrigation system itself (i.e., large-scale) to the *manner* in which irrigation activities are organized. The hypothesis then becomes: If irrigation is regulated centrally in arid or semiarid environments, then there is a corresponding increase in centralized political power in other areas of social life. The extent of political power varies directly with the extent of the irrigation system and its importance to the local economy."

Jonathan Mabry (2000) has expressed his opinion in his paper title "Wittfogel was half right." Mabry thinks that Wittfogel was right when he said that large-scale irrigation systems can only be managed by hierarchical administrative institutions; he thinks that Wittfogel was wrong when he focused on the scale of the irrigation system. Mabry, like Mitchell (1976) and some contributors in this volume, thinks Wittfogel should have focused on the *number of irrigators* that had to be coordinated. This focus on the total number of irrigators—especially on the way that labor was organized and coordinated—has emerged as a key ingredient in recent discussions. Two other related variables are being evaluated—the number and arrangement of managers who coordinated those irrigators, and the number of decision makers and policy-making levels over time.

Even though it was once trendy to criticize or reject all of Wittfogel, recent scholars such as David Price (1994) and Mabry (2000) are finding much to applaud. Mabry (2000:285), in fact, says that he is joining a "growing number of voices that call for separating the baby from the bathwater." Mabry discusses how Wittfogel came to be ignored by some and refuted by others who simplified his views to create a straw man. Those who have taken the time to read Wittfogel's original works (1929, 1931, 1935, 1938, 1953, 1957, 1960) have gained new respect for his data and many of the ideas he developed (Price 1994). This trend reminds us that we should always read the original works in their entirety, remembering that they were written long before we had amassed the archaeological data that can be brought to bear on the topic of water management today.

DID CENTRALIZATION PRECEDE OR FOLLOW CONSTRUCTION OF LARGE-SCALE HYDRAULIC SYSTEMS?

While Wittfogel stressed the linkage between large-scale irrigation systems and state bureaucracy, it remained for archaeologists to determine whether (1) investments in large-scale irrigation led to greater centralization, or (2) preexisting centralized bureaucracies were needed to plan and construct such systems. More than 30 years ago, Robert McC. Adams (1974:5) had decided that "truly large-scale irrigation works depended on state initiatives not only for their construction but for their subsequent maintenance." Eva Hunt and Robert Hunt (1974:153) agreed with Adams, adding that "the power elites are crucial for conflict resolution over water. The persons occupying water-control roles are intimately connected to the distribution of power in the society." Hunt and Hunt (1974:154) went on to argue that without unified and centralized decision making, high levels of social disruption and conflict ensue, and such conflict would tend to paralyze the socioeconomic system.

The argument that state control is needed to suppress conflict and ensure the smooth running of the system is a recurrent theme of some importance. Many anthropologists have looked at systems that operated over long spans of time. Those studying small-scale irrigation systems have emphasized the role of consensus and voluntarism in decision making at the local level (e.g., Lees 1970; Mabry 1996, 2000). It is mainly those who have studied large-scale systems who stress the need for prior centralized authority (Adams 1974).

DOES SCALE MATTER?

Scale matters, but *scale* must be defined. What is described as "large-scale" in one region may be "small-scale" in another. And scale is clearly only one of several key variables. Wittfogel focused on "top-down" state-managed systems, but researchers like Clark Erickson (2000, this volume) and Gray Graffam (1992) have emphasized the need for a balanced approach, devoting just as much time to "bottom-up" and "locally managed" decentralized systems.

Bottom-up approaches are usually based on modern ethnographic or ethnohistoric data, while top-down approaches are often based on prehistoric data. This distinction is due in part to the fine-grained and temporally restricted analysis that ethnographers can achieve by observing and interviewing living informants and contemporary farmers. Ethnohistorians have eyewitness accounts of similar behaviors. In contrast, archaeologists focus on multigenerational units of time and work with a body of data that generally precludes observation of individual decision making. A good example of an ethnohistoric bottom-up study is that of Patricia Netherly (1984:229) on Peru's north coast. Although she expected to find evidence for state control of the irrigation systems in the Chimu and Chimu-Inca areas, "it soon became clear that political and economic institutions had not been managed by state bureaucracies and that, in general, hydraulic management was carried out on a number of different levels that tended to be lower rather than higher on a hierarchic scale."

Today we recognize that terms like *large-scale* and *small-scale* need to be defined more precisely to facilitate cross-cultural studies. Some quantification would enhance comparative studies of agricultural strategies and help define the continuum of scale.

The difficult task will be determining how many irrigators and managers were needed to keep various systems up and running smoothly. It is probably the case that two hydraulic systems built on the same scale sometimes utilized labor forces of different sizes and organized them in different ways. And although we can recover canals, terraces, cisterns, and aqueducts, it is much more difficult to recover data about people, social units, and the strategies they followed: coordinating workers to clean canals, performing agricultural rituals, repairing and expanding irrigation canals, constructing check dams, terraces, and the like.

IS IT USEFUL TO THINK IN TERMS OF DICHOTOMIES?

Dichotomies tend to structure research in new fields of study; certainly they structured much of our early thinking about water management. Examples include the dichotomies "top down" vs. "bottom up," "state-managed" vs. "jointly managed," "consensual" vs. "nonconsensual," "small-scale" vs. "large-scale," and "intensification" vs. "disintensification" (e.g., Brookfield 1972; Connelly 1994; Kelly 1983; Janusek and Kolata 2004; Mabry 2000; Price 1994; Wittfogel 1957:3).

As the field of hydraulic studies continues to grow and mature, however, many archaeologists (both in this volume and elsewhere in the literature) seem to want to go beyond dichotomies. To many scholars they seem too simplistic or restrictive to fit "real-world" case studies. Some scholars have therefore turned to (1) continua of scale, (2) a wider range of intermediate economic and strategic policies, (3) a complex mix of technologies that amplify each other's effects over time, and (4) long-term processes in which different variables predominate at different times. The old accretion models (in which technologies were added one by one) and monolithic models (in which only one strategy was employed at a time) do not seem to fit most of the archaeological cases we know. The coexistence of multiple strategies, multiple technologies, and multiple work groups makes our task of modeling these systems much more difficult than we formerly thought it would be.

A salient example of the problem of simple dichotomies would be the massive agricultural terrace systems of Bali, as described by J. Stephen Lansing (1987, 1993; see also Lansing et al. 1998). It is virtually impossible to understand this system by applying the dichotomous framework of "bottom up" or "top down." Rather, such a system represents an example of a phenomenon known as "emergent complexity." Emergent complexity refers to the development of a self-organizing complex system out of the interaction of numerous individual actors.[2] In the Balinese water-temple example Lansing and his colleagues illustrate how the self-interested decisions of thousands of farmers will give rise to cooperation (a form of reciprocal altruism on a large scale over many decades) that resulted in a very complex regional irrigation system.[3] In fact, the system approached theoretically optimal patterns of fallowing and irrigation timing. It achieved this without any conscious planning but rather through numerous interactions and trial-and-error learning over several generations (Lansing 2003; Lansing et al. 1998).

Since leadership positions exist in the Balinese case (the water-temple priests), one could simplistically see these water-temple priests as "managing" the system. However, it is more profitable to view these priests as a necessary part or component of a self-organized system that evolved over centuries by trial-and-error. Some kind of mediating institution is necessary to facilitate communication between individual actors and communities (known as *subaks*), but it was the interaction of those actors (not the temple authorities themselves) that ultimately created the system. Extrapolating from this example, one could imagine a state either co-opting an extant self-regulated system or providing the context for such a system to develop. In neither case can we speak of a top-down or bottom-up system because the system shares features of both in a complex relationship that most certainly shifted and changed over time.

In a similar vein many scholars would now argue that hydraulic and other agricultural systems did not follow a linear progression from simple to complex, small-scale to large-scale, and extensive to intensive. Our historical trajectories are also known to include genetic drift and mutation in plant species, the cumulative effects of intensification on soils and nutrients, and the long-term effects of "landesque capital intensification"—that is, labor invested in permanent landscape modifications such as terraces, canals, raised fields, and cisterns. Just as these multiple technologies and scales affected the system over time, so also did changing political and economic policies.

Decades ago it was common for scholars to concentrate on the natural environment while underemphasizing the sociopolitical environment in which indigenous agricultural practices and economic strategies were embedded. Today we know that "man-man" relations were at least as important as "man-land" relations, and many more variables—social, economic, and political, as well as geologic, hydrologic, and physiographic—must be incorporated to model both successful and failed agricultural strategies.

ENDURING ISSUES AND NEW DIRECTIONS

As social scientists, most of us look for patterns and trends. Although we recognize that the culture and specific trajectory of each region's hydraulic system might be distinctive, we also look for general trends and similarities shared by several regions. The number of general patterns and trends turns out to be much smaller than the number of specific cases. Such patterns provide insights into the way different societies, having no connection to each other, resolved similar problems in similar ways.

Our interest in hydraulics and agricultural systems naturally leads us to pay attention to matter-energy transactions, but we have also found that sociopolitical information and ritual transactions can be just as important. Technology and labor organization are key, but so is the indigenous framework. An example of information and ritual transaction comes from Oaxaca, Mexico, where a Zapotec farmer makes predictions about the adequacy of the year's precipitation based on the volume of spring rainfall. If he predicts inadequate rain, the farmer offers his own blood to *Cociyo* (Lightning) at a public building, asking *Cociyo* to split the clouds to produce more rain; the farmer then increases his planting on land irrigated by wells and canals. Predictions of adequate rainfall encourage the farmer to take a risk, that is, to cultivate more nonirrigated land.

Other sophisticated non-Western systems of managing risk abound in ethnographic studies of farmers. In the highlands of Peru people have learned over generations to cultivate many small patches of land in different microniches. Local ecological conditions vary by altitude in the Andes. In response Andean peoples strategically adopted practices that ensure ownership or usufruct rights over land at different altitudes, thereby spreading out or buffering the risk from frost and rainfall variation.

Gambling and risk taking are, of course, relative. What we think of as risk may not be the way the indigenous people perceived things. And risk can operate on different scales—at the level of the family, the community, or the region. We often consider large-scale systems to be more vulnerable than small-scale systems, but that assumption may not always be accurate. How did each family, corporate unit, village, or polity perceive risk? How were their short-term and long-term strategies affected by that perception? Was their tolerance for risk different from that of modern ethnic groups in the same region? Or was risk of secondary importance when the prospects of political control were high?

The authors of this book take a new look at intensification, technology, water rights, strategies, agricultural rites, man-man relations, and man-land relations. We believe that the volume shows that progress is being made in understanding ancient water management. Perhaps the future will add case studies to the theoretical armature we need before we can fully model the relations among sociopolitical groups, regional economies, and landscapes.

NOTES

1. Population projections are very difficult, as described in Bongaarts and Bulatao (2000). However, there is little question that the population of the world will increase by at least a factor of 50 percent in the next century, greatly depleting the freshwater resource base.

2. Emergent complexity is also a central feature of evolutionary game theory and other work in the evolution of cooperation among adaptive agents (e.g., Andreoni 1995; Axelrod and Hamilton 1981; Boyd and Richerson 1992; Gintis et al. 2001).

3. But see Hauser-Schaublin (2003) for a critique of this interpretation.

REFERENCES

Adams, R. McC.

1974 Historic patterns of Mesopotamian irrigation agriculture. In *Irrigation's Impact on Society*, edited by T. E. Downing and M. Gibson, 1–6. University of Arizona Press, Tucson.

Adams, W. M., and D. M. Anderson

1988 Irrigation before development: Indigenous and induced change in agricultural water management in East Africa. *African Affairs* 87:519–535.

Andreoni, J.

1995 Cooperation in public-goods experiments—kindness or confusion? *American Economic Review* 85:891–904.

Axelrod, R., and W. D. Hamilton

1981 The evolution of cooperation. *Science* 211:1390–1396.

Bongaarts, J., and R. A. Bulatao (editors)

2000 *Beyond Six Billion: Forecasting the World's Population*. National Research Council, National Academies Press, Washington, DC.

Boyd, R., and P. J. Richerson

1992 Punishment allows the evolution of cooperation (or anything else) in sizable groups. *Ethology and Sociobiology* 13:171–195.

Brookfield, H. C.

1972 Intensification and disintensification in Pacific agriculture: A theoretical approach. *Pacific Viewpoint* 13:30–48.

1986 Intensification intensified: Review of prehistoric intensive agriculture in the tropics. *Archaeology in Oceania* 21(3):177–180.

Connelly, W. T.
1994 Population pressure, labor availability, and agricultural disintensification: The decline of farming on Rusinga Island, Kenya. *Human Ecology* 22(2):145–170.

Erickson, C.
2000 The Lake Titicaca Basin: A pre-columbian built landscape. In *Imperfect Balance: Landscape Transformations in the Precolumbian Americas*, edited by D. Lentz, 311–356. Columbia University Press, New York.

Flannery, K. V., and J. Marcus
2005 *Excavations at San José Mogote 1: The Household Archaeology*. University of Michigan Museum of Anthropology, Memoir 40. Ann Arbor.

Gintis, H., E. A. Smith, and S. Bowles
2001 Costly signaling and cooperation. *Journal of Theoretical Biology* 213:103–119.

Gleick, P. H.
1998 *The World's Water: The Biennial Report on Freshwater Resources*. Island Press, Washington, DC.

Graffam, G. C.
1992 Beyond state collapse: Rural history, raised fields, and pastoralism in the south Andes. *American Anthropologist* 94(4):882–904.

Guyer, J., and E. Lambin
1993 Land use in an urban hinterland: Ethnography and remote sensing in the study of African intensification. *American Anthropologist* 95(4):839–859.

Hakansson, N. T.
1998 Rulers and rainmakers in precolonial South Pare, Tanzania: Exchange and ritual experts in political centralization. *Ethnology* 37:263–283.

Harrower, M. J.
2003 Environmental versus social contingencies and the origins of irrigation farming in highland southwest Arabia. Proposal submitted to the National Science Foundation. Manuscript in possession of the authors.

Hauser-Schaublin, B.
2003 The precolonial Balinese state reconsidered: A critical evaluation of theory construction on the relationship between irrigation, the state, and ritual. *Current Anthropology* 44(2):153–181.

Hunt, E., and R. Hunt
1974 Irrigation, conflict, and politics: A Mexican case. In *Irrigation's Impact on Society*, edited by T. E. Downing and M. Gibson, 129–157. University of Arizona Press, Tucson.

Hunt, R. C.
2000 Labor productivity and agricultural development: Boserup revisited. *Human Ecology* 28(2):251–277.

Janusek, J. W., and A. L. Kolata
2004 Top-down or bottom-up: Rural settlement and raised field agriculture

in the Lake Titicaca Basin, Bolivia. *Journal of Anthropological Archaeology* 23:404–430.

Kahrl, W. L.

1982 *Water and Power: The Conflict Over Los Angeles' Water Supply in the Owens Valley*. University of California Press, Berkeley.

Kelly, W. W.

1983 Concepts in the anthropological study of irrigation. *American Anthropologist* 85(4):880–886.

Lansing, J. S.

1987 Balinese "water temples" and the management of irrigation. *American Anthropologist* 89(2):326–341.

1993 Emergent properties of Balinese water temple networks: Coadaptation on a rugged fitness landscape. *American Anthropologist* 95(1):97–114.

2003 Complex adaptive systems. *Annual Review of Anthropology* 32:183–204.

Lansing, J. S., J. Kremer, and B. Smuts

1998 System-dependent selection, ecological feedback, and the emergence of functional structure in ecosystems. *Journal of Theoretical Biology* 192:377–391.

Lees, S. H.

1970 Socio-Political Aspects of Canal Irrigation in the Valley of Oaxaca, Mexico. PhD dissertation, Department of Anthropology, University of Michigan, Ann Arbor.

Mabry, J. B.

1996 The ethnology of local irrigation. In *Canals and Communities: Small-Scale Irrigation Systems*, edited by J. B. Mabry, 3–30. University of Arizona Press, Tucson.

2000 Wittfogel was half right: The ethnology of consensual and nonconsensual hierarchies in irrigation management. In *Hierarchies in Action: Cui Bono?* edited by M. W. Diehl, 284–294. Center for Archaeological Investigation, Southern Illinois University, Carbondale.

Mitchell, W. P.

1976 Irrigation and community in the central Peruvian highlands. *American Anthropologist* 78(1):25–44.

Mitchell, W. P., and D. Guillet (editors)

1993 *Irrigation at High Altitudes: The Social Organization of Water Control Systems in the Andes*. Society for Latin American Anthropology, American Anthropological Association, Arlington, Virginia.

Netherly, P.

1984 The management of late Andean irrigation systems on the north coast of Peru. *American Antiquity* 49(2):227–254.

Postel, S.

1997 *Last Oasis: Facing Water Scarcity*. Norton, New York.

Price, D. H.
1994 Wittfogel's neglected hydraulic/hydroagricultural distinction. *Journal of Anthropological Research* 50:187–202.

Raskin, P., H. Hansen, and R. Margolis
1995 *Water and Sustainability: A Global Outlook*. Polestar Series report #4. Stockholm Environmental Institute, Stockholm.

Scarborough, V. L.
2003 *The Flow of Power: Ancient Water Systems and Landscapes*. School of American Research Press, Santa Fe, New Mexico.

Sheridan, M. J.
2002 An irrigation intake is like a uterus: Culture and agriculture in precolonial North Pare, Tanzania. *American Anthropologist* 104(1):79–92.

Steward, J. H.
1936 The economic and social bases of primitive bands. In *Essays in Anthropology Presented to Alfred L. Kroeber*, edited by R. H. Lowie, 331–350. University of California Press, Berkeley.
1949 Cultural causality and law: A trial formulation of the development of early civilizations. *American Anthropologist* 51:1–27.
1953 Evolution and process. In *Anthropology Today*, edited by S. Tax, 313–326. University of Chicago Press, Chicago.
1955 *Irrigation Civilizations: A Comparative Study*. Pan American Union, Washington, DC.

Trawick, P. B.
2001a Successfully governing the commons: Principles of social organization in an Andean irrigation system. *Human Ecology* 29(1):1–25.
2001b The moral economy of water: Equity and antiquity in the Andean commons. *American Anthropologist* 103(2):361–379.

Treacy, J.
1994 *Las chacras de Coporaque: Andenería y riego en el valle del Colca*. Instituto de Estudios Peruanos, Lima.

Uphoff, N.
1986 *Improving International Irrigation Management with Farmer Participation: Getting the Process Right*. Westview Press, Boulder, Colorado.

Wittfogel, K. A.
1929 Geopolitik, geographischer materialismus und Marxismus. *Unter dem banner der Marxismus* 3:17–51; 485–522; 698–735.
1931 *Wirtschaft und gesellschaft cinas*. C. L. Hirschfeld, Leipzig.
1935 The foundations and stages of Chinese economic history. *Zeitschrift für sozialforschung* (Paris) 4:26–60.
1938 Die theorie der orientalischen gesellschaft. *Zeitschrift für sozialforschung* (Paris) 7:90–122.

1953 The ruling bureaucracy of Oriental despotism: A phenomenon that paralyzed Marx. *Review of Politics* 15:350–359.

1957 *Oriental Despotism: A Comparative Study of Total Power*. Yale University Press, New Haven, Connecticut.

1960 A stronger oriental despotism. *China Quarterly* 1:29–34.

Wunderlich, W. O., and J. E. Prins (editors)

1987 *Water for the Future: Water Resources Developments in Perspective*. A. A. Balkema, Rotterdam.

Zuidema, R. T.

1986 Inka dynasty and irrigation: Another look at Andean concepts of history. In *Anthropological History of Andean Polities*, edited by J. Murra, N. Wachtel, and J. Revel, 177–200. Cambridge University Press, New York.

Middle East

Intensified Large-Scale Irrigation as an Aspect of Imperial Policy

Strategies of Statecraft on the Late Sasanian Mesopotamian Plain

ROBERT McC. ADAMS

Elementary forms of supplemental watering may be almost as old as consciously conducted, archaeologically recognizable agriculture itself. Appearing only millennia later, and at the other extreme of scale, complexity, and structural centrality within the societies in question, are giant canal systems that we think of as primary characteristics of the great river valley civilizations of the Old World. How unifying and dominant as an explanatory factor behind that grand succession are an enormous variety of environmental enabling factors, opportunities, and constraints? In what sense, if at all, is it useful to think of the many manifestations of all or parts of the long transition as a unitary process of incremental changes in a common direction? At what point in the course of that transition do the diverse requirements of truly large-scale irrigation become so compelling that societies relying primarily on this form of subsistence can profitably be viewed as constituting a single class? Can increasing scale and complexity simply be equated with improvement and "progress," or if they result in increasing exposure to unforeseen risks and unsustainable long-term managerial requirements, can they sometimes take on an almost pathological character? What can be done to distinguish useful clusters of commonalities within the huge, disorderly aggregate that is our subject matter, and what, if anything, can we discover of the causative forces that in at least some circumstances contributed to disjunctive changes within it? These and related concerns are broadly shared by the participants in this symposium, but we can approach them only from our different—locally and historically specific—perspectives. Mine has long concerned the watersheds of the Tigris and Euphrates rivers at the eastern end of the ancient Near Eastern

"Fertile Crescent." And here I deal with the character and causation of the largest irrigation systems ever constructed there.

Broadly speaking, after its fourth millennium BC inception western Asiatic civilization can be divided into three successive phases. In the first, subregional systems centered on individual city-states contested with one another almost continuously, with one or another periodically achieving larger but tenuous temporary dominance and political control. With regard to the long-prevailing assumption that irrigation was a prerequisite for the development of civilized, urban life in presumably arid or semiarid areas, an important recent analysis of the southern Mesopotamian plain based largely on Corona satellite imagery strikingly contradicts it for precisely the region where this was first achieved. Under radically different hydrological, climatic, and geomorphological conditions that prevailed at least through the fourth millennium, widespread marsh environments encouraged richly diversified riparian and estuarine subsistence systems in which water management technologies were prevailingly small-scale and localized (Pournelle 2003). Farther upstream, in the Tigris-Euphrates-Karun drainage basin, and in the western part of the so-called Fertile Crescent, more ambitious but still relatively small-scale irrigation regimes had already appeared earlier.

Macroscale, regional economic integration does not seem to have been a recognized objective throughout this initial phase. During it there was, however, a lengthy, irregular environmental transition in the direction of present conditions. Coinciding with this transition were many additions to the scale, technological inventory, and probable systemic character of water management during the third millennium. At around the turn of the second millennium some cuneiform records of dam, reservoir, and canal construction (e.g., Walters 1970) suggest that projects of quite substantial scale sometimes were being undertaken. The areal extent they served is unknown, however, and there is no evidence that they were generally able to outlast the continuing political vicissitudes that were so characteristic of the time.

The second phase, its beginnings associated with the rise of Assyrian power late in the second millennium, saw larger, often panregional and increasingly durable domains take shape—essentially, a transition from states to empires. This transition occurred initially in a northerly region of predominantly nonirrigation-dependent agriculture. It was accompanied by forced, large-scale interregional population transfers and by enlarged royal aspirations for the display of power and opulence, all of which led to canalization associated with imperial capitals that were sometimes of considerable scale. Since these systems were closely linked to the requirements of the largest Assyrian centers, they are unlikely to have survived for long after the fall of Assyrian power and the destruction of those cities at around the end of the eighth century

BC. But a long series of territorially extensive empires of a generally Assyrian pattern succeeded it—Neo-Babylonian, Achaemenid, Seleucid, Parthian, and Sasanian—bringing irrigation systems on a progressively larger scale not only to the Mesopotamian plain but to many surrounding regions.

The second phase may be considered to have come to an end with the immense transformation that rapidly emerged out of the rise and spread of Islam as a pancontinental, world religion but soon politically fragmented imperium, extending from the Iberian peninsula deep into central Asia. A new pattern of urban primacy arose, involving great expansion of a handful of major urban centers, extensive population movements, and a new emphasis on commerce and commercial agriculture in politically dominant areas (Wheatley 2001). This third distinguishable phase lasted at least until the advent of modernity. But already before the end of the second phase the demands of virtually continuous military campaigning at great distances had led to escalating state demands for revenue. These demands were concurrent with—and I will argue sometimes led directly to—major, state-directed programs of economic investment in agricultural intensification.

In southern Mesopotamia, where hydrological regimes and alluvial topography made intensification feasible, irrigation on a vast scale was an understandable, if not necessarily inevitable, outcome. And there we can now see that such systems were introduced apparently not in a process of organic, incremental growth but as a conscious, "top-down" policy choice arising from considerations of politico-military strategy. They were, in short, neither a consequence of autonomously growing technological scale and complexity that resulted in self-sustaining, spreading Oriental despotism, as Karl Wittfogel (1957) once forcefully argued, nor embodiments of some demographically or environmentally imposed course of action.

In identifying a series of three phases I speak, of course, in terms of an irregular succession of loosely interrelated tendencies. There were lengthy, discontinuous transitions between the clearest manifestations of them rather than a series of abrupt and irreversible turning points. But these are cumulative, metahistoric trends or patterns in Near Eastern history that deserve our attention and that may well have parallels (or at least resonances) in other world regions.

Seen in this light, it is disappointing that archaeologists and historians have generally given disproportionate weight to the first and earlier part of the second phase. The availability of the almost incomprehensibly vast resources of cuneiform documents is of course an explanation and substantial (but incomplete) justification for this. For archaeologists, however, the shoe is on the other foot. The resources of the later periods are also rich, almost incomprehensibly vast, and, with their recency and correspondingly reduced overburden of later deposits, much more accessible. Major urban centers, sometimes irregularly sprawling but in other cases stamped out in almost identical, geometric modules, made their appearance. Recoverable town and building plans, supplemented with surface collections that can provide clues as to status and function of population segments, can be expected to open up entirely new horizons of socioeconomic and sociocultural knowledge. Yet these possibilities, rapidly being demonstrated elsewhere, are still almost entirely neglected for the later periods in Mesopotamian history.

These background considerations provide the context for concentrating for present purposes on the Sasanians (AD 226–637), who may most fully represent the central tendency of the second phase.[1] No contrast, favorable or unfavorable, is intended with their predecessors during several preceding centuries or with their early Islamic successors, in terms of the magnitude and quality of Sasanian technological and economic achievements. But there is a further, external and pragmatic, reason for concentrating on the Sasanian period. In much of the Mesopotamian plain, due to extensive surface disturbances of later centuries and especially the modern period, Sasanian surface remains are most strikingly widespread and (potentially) accessible for multidisciplinary study.

The outer boundaries of the aggregate area within the Mesopotamian plain that has to receive some measure of archaeological survey have an important bearing on this discussion. To begin with, they are an artifice of the time, personnel, and funds available during the period in which they were conducted (Figure 2.1). Neglecting for present purposes the relatively well-studied upper plains of Khuzistan to the east across the Iranian border, only the west-central portion of the Mesopotamian alluvium proper, plus the alluvial fan of the tributary Diyala River east of the Tigris, have been comprehensively (not in any case exhaustively!) examined. A larger surrounding region to the east and west is essentially unsurveyed. If and when conditions again permit extensive field surveys in Iraq following the nation's recovery of its sovereignty and the end of the Coalition's occupation of the country, further research along these lines (but with much advanced technologies and methodology) will undoubtedly be quickly resumed. But in the meantime we must be conscious of the fact that we are not dealing with a representative sample and that the impediments to generalization introduced by the geographic limitations of our sample may vary greatly from period to period. I will return to some of the implications of this problem at the end of this chapter.

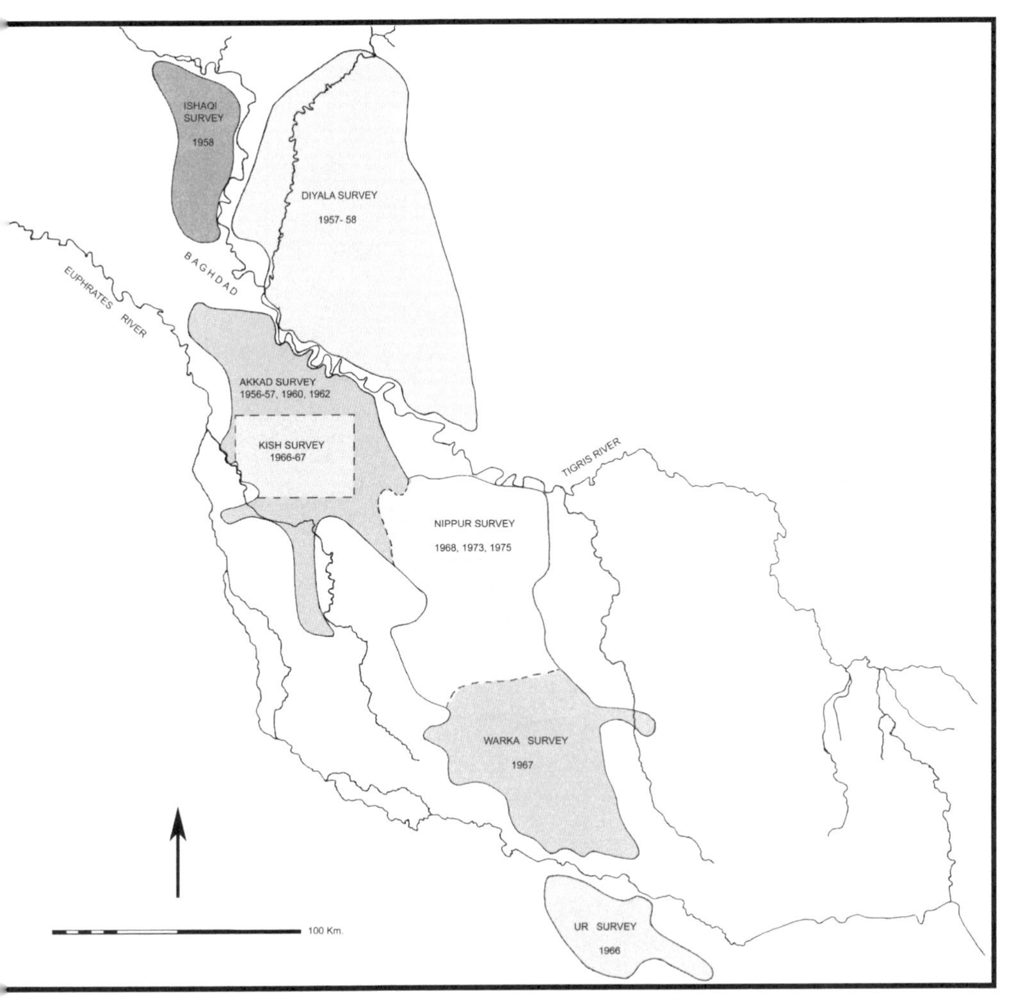

'igure 2.1. Geographic coverage of archaeological surveys in southern Iraq.

Quickly overrun by the Arab conquest around AD 637, the Sasanian Empire left behind little more than some outstanding architectural and artistic monuments and an evident weakness and disorganization at the time of its demise that paradoxically accompanied its long duration and impressive size. It is this terminal period of dissolution that is historically best documented, very largely in the accounts of its victorious successors. Periodic earlier successes against the Byzantines and other adversaries similarly are best known to history from the testimony of their enemies. A relatively large responsibility therefore falls on archaeology to amplify the record and better elucidate both the strengths and limitations of Sasanian achievements.

The conduct of irrigation agriculture under the Sasanians offers exceptional opportunities for pursuing this objective. Irrigation suggests something close to at least a transitory realization of an ambitious vision to maximize the use of the available land and water resources, deploying unprecedented massive investment capabilities and impressive engineering sophistication toward this end. Khusraw Anushirwan (531–579), perhaps the greatest of its rulers (or at least often so perceived by later Arab chroniclers), claimed personal responsibility for the larger strategy behind this in a form later preserved for us by Mas'udi: "Royal power rests upon the army, and army upon money, and money upon the land-tax (kharaj), and the land-tax upon agriculture, and agriculture upon just administration, and just administration upon the integrity of government officials, and the integrity of government officials upon the reliability of the vizier, and the pinnacle of all of these is the vigilance of the king in resisting his own inclinations, and his capability so to guide them that he rules them and they do not rule him" (Altheim and Stiehl 1954:46).

Confronting the Sasanian dynasty during the long course of its rivalry with the Byzantine Empire was a major challenge posed by the relative inadequacy of its resource base. As James Johnson (1995:169) has emphasized in a provocative paper, "the Sasanians disposed of no more than half of the material resources [of arable lands and manpower] available to their Roman rivals." Ultimately, means had to be sought to overcome this disparity in centralized fiscal, administrative, and political measures.

Traditionally, Sasanian armies taking part in imperial campaigns were assembled from forces drawn up and largely supported by the powerful, restive nobility. This procedure had obvious limitations that greatly intensified the problem, for even under able royal leadership military command and control were subject to negotiation and delay. Greater reliance on mercenaries, who would be loyal to the crown as their paymaster, was a clearly preferable alternative, and these mercenaries were available from opposite ends of the realm in Yemen and the northern Caucasus. But of course their procurement was possible only at escalating cost to the crown—whose fiscal resources had to be extracted from the same quasi-independent "feudal" lords.

Anushirwan found a solution to this conundrum in the suitability of the Mesopotamian plain for greatly intensified irrigation agriculture. That process of integrated development simultaneously served to direct the major part of the flow of the resultant surpluses into the royal treasury. The principal underutilized resource available for this purpose was the Tigris River, the larger but also the more dangerous and difficult to control of Mesopotamia's Twin Rivers (Adams 1981:3–4). Given the virtual absence of any knowledge at the time of the long-term dangers of hydrodynamic variability and saline groundwater, the challenge of more fully employing its waters was addressed with great initial success.

Massive labor resources were clearly a first essential for Anushirwan's chosen objective. Captives (and food stocks to support them?) from his highly successful military campaigns in the west and elsewhere must have been key in filling this need. But also integral to that success were the impressive skills in land planning and the engineering of public works he could call on. These are repeatedly evident in a characteristic readiness to transform entire landscape and drainage configurations. In the long run, to be sure, the systems thus introduced still remained vulnerable to the complex instabilities of agricultural ecology in the region, as well as to those of ancient systems of imperial rulership. In fact, the very scale of some of the largest of the new undertakings imposed management requirements that could not be consistently met. This recurring problem contributed heavily to their serious, periodic failures and may bear no small responsibility for the ultimate ineffectiveness of the dynasty's resistance to the Arab onslaught.

Integral to Anushirwan's physical improvement of the agricultural base was an accompanying cadastral survey and tax reform. While these are traditionally associated with his name, it appears that his predecessor, Qubadh I (488–496, 499–531), initiated the process and that its full implementation may have involved one or more of his progressively weaker successors. But at any rate the details of the reform clearly establish the fiscal as well as political imperatives behind it. To draw selectively from the extended analysis of Zeev Rubin:

> The old Iranian aristocracy was its main victim. . . . It seems to be perfectly true that the new system brought about a significant increase in the royal income from the land tax. . . . [The new system of taxation made] it possible to anticipate incomes and budget expenses. . . . The overall result of this should have been to maintain a system of small farms that might be easily taxed, and to prevent the concentration of land in huge estates whose owners might become excessively powerful, accumulating privileges and immunities, and obstruct effective taxation. (Rubin 1995:229, 248, 291–292)

The part of the Mesopotamian plain receiving the major thrust of Sasanian development efforts happens to coincide to a considerable extent with the areas covered by existing archaeological surveys (Figure 2.2). To be sure, in the area covered by the Akkad survey, the first of the series in 1956–1957, few Sasanian sites are shown. At that time, however, ceramic "index fossils" suitable for dating surface collections from sites later than the mid-first millennium BC were poorly understood, and attention was focused almost exclusively on earlier periods. By far the greater proportion of sites encountered were classified simply as (Seleucid-)Parthian-Sasanian and not recorded. When work resumed six months later on the Diyala plains east of Baghdad, and in all subsequent campaigns, this omission was carefully rectified.

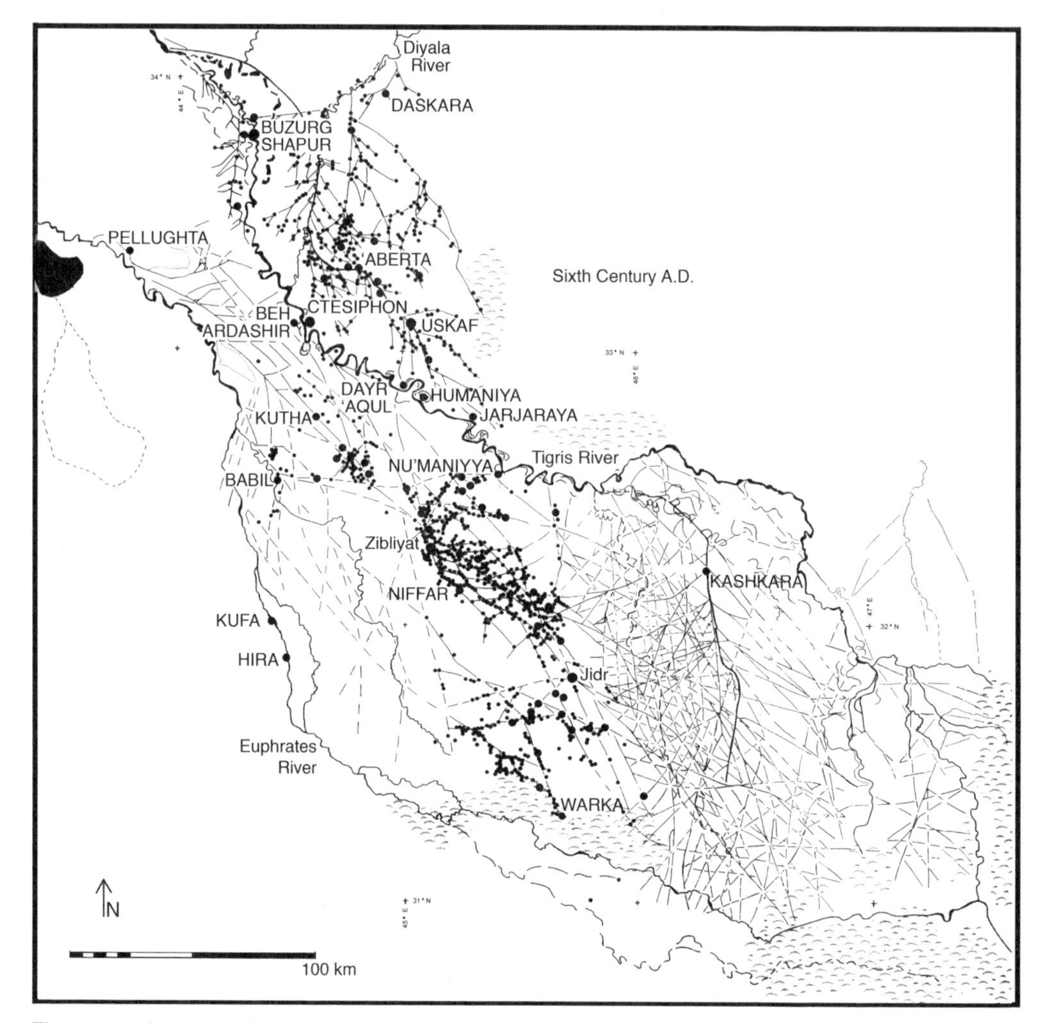

Figure 2.2. Sasanian settlement and irrigation systems, dated mainly from surface collections.

Largest and best known of the major Mesopotamian projects is the Nahrwan Canal system. The Diyala River's alluvial fan had long been settled and under the preceding Parthian dynasty had experienced great population and urban growth to an extent that must have seriously tested the limited and variable supplies of water available from the Diyala River alone (Adams 1965:63). Anushirwan decisively overcame these limits by drawing water directly from the Tigris River just above the upper end of the alluvium in a great canal sufficient alone to sustain irrigation on some 4,000 or more square kilometers. The ambitiousness of the design of this effort is remarkable, even by modern standards, for it had to emanate from a dangerously flood-prone source, follow a straight, deeply excavated course through 100 km of headland, and cross two intervening rivers before reaching the area it was intended to serve. Its carefully engineered weirs and massive branch-canal off-takes, repaired or renewed to remain in operation for several centuries of the succeeding early Islamic period, have been mapped and illustrated in considerable detail (Adams 1965). For an understanding of the original Sasanian design, however, the Nahrwan has the drawback that most of its original features are obscured by five or more centuries of later use and protracted deterioration before it was finally abandoned.

The course (or more accurately, numerous courses) of the lower Tigris prior to the late Sasanian period is a subject to which considerable study is being devoted with the aid of satellite imagery. The details of their changing configurations through time cannot be dealt with here, but it may be enough to note that for many millennia the greater part of them lay well to the west and south of the river's modern position. The late Sasanian course may have approximated the modern one as far downstream as modern Kut but then diverged to the southeast, following clear traces of a partly meandering and partly straight, evidently constructed, bed. Below Kut the modern course of the Tigris lies considerably north, in a bed that is demonstrably recent since it lacks significant adjoining settlement or the wide levee that would have encompassed meander traces.

The more southerly course constituted another major, new, integrated irrigation system, not far short of the scale of the Nahrwan system. It entailed, as Michael Morony notes, "the expansion of the district of Kaskar into an important province in the quarter of the West in the late Sasanian period [and] reflected significant changes in the hydrology of the lower Tigris and in the local irrigation system" (1984:155). Part of the area it served may have been very lightly settled previously, although in the absence of surveys this cannot yet be known conclusively. But farther south its canal tails intersected extensively with lands of the ancient kingdom of Lagash, and there it served as an important supplement to irrigation supplies drawn from Euphrates branches far to the west and northwest.

The general pattern of late Sasanian irrigation development as illustrated by the layout in this province can be provisionally reconstructed in general outline even in the absence of archaeological surveys. Especially helpful in this regard are the accounts of Islamic geographers (al-'Ali 1970–1971) and others and, to a lesser extent, the records of inspectors employed by the Iraqi Department of Antiquities even in the absence of archaeological surveys. Although often complicated by underlying Parthian antecedents, it basically consisted of two associated elements. First, there were long, intersecting, branching networks of primary distributary canals stemming from the main Tigris artery, each of them essentially dendritic or treelike in form if considered in association with its secondary off-takes. Second, provision was made for irrigation water to be supplied from two alternate sources: what we may label the older "Euphrates" network to the northwest and the new Tigris network stemming from the expansion of the Sasanian province of Kaskar to the northeast. Immensely costly in construction and annual maintenance, these duplicate systems can only have been intended to ensure continuity of flow, and hence of agricultural output and consequent tax income, in the event of various forms of blockage or reduction in either of the two major sources. Strategic considerations, it seems, dictated the payment of a heavy premium—greater assurance of constancy of state income from land-tax receipts.

But the development of this new system presently entailed still other difficulties and costs. The complex redistribution and shift of Tigris flow probably could not have been adequately planned or controlled with available means and knowledge, especially during periodic inevitable intervals of dynastic weakness. Major floods were a special danger and are known to have resulted in large-scale expansions of marsh, beginning in the time of Bahram V (AD 420–438) and occurring again under Qubadh I. According to Mas'udi these resulted in a reduction in the number of *kuwar* (intermediate administrative districts) in the Sawad from 12 to 10 and the number of *tasasij* (still smaller units) from 60 to 48. The long-term effect, as Morony notes, was an overall decline in the importance of the region. Reclamation awaited the accession of Anushirwan, who also enlarged Kaskar province by adding to it the *tassuj* of Zandaward, apparently in the interest of the Sasanian royal family as a crown province (Morony 1984:155–157). Adding the (only satellite-observed) "footprint" of this irrigation system to that already known in the adjoining region to the west (Figure 2.2), a basic characteristic of the combined systems becomes immediately evident: a fairly regular, roughly evenly spaced and rectangular grid of smaller, shorter canals that must have been aimed at providing more uniform allocations of water based on potentially cultivable areas.

The significance of the westward extension of Kaskar to include the region bounded more or less by Warka (ancient Uruk), Niffar (ancient Nippur), and Zawabi (which I would like to identify with the very extensive ruins northwest of Niffar that are now known as Zibliyat)[2] in the late Sasanian period lies in the fact that this region had long been occupied by the lower end of a massive irrigation system developed in this period with water drawn primarily from the Babylon branch of the Euphrates. Through this expansion the late Sasanian province of Kaskar newly combined two, essentially distinct, hydrographic systems: the great trunk canal drawn from the Euphrates in its western part and the lower course of the rerouted Tigris in its eastern part. Intersecting branch canals fanned out over this region from both systems.

Most irrigation systems not only are endlessly, incrementally modified in the course of regular desilting and maintenance operations but also are repeatedly subject to modifications on a larger scale. Frequently they newly incorporate remnants of extensive older systems conforming to entirely different basic patterns, and they are in turn incorporated into dissimilar later ones. Thus the designed character of ancient schemes of irrigation and resettlement at the time of their first development is normally difficult to discern beneath the blurred succession of its evolving forms. What can be readily identified are only the final, fragmentary remains of extended processes in time. But fortunately, a convergence of special circumstances has provided a means of at least partly overcoming these obstacles in one instance in southern Iraq.

The limited segment of a canal-and-settlement system shown in Figure 2.3, only some 35 km in length, was an integral part of Anushirwan's new westward extension. It is located about 40 km north of the great ancient city of Uruk (modern Warka). The latter retained a garrison until the arrival of the Arab conquerors, although by this time that city itself was largely in ruins, and the surviving settlement lay southeast of the old walled center, in a region otherwise given over to marshes (Finster and Schmidt 1976:164–166). Unfortunately, both the major upstream and downstream continuations of the central canal that is shown are obscured by overlying deposits and later canal levees. This suggests that there has been substantial wind deflation here, which also must account for the extraordinary detail of preservation of minor, contemporary surface features.

Although only those sites of Sasanian date are of primary interest for present purposes, all sites in the vicinity that were identified in the original archaeological survey are also shown in their generally scattered locations on this map. Adjoining numbers assign them to their appropriate millennium (or millennia) BC, with the letter *P* denoting an occupation of the preceding Parthian period (ca. 120 BC–AD 220). Readily apparent is the disproportionately greater density and linearity of the Sasanian settlement system, and its apparently more extensive exploitation of the surrounding landscape, as compared with that of all preceding periods in this immediate region.

More relevant for the theme of irrigation are the numerous alternative or supplementary sources of water beyond the major one to the west, reflecting the new inputs of water from the Kaskar system with the Tigris as its source. Given the regularity of spacing of these numerous intake canals, one strongly suspects that they were an element of the original, risk-avoidance design of the system as a whole. Once again we see here that the system consisted of a complex but fairly regular arrangement of descending, dendritic fans. Its linearity testifies to the employment of experienced surveyors in the execution of the plan as well as in its design. What became the main branch, along which almost all of the settlements were concentrated, was not planned originally to serve that function but instead adopted that character only when the settle-

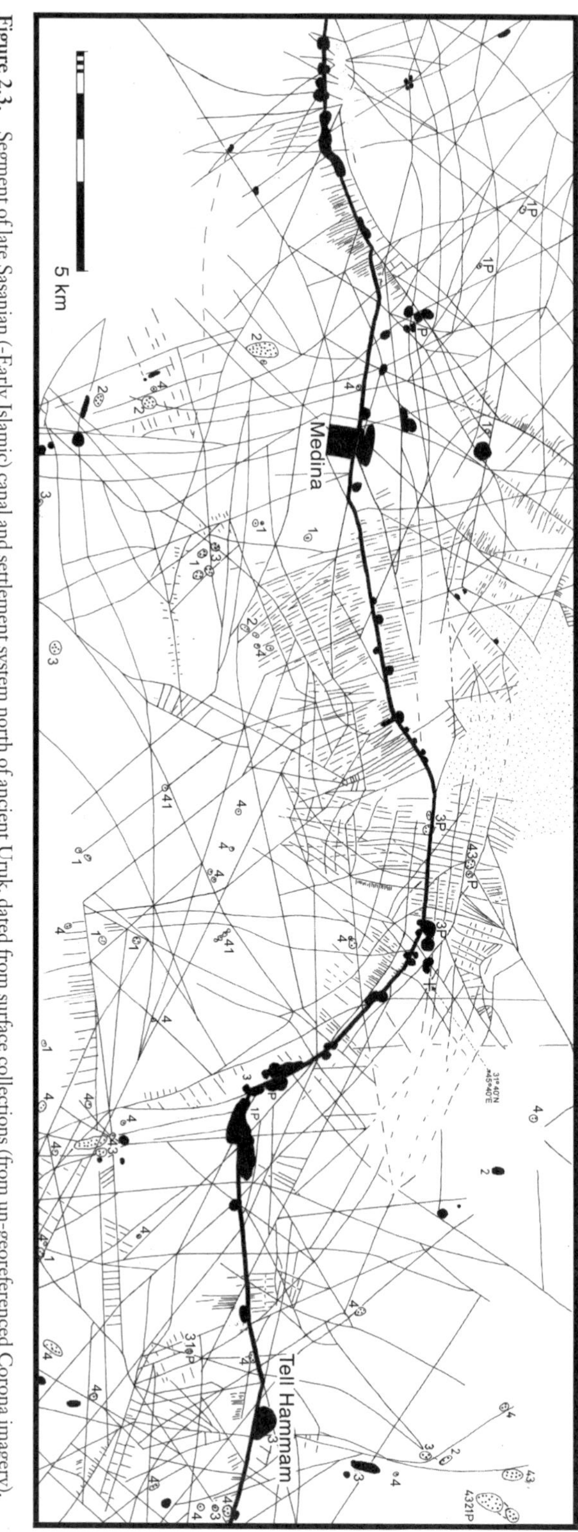

Figure 2.3. Segment of late Sasanian (-Early Islamic) canal and settlement system north of ancient Uruk, dated from surface collections (from un-georeferenced Corona imagery).

ments themselves came into existence. This suggests that the design, and most of the installation of its larger component elements, probably was carried out by a separate, possibly captive, labor force before the new agricultural population had taken up residence.

Other, smaller-scale features of the system are wonderfully preserved in the satellite imagery. Small, rectangular, bounded fields, on the order of 1–2 ha in size, are concentrated well away from the larger centers of population. This would be strange if they were intended for orchards or private garden cultivation and may suggest that they had specialized uses more directly in the interest of the state. There is also a more widespread pattern of long, parallel, fairly close-spaced channels conforming with irrigation and land-use practices that had prevailed for cereal cultivation in the southern part of the Mesopotamian plain for several millennia at least. The latter component appears somewhat opportunistic, the work of individual cultivators or groups of cultivators once they were resident. The rectangular plots, on the other hand, appear in carefully ordered, linear, fairly uniform series. (This may well indicate that they were part of the original design of the system.)

A prominent square enclosure bearing the modern local name simply of *Medina* (city) will be noted in the left central portion of the map. This was found to have the visible, low remains of heavy, buttressed mud-brick walls and gateways, while the interior depression contained nothing more than a moderate concentration of surface ceramics in its southeast quadrant and extremely sparse debris elsewhere. No similar enclosures have yet been identified from earlier periods, including the great era of antique urbanization in the late fourth, third, and second millennia BC, although several others of Sasanian date are described in reports of archaeological surveys of central and southern Mesopotamia. All may be harbingers of a new, fully imperial level of specialization and control—perhaps garrisoned strong points to serve as reception depots for in-kind tax collections. Finally, a string of several adjoining mounds at the western end of this subsystem consist largely of glass wasters and slag. Thus this was a substantial center of glass-vessel production, as well as a region of intensive agriculture. Its occupation continued for some time, probably into the second century of the early Islamic period.[3]

Another glimpse of Sasanian irrigation that is generally congruent with the impressions left by this one comes not from an immediately adjoining region but from a northeastern periphery of the Mesopotamian plain proper. Flanking the alluvium in this direction, on the left bank of the modern Tigris, the gradually rising land gives way to low, rocky outcrops and then to parallel ridges of the Zagros Mountains. Only the lowermost region, in the vicinity of the river, can be irrigated with (today mostly pump-fed) canals from that source, and both modern settlements and cartographic representations of earlier ones are generally small and sparse. But the best-known exception (from the limited available evidence, in the absence, again, of archaeological surveys there) appears from documentary references to center on the Sasanian period. It is associated in Sasanian and early Islamic tax and other records with two districts, Baduraya and Bakusaya. Larger-scale Corona satellite photographic images have not been consulted for this marginal district, but in the apparent absence of intensive land use on any significant scale in any more recent period 1:250,000 Landsat images have proved very informative (Figure 2.4).

Numerous runoff channels emanating from between the higher Zagros ridges provide surface water to these slopes, surely somewhat irregularly. In early antiquity they served as an ancient corridor for long-distance lateral movement of pastoral groups migrating with their animals. But on present evidence it was primarily in Sasanian times that great efforts were made to capture the widely dispersed runoff for irrigation and thus to farm the entire arable surface.

The canals making up this network are fairly simple, single-stranded, and arrayed in quadrilateral if not rectilinear patterns. There is little to suggest successive modifications or a long buildup of spoil banks. Thus there must have been a relatively limited period of canal construction and, presumably, operation in the irrigated area they served. The relatively faint lines left on the images by these canals also suggest that they were designed to carry modest, not wholly dependable, supplies of water. Once again, this reflects a readiness to expend great efforts in maximizing agricultural yields from a marginally productive area.

These two examples should be seen as tesserae in a very large-scale, Mesopotamian alluvium-wide mosaic. Together with the giant Nahrwan system of largely Sasanian origin on the Diyala plains, the prominence of Sasanian remains in the central region of the plain between the Tigris and Euphrates where archaeological surveys have been concentrated, and the great canal system serving the Sasanian city of Kaskar and its environs that was discussed earlier, a fairly consistent pattern can be reconstructed. At its heart was a massive Sasanian effort at agricultural intensification based on state-initiated irrigation construction and maintenance. And although no pronouncement of a motive for this effort has survived, it can with high probability be associated with the crown's efforts to centralize its powers and augment its resources in order to pursue more effectively its vital military struggles with Byzantine and other enemies.

As a final note, however, there are substantive, methodological, and geographic limitations to the presently available evidence on which some of the essential findings of this study are based, and they must not be overlooked. Settlement pattern surveys are the primary source of the archaeological

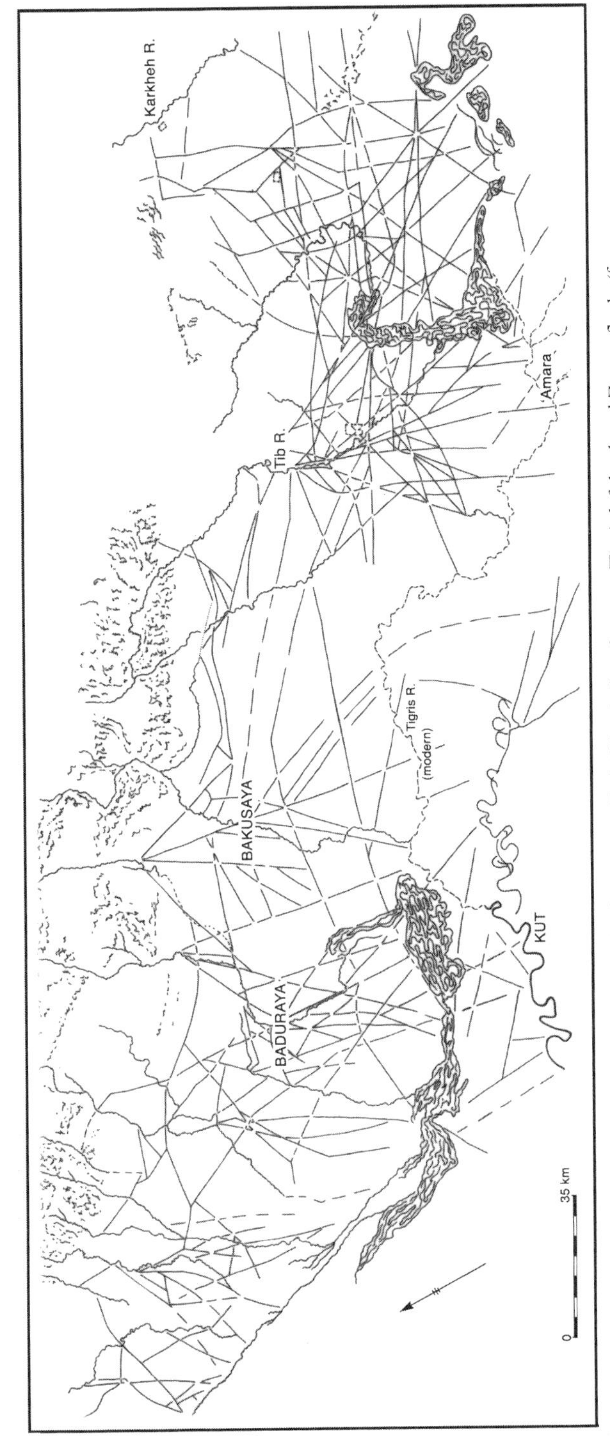

Figure 2.4. Peripheral, unsurveyed canal traces, probably of Sasanian – Early Islamic date, between Tigris left bank and Zagros flanks (from LANDSAT imagery).

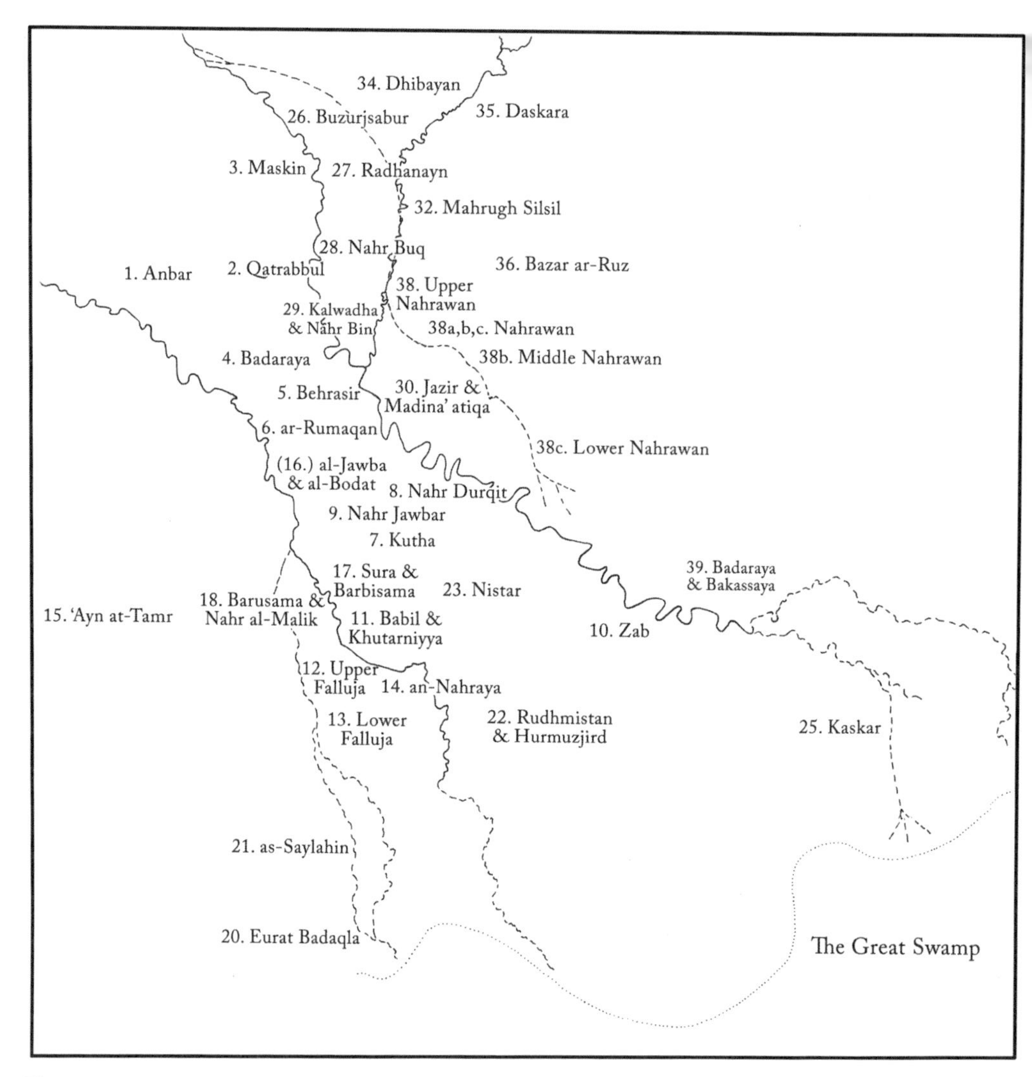

Figure 2.5. Provisional locations of Early 'Abbasid tax collection districts, with their reported tax yields and numbers of towns (after de Goeje 1889 and in consultation (1958) with Fuad Safar).

1. Anbar
Towns 5
Wheat 2,300
Barley 1,400
Silver 150,000

2. Qatrabbul
Towns 10
Wheat 2,000
Barley 1,000
Silver 300,000

3. Maskin
Towns 6
Wheat 3,000
Barley 2,000
Silver 150,000

4. Badaraya
Towns 14
Wheat 3,500
Barley 2,000
Silver 2,000,000

5. Behrasir
Towns 10
Wheat 1,900
Barley 1,700
Silver 150,000

6. ar-Rumaqan
Towns 10
Wheat 3,300
Barley 3,050
Silver 250,000

7. Kutha
Towns 9
Wheat 3,000
Barley 2,000
Silver 150,000

8. Nahr Durqit
Towns 8
Wheat 2,000
Barley 2,000
Silver 200,000

9. Nahr Jawbar
Towns 8
Wheat 1,700
Barley 6,000
Silver 150,000

10. Zab
Towns 12
Wheat 1,400
Barley 7,200
Silver 250,000

11. Babil & Khutarniyya
Towns 16
Wheat 3,000
Barley 5,000
Silver 350,000

12. Upper Falluja
Towns 15
Wheat 500
Barley 500
Silver 70,000

13. Lower Falluja
Towns 6
Wheat 2,000
Barley 3,000
Silver 280,000

14. an-Nahrayn
Towns 3
Wheat 300
Barley 400
Silver 45,000

15. ‘Ayn at-Tamr
Towns 3
Wheat 300
Barley 400
Silver 45,000

(16). al-Jawba & al-Bodat
Towns 8
Wheat 1,200
Barley 1,600
Silver 150,000

17. Sura & Barbisama
Towns 10
Wheat 700
Barley/Rice 2,400
Silver 100,000

18. Barusma & Nahr al-Malik
Towns 10
Wheat 1,500
Barley 4,500
Silver 150,000

20. Furat Badaqia
Towns 16
Wheat 2,000
Barley/Rice 2,500
Silver 900,000

30. Jazir & Madina ‘atiqa
Towns 7
Wheat 1,100
Barley 1,500
Silver 140, 000

32. Mahrudh & Silsil
Wheat 2,000
Barley 2,500
Silver 250,000

34. Dhibayn
Towns 4
Wheat 700
Barley 1,300
Silver 40,000

35. Daskara
Towns 7
Wheat 2,000
Barley 2,000
Silver 70, 000

36. Baraz ar-Ruz
Towns 7
Wheat 3,000
Barley 5,500
Silver 120, 000

38a. Upper Nahrawan
Wheat 2,700
Barley 1,800
Silver 350, 000

38a, b, & c Nahrawan
Towns 21

38b. Middle Nahrawan
Wheat 1,000
Barley 500
Silver 100, 000

38c. Lower Nahrawan
Wheat 1,000
Barley 1,200
Silver 150, 000

39. Badaraya & Bakasaya
Towns 7
Wheat 4,700
Barley 5,000
Silver 330,000

Figure 2.5.

evidence that has been employed, since excavations are distinguished both by their rarity and by generally out-of-date approaches and methodologies. Settlement pattern surveys in the Mesopotamian plain have very substantial limitations of their own; I can do no better here than to refer to Tony Wilkinson's (2000) thoughtful and comprehensive review of the most salient of these problems.

But the geographic limitations need further discussion. Very largely missing are archaeological data, survey and otherwise, from the eastern, western, and southern parts of the alluvium (except for the basin around ancient Eridu, which seems to have been largely under water and almost completely abandoned in both Sasanian and early-middle Islamic times). Morony has argued effectively against any presumption of Sasanian primacy at the apogee of their power under Anushirwan, as contrasted with their early Islamic successors. On the basis of literary and historical sources he favors, instead, a mere shift in the focus of intensity of reclamation and development after the collapse of Sasanian power in AD 637 (Morony 1994:224–225). His further characterization of the likely context of that shift is equally important:

> Apart from the fact that there is no direct evidence for the migration of the agrarian population of Iraq to the new cities founded by Muslims in the seventh and eighth centuries, and that we are told that the population of Samarra came from more distant parts of the Abbasid empire, it is worth emphasizing that shifts in population are not necessarily to be explained by migration from the places where it decreased to places where it increased. Especially when shifts occur over several generations or even centuries, such change is just as likely to be brought about by a drastic reduction in fertility and an increase in mortality caused by oppressive conditions, disease, the lack of opportunities, and increasing levels of violence. . . . Conversely, after their initial settlement, mostly by people from outside Iraq, the burgeoning population of the newly founded cities in early Islamic Iraq is likely to have been due as much to procreation as to immigration, and to the reorientation of the regional economy by these urban manufacturing and commercial centers. Shifts in the countryside are just as likely to have been encouraged by the protection, patronage and development of favoured alternative sites. (Morony 1994:226)

Highly relevant to this discussion is a further line of geographic and settlement evidence reflecting spatial distributions and activities but on a textual rather than archaeological basis. Early 'Abbasid tax registers dating to the turn of the ninth century have long been known, 160 years or so after the triumph of Islam (de Goeje 1889:4–10). Figure 2.5 approximately locates these districts and records the number of their market towns and their tax yields in wheat, barley, and silver. The given locations are a product of my lengthy discussions many years ago with the late Fuad Safar, long the deputy director general and director of research of the Iraq Department of Antiquities, excavator of Kaskar/Wasit (Safar 1945), and a widely acknowledged authority on the subject. This compilation, despite the informality of its source and the admitted uncertainties of some given locations, is a suggestive source of information on many subjects. It provides, for example, a graphic representation of the apparently very limited degree to which the center of gravity of the Mesopotamian alluvium had moved west more than a century and a half after the Sasanian dynasty was overwhelmed by the Arabs.

No less significant is the vast lacuna that is evident in recorded towns and tax collections in the entire southern portion of the region—the "Great Swamp." It would be a serious error to regard the absence of this particular form of evidence as reflecting an absence, or even diminution, of human occupation at the time of these records. To the contrary, the lengthy, well-described struggle that began some seven decades after the time of these records and for a time threatened to bring down the Caliphate before the final suppression of the "Revolt of the Zanj" makes clear in abundant detail not only the military prowess but the resources, (particularly aquatic) mobility, and resilience of the very considerable population of precisely this region (al-Tabarī 1992).

This is not the place to pursue this very worthwhile subject further, but the very characterization of what was clearly a widespread regional upheaval with an ethnic term narrowing its participants to east African agricultural slave laborers is obviously a considerable distortion. The undoubted presence of that population element already speaks to the presence of large estates with the resources to import and manage slaves but the capacity to evade accountability for them. And while slaves taking up arms may well have constituted key cadres with nothing to lose, al-Tabarī 's graphic and voluminous account leaves no room to doubt the full participation of other, very disparate population elements. The key probably lies in the marshy environment to which they were all so well adapted but that both the Caliphate's tax collectors and its military formations found so difficult to penetrate.

There is thus much work yet to be done before the full extent and significance of the Sasanian achievement in irrigation agriculture can be properly assessed. But for the comparative study of premodern irrigation systems, at least one salient aspect of its theoretical significance appears to lie in the centralized, politico-military stimulus that was sometimes responsible for initiating them. Large-scale irrigation, as primarily a recourse of statecraft and military strategy and only in a longer-term secondary way as a possible response also to demographic pressures and technological/managerial advances, offers a vast and fertile field of new, empirical investigation that awaits a new generation of investigators.[4]

NOTES

1. I am indebted to Joyce Marcus for very useful critical comments on an earlier draft of this chapter.

2. Morony (1994:224) is under the misapprehension that a radiocarbon determination of AD 848 obtained from this great center comprising some 1.6 km^2 of ruins makes it early Islamic rather than overwhelmingly Sasanian in date. That radiocarbon sample, however, comes from a prominent ruined tower that is built of later, 30–30-cm bricks. The tower, which gives its name to the site, is obviously of later construction than all of the remainder, which is characterized by clearly Sasanian 34–34-cm bricks (Adams 1981:259).

3. Chemical analyses of 24 samples of glass fragments from these sites have recently been published (Brill 1999:1:82; 2:152–153).

4. Financial assistance from the Smithsonian Institution, the University of California at San Diego, and National Geographic Society Research Committee in the preparation of this chapter is gratefully acknowledged.

REFERENCES

Adams, R. McC.
1965 *Land behind Baghdad: A History of Settlement on the Diyala Plains.* University of Chicago Press, Chicago.
1981 *Heartland of Cities: Surveys of Ancient Settlement and Land Use on the Central Floodplain of the Euphrates.* University of Chicago Press, Chicago.

al-'Ali, Salih A.
1970–1971 Wasit region: A topographical study based on literary sources. Translated by Yasin al-Khalesi. *Sumer* 26:237–262 and 27:153–183.

al-Tabarī
1992 *The Revolt of the Zanj.* The History of al-Tabarī, vol. 36. Translated and annotated by David Waines. State University of New York Press, Albany.

Altheim, F., and R. Stiehl
1954 *Ein asiatischer Staat: Feudalismus unter den Sasaniden und ihren Nachbarm*, vol. 1. Limes-Verlag, Wiesbaden.

Brill, R. H.
1999 *Chemical Analyses of Early Glass.* 2 vols. Corning Museum of Glass, Corning, New York.

de Goeje, M. J.
1889 *Kitab al-Masalik wa'l-Mamalik, auctore Abu'l-Kasim Obaidallahibn Abadallah Ibn Khordadhbah.* Bibliotheca Geographorum Arabicorum, no. 6. Leiden.

Finster, B., and J. Schmidt
1976 *Sasanidische und fruehislamische ruinen in Iraq.* Baghdader Mitteilungen 8. DAI

Abteilung Baghdad, Baghdad.

Johnson, J. H.

1995 The two great powers in late antiquity: A comparison. In *The Byzantine and Early Islamic Near East III: States, Resources, and Armies*, edited by A. Cameron, 157–226. Papers of the Third Workshop on Late Antiquity and Early Islam. Darwin Press, Princeton, New Jersey.

Morony, M.

1984 *Iraq after the Muslim Conquest*. Princeton University Press, Princeton, New Jersey.

1994 Land use and settlement patterns in Late Sasanian and Early Islamic Iraq. In *The Byzantine and Early Islamic Near East, II: Land Use and Settlement Patterns*, edited by G. R. D. King and A. Cameron, 221–229. Darwin Press, Princeton, New Jersey.

Pournelle, J. R.

2003 Marshland of Cities: Deltaic Landscapes and the Evolution of Early Mesopotamian Civilization. PhD dissertation, Department of Anthropology, University of California, San Diego.

Rubin, Z.

1995 The reforms of Khusro Anushirwan. In *The Byzantine and Early Islamic Near East III: States, Resources, and Armies*, edited by A. Cameron, 227–296. Papers of the Third Workshop on Late Antiquity and Early Islam. Darwin Press, Princeton, New Jersey.

Safar, F.

1945 *Wasit: The Sixth Season's Excavations*. Imprimerie de l'Institut Français d'Archeologie Orientale, Cairo.

Walters, S. D.

1970 *Water for Larsa: An Old Babylonian Archive Dealing with Irrigation*. Yale University Press, New Haven, Connecticut.

Wheatley, P.

2001 *The Places Where Men Pray Together: Cities in Islamic Lands, Seventh through Tenth Centuries*. University of Chicago Press, Chicago.

Wilkinson, T. J.

2000 Regional approaches to Mesopotamian archaeology: The contribution of archaeological surveys. *Journal of Archaeological Research* 8:219–267.

Wittfogel, K. A.

1957 *Oriental Despotism: A Comparative Study of Total Power*. Yale University Press, New Haven, Connecticut.

From Highland to Desert

The Organization of Landscape and Irrigation in Southern Arabia

TONY WILKINSON

Major irrigation systems are frequently perceived to have developed in the context of state-level political control, under the jurisdiction of powerful kings, or as a result of large-scale private investments. For hydraulic works in general, Karl Wittfogel argued that the construction and operation of large hydraulic systems not only was associated with politically centralized, that is state-controlled, management but also ultimately caused the state to develop (see, e.g., Sidky 1996). Robert Hunt has noted, however, that large irrigation systems (up to 458,000 ha) can be successfully run by local communities (Hunt 1988; also Erickson, this volume). Because Wittfogel's writings are frequently misinterpreted and also couched in biased political language, both his writings and scholars who attempt to refute him tend to muddy rather than clarify our understanding of social processes and irrigation systems (Price 1994). Here I present evidence for a series of water-gathering technologies that have been in use in Yemen for millennia, and I discuss them within their hydrological and social contexts. Overall, it appears that the largest-scale systems were most vulnerable to catastrophic failures, whereas smaller-scale systems that were administered at the household level were less vulnerable to major failures. Where household-scale and large-scale systems coexisted, as in the highlands of Yemen, the household-scale systems acted as a safety net by allowing agriculture to continue after the failure of major dam-irrigation systems.

In the Middle East large-scale ancient irrigation works are most frequently associated with southern Mesopotamia and neighboring parts of western Iran. Although less well known, the irrigation systems of southwest Arabia provide an ideal laboratory for understanding the relationship between social organization and ancient irrigation systems, in part because many traditional irrigation practices have continued up to the present day. By examining a range of land-use systems in highland Yemen, I examine how local rules of organization, acting at a variety of scales, give the impression that there has been some "guiding hand" behind the organization of irrigation systems. However, more detailed examination suggests that all the systems under examination operated independently and according to local rules of behavior. In the highlands of Yemen, around Dhamar, during the late first millennium BC and first millennium AD, the patterning of land-use systems included both rain-fed terraced agriculture and irrigated fields, and each agricultural niche appears to have been well adapted to the local environment. Archaeological and paleoenvironmental studies demonstrate that in any one location such land-use systems varied through time, indicating that the local communities adopted very different land-use strategies at different times. We should not therefore assume that there was necessarily a one-to-one relationship between the environment and the system of land use that developed.

On the desert fringes (termed the Sayhad) the massive dam and irrigation complex at Marib was organized by kings whose rule held sway over relatively large areas, although irrigation was not necessarily administered under state control in the bureaucratic sense. In fact in southern Arabia the distinction between state and tribe is a slippery one. Thus, according to Paul Dresch (1990:254) north of the capital of San'a tribes are territorial entities of farmers, and it is therefore inappropriate to equate tribes with nomads. Similarly, in parts of the western highlands tribalism is weak, whereas in the terraced areas of the southern highlands ("Lower Yemen") the social geography is dominated by relations between landlord and peasant. What Dresch refers to as "statelike tendencies" can be recognized, however, for example in the case of the Zaydi, whose first imam was installed in the late ninth century AD. Owing to this flux between territorial tribes, pastoral nomadic tribes, areas of nontribal cultivators, and statelike traditions, one must be wary of imposing Westernizing terms without justification.

The operation of local rules of agricultural organization can result in a deceptive degree of apparent landscape organization. For example, in southwest Arabia at the present time rain-fed terraced fields predominate in the moist highlands, slightly larger runoff cultivation systems are typical of the semiarid parts of the highlands and their river basins, and flood runoff systems of maximum scale are the norm farther downstream around the desert margins. Such a continuum of land-use practices occurs in the Red Sea coastal plains of Saudi Arabia (Abdulfattah 1981), as well as in Yemen, and field evidence suggests that during the first millennia BC and AD similar patterns were in operation. At this time it appears that terraced farming was already a well-established technology in the moist highlands, whereas in the neighboring oases of the Sayhad that fringe the interior deserts, massive flood runoff systems were associated with large dams or other water deflection devices (Figures 3.1 and 3.2). In between the moist highlands and the flood systems of the arid fringe, localized runoff irrigation systems are not so well established. Nevertheless, these can be inferred from the evidence of the sustained use of valley-floor terraces, and such systems can now be traced back to c. 4000 BC or thereabouts (see below). Each of the above-mentioned land-use systems appears to have been well configured and adjusted to its local environment so as to give the impression that each local configuration was in equilibrium with the environment. In reality, however, each locality had a range of potential land uses through time. Only in the largest systems, such as around Sabaean (first millennium BC) sites such as Marib, was any large-scale coordination required.

THE SPATIAL PATTERNING OF IRRIGATION IN THE YEMEN HIGHLANDS

The often spectacular landscapes of Yemen are, in part, the result of human actions that have operated over many thousands of years to create land-use systems that could support the local population within a wide range of often marginal environments. These environments range from arid lowlands fringing some of the driest deserts in the Near East to relatively verdant highlands at elevations of more than 2,500 m above sea level. In many mountainous parts of the country, where the amount of cultivable land is rather limited, the construction of terraced fields has overcome this limitation by increasing the cultivated area so that it was sufficient to support the population. Nevertheless, the archaeological evidence suggests that when terraced farming was introduced (probably in the fourth or third millennium BC) population densities were not high, so one cannot argue that population pressure forced the development of terraced agriculture (Wilkinson 2003:191). In the driest locations fringing the desert sophisticated flood irrigation systems have been in operation for at least 4,000 years (Brunner 1999:38–39), and massive systems of irrigated fields, water-control structures, and in some cases accompanying texts in epigraphic Old South Arabian provide compelling evidence for the existence of a complex and hierarchical level of organization (Bowen 1958; Brunner 1983; Caton-Thompson and Gardner 1939; Hehmeyer and Schmidt 1991; Wagner 1993).

Rather than treat the entire range of irrigation systems throughout Yemen, this study focuses on one area dominated by a single large wadi system, that of the northeast-flowing Wadi Dhanah, together with the neighboring Tihama plain to the west (Figure 3.1). The Wadi Dhanah system, which drains a

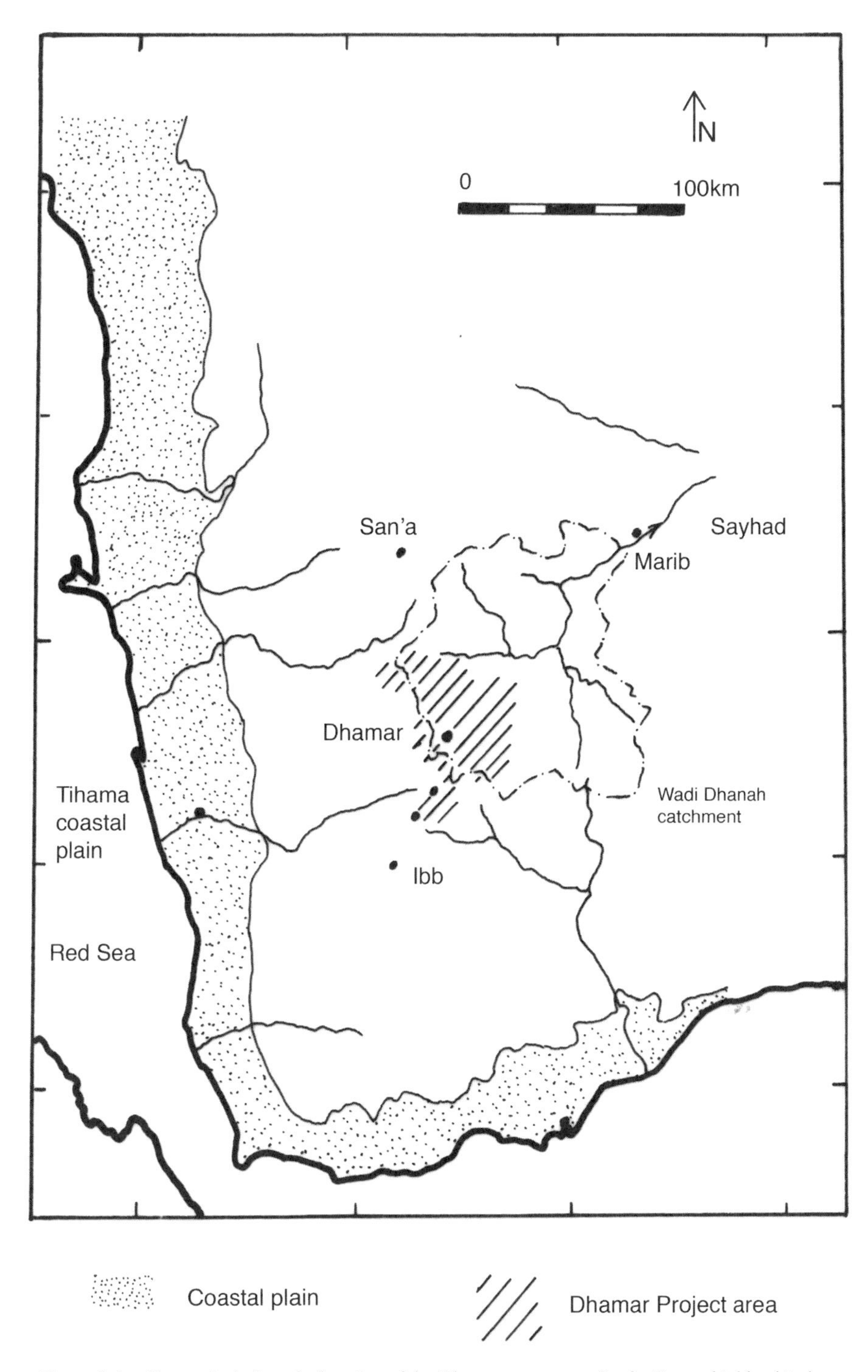

Figure 3.1. Yemen, including the location of the Dhamar survey area (in the Yemen highlands), the flood-irrigated areas of Marib, and the Tihama coastal plain.

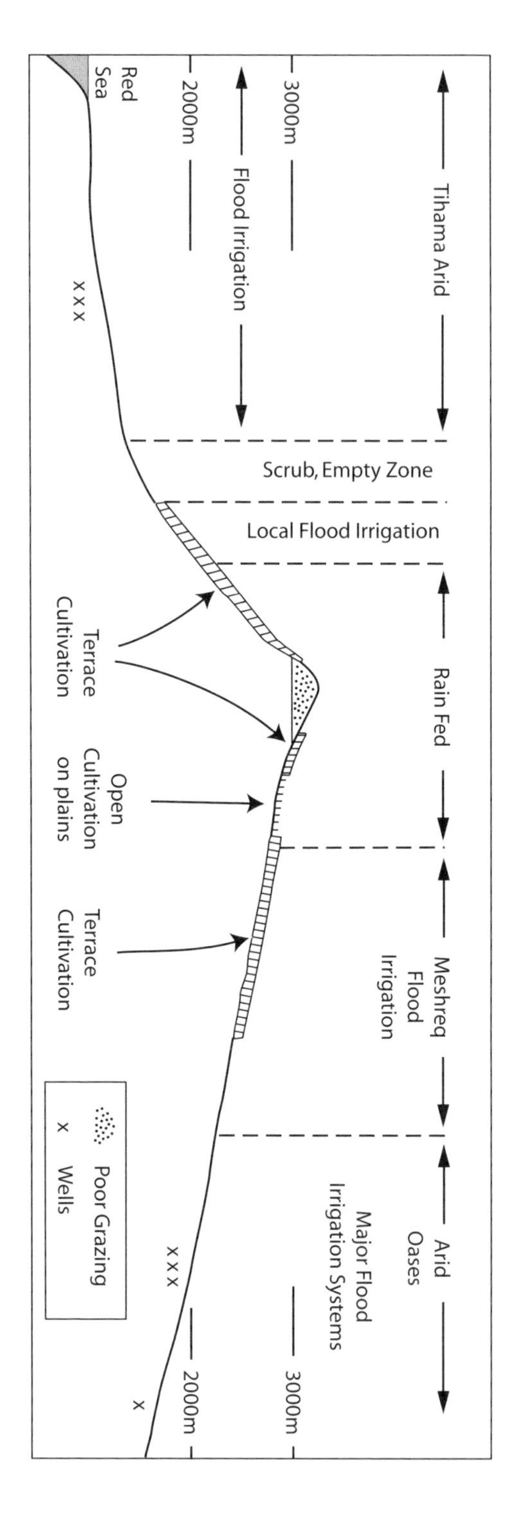

Figure 3.2. Cross section of the Arabian mountains showing land use and irrigated zones. Drawing by Peggy Sanders.

roughly 10,000-km^2 catchment of the Yemen highlands, is of particular relevance because it contains in its lower reaches the remains of the massive Marib Dam. In the more verdant highlands to the southwest, where irrigation was either unnecessary or was supplementary to rainfall, the Himyarite state developed in the first century BC. The relatively high rainfall (300–800 mm per annum) enabled rain-fed cultivation to support the bulk of the population, but in addition monumental dams were constructed. Although by no means as impressive as the dam at Marib, these structures provided supplementary water for modest-sized Himyarite towns, villages, and estates that dotted the highlands during the late first millennium BC and early first millennium AD.

If we take a representative swathe of land from the highlands to the neighboring desert fringes to either the northeast or the southwest, the following three systems of land use can be seen to have been in operation in various degrees during recent times (based on Beck 1990).

1. In the highlands and moister western districts (rainfall c. 500–900 mm per annum: Figure 3.3) extensive staircases of terraced fields predominate today (Figure 3.4). Terraced cultivation can be traced back archaeologically for some 4,000 to 5,000 years (Wilkinson 1999), and well-made monumental terraces in the Himyarite style (first century BC to seventh century AD) are common. It is therefore clear that terraced agriculture has formed a significant part of the local economy for at least the last 2,000 years or so. The terraced fields, with their associated runoff cultivation and small-scale floodwater farming systems, could have been constructed and maintained by the family or small cooperative groups of families, and crops were either sustained by the rain that fell on them or by a bonus of excess water from upslope or up-valley. Although some traditional systems of terraced fields to the north of San'a have been shown to have been constructed by importing soil from elsewhere (Varisco 1991:168), sections through eroded fields in the Dhamar area demonstrate that many highland fields developed incrementally as a result of the aggradation of sediments washed from upslope. This situation contrasts with parts of Palestine, where large systems of terraced fields are thought to have been constructed as a single unit by lineage groups or by families working cooperatively with other social groups (Hopkins 1985:269). In the Dhamar area, terraced fields could have been built and maintained by single or extended families. Such a system was flexible, resilient, and extendable as long as there was more land that could be made into terraces.

2. In drier parts of the highlands, where mean annual rainfall today falls in the range of 200 to 350 mm per annum, terraced cultivation is also practiced, but to achieve the appropriate level of soil moisture, it is necessary to gather additional moisture from neighboring slopes or up-valley (Figure 3.5) so that a designated runoff catchment area provides supplementary water to the field onto which the

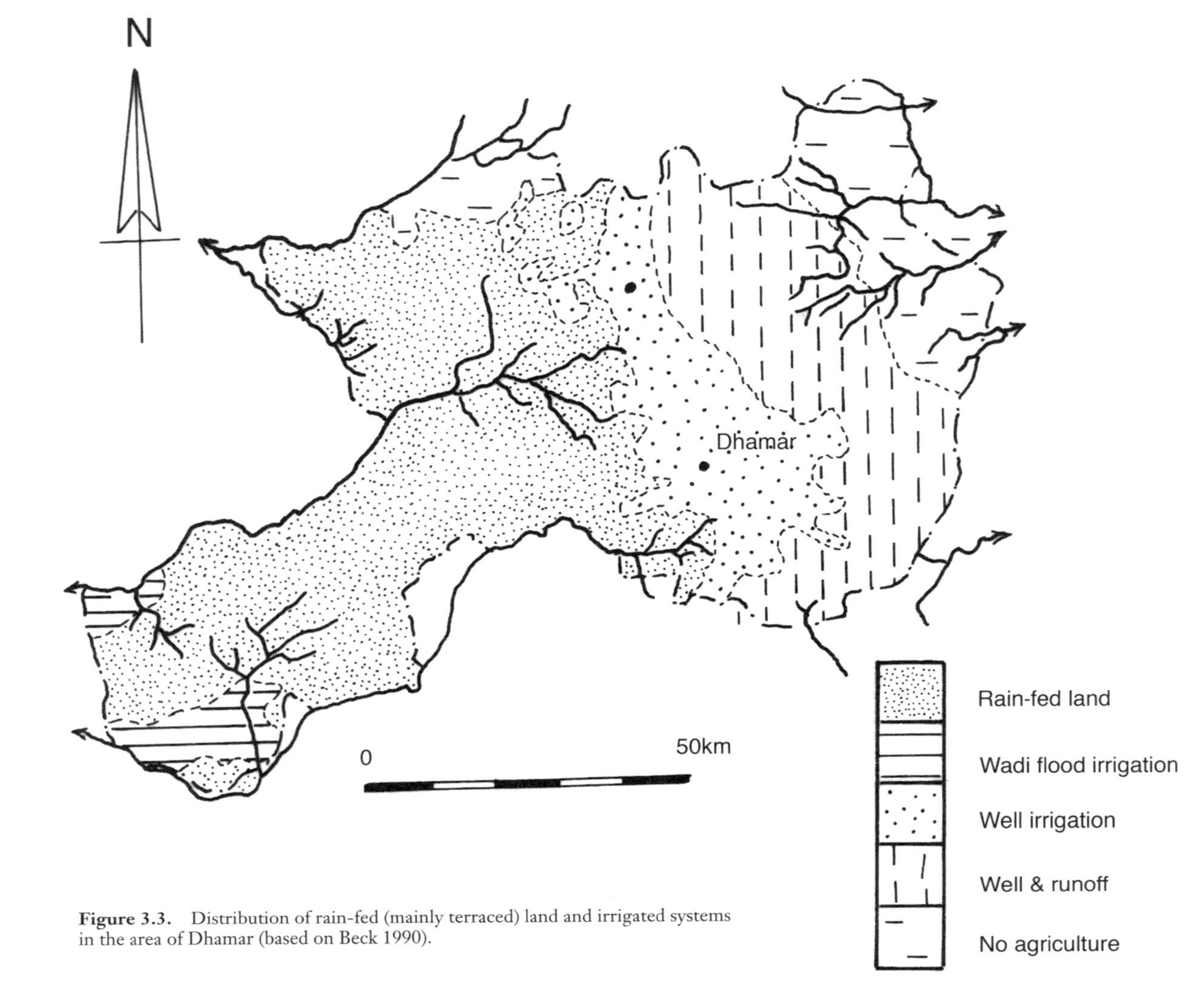

Figure 3.3. Distribution of rain-fed (mainly terraced) land and irrigated systems in the area of Dhamar (based on Beck 1990).

Figure 3.4. Landscape of terraced fields in highland Yemen. View taken at nearly 3,000 m above sea level to the southeast of Yarim. Photo by T. J. Wilkinson.

Figure 3.5. Modern runoff-irrigated land within valley floor near Nunah (Figure 3.10). Note that the upland (with a small archaeological site) forms the catchment area for the runoff, which is led to fields in the valley floor by means of the linear walls (*saqiyah*) indicated. Photo by T. J. Wilkinson.

water runs. Such systems resemble the Roman/Byzantine runoff agriculture on the fringes of the Negev desert (Evenari et al. 1982), and in highland Yemen the runoff systems that continue in use to this day (Eger 1987) provide an excellent model for these Negev systems.

3. In theory, in progressively drier areas the proportion of runoff to run-on land should increase until the area of runoff becomes so large that it operates at the scale of a wadi catchment. In the very driest areas floods from large wadi catchments are harnessed and deflected by dams or earthen bunds onto the fields downstream. An excellent example of such a traditional system in operation in southwest Saudi Arabia has been described by Kamal Abdulfattah (1981), who shows how wadi (or *sayl*) floods are captured by a temporary dam or barrage (*aqm*) and are fed onto the neighboring fields where the waters are progressively distributed among numerous farmers. The *aqm* of the Tihama coastal plain of Saudi Arabia must be maintained by all who hold lands that receive irrigation water, and these landholders contribute cash or labor for maintenance in proportion to the extent of their holdings or some equivalent measure (Serjeant 1988:145). Similar diversion structures in Yemen have been recorded by Serjeant (1988), as well as by a number of irrigation consultants (Makin 1977; Pratt 1977). In such traditional systems water is organized by the community and administered by a local administrator (*wakil*) so that upstream landholders get first priority to use the floodwaters, and downstream users must wait until those upstream have received their allocation. Such traditional systems have operated for centuries in the western mountains fringing the Tihama plain (Figure 3.1) and also functioned during the first millennium BC and AD at Marib (Breton 1998:13–18), around Shabwa (Gentelle 1991), in the Wadi Beyhan (Bowen 1958; Breton et al. 1998), Hadhramaut (Caton-Thompson and Gardner 1939), and Wadi Marha (Brunner 1997). In each case the system appears to have been managed by the local king, chief, or head of community.

In general, the ratio of runoff catchment to receiving field area is greater where runoff catchments are created outside the fields themselves (external catchments) than those within the field areas (internal catchments). This ratio also increases from wetter to drier areas (Table 3.1; Pacey and Cullis 1986:145) to attain its maximum where entire wadi basins gather water in the form of flood (*sayl*) irrigation. Needless to say there is a wide range of scatter in the ratios (Figure 3.6), and estimates for larger basins incorporate some oversimplifications, in part because the larger catchments include large areas of rangeland, and their wadis irrigated more than just a single cultivated enclave. Overall, it appears that as rainfall declines larger areas of landscape are required to service the progressively smaller areas of directly productive land (Figure 3.6).

Table 3.1. Ratios of catchment to cultivated areas for a range of environments.

Mean Annual Rainfall	Catchment: Field Ratios (Within-Field Catchments)	Catchment: Field Ratios (External Catchments)	Area Ratios: Flood Irrigation
300–600 mm			
Kenya, Baringo area	1:1, 2:1	5:1, 20:1	
Kenya, N. Turkana	—	10:1, 20:1	
Texas, USA	2:1	—	
Arizona, USA	3:1, 5:1	—	
150–300 mm			
Tunisia	2:1	10:1	
Arizona, USA	12:1	33:1	
Yemen, 'Amran		3:1, 20:1	
India, Rajastan	—	11:1, 15:1	
150 mm			
Israel, Negev			
a) runoff farms		17:1, 30:1	
b) microcatchments	10:1		
c) contour strips	4:1, 20:1		
Yemen, Marib			100:1, 125:1
Yemen, Rima'			29:1, 36:1, 41:1

Note: Table adapted from Pacey and Cullis (1986:Table 6.4), with added data from Yemen. For sources see Pacey and Cullis (1986).
[a]Eger (1987).
[b]Based on Brunner and Haefner (1986) and Hehmeyer (1989).
[c]Makin (1977).

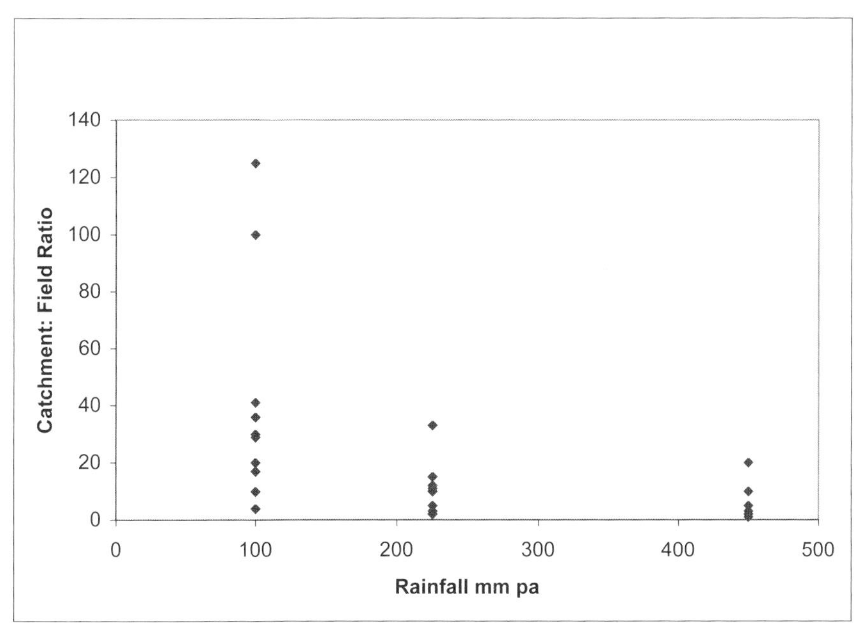

Figure 3.6. Scatter diagram showing the increased ratio of runoff catchment to fields in drier regions of the tropical zone (based on Table 3.1).

THE SCALE OF IRRIGATION SYSTEMS AND THEIR ADMINISTRATION

System scale is one way of estimating the significance of irrigation systems because large systems have the capacity to support larger populations. Moreover, it is frequently thought that larger systems require a state-level political control to administer them, or some form of powerful authority is required to build and coordinate such systems. Here irrigation systems in the Yemen highlands and adjacent areas are discussed in terms of increasing scale from the small-scale systems in the highlands to the much larger systems of the desert fringe.

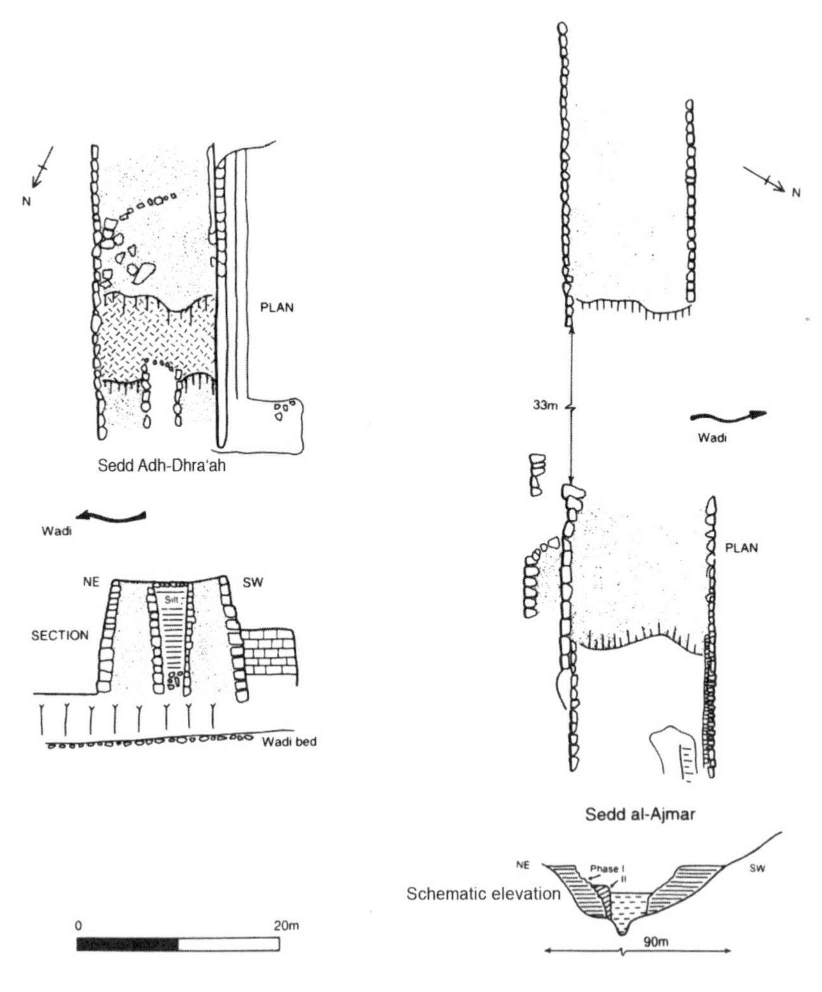

Figure 3.7. Plans and sections of two major Himyarite dams: Sedd al-Ajmar (to right) and Sedd Adh-Dhra'ah (to left) (from Gibson and Wilkinson 1995).

Dam Irrigation in the Highlands around Dhamar

During the first millennium BC agricultural production in the Yemen highlands around Dhamar was based on terraced agriculture supplemented by irrigation water supplied by means of dams. These monumental stone-built dams collected and distributed flood runoff water from valley catchments and conducted the intercepted water via a system of distribution canals to irrigate fields in valleys downstream or in the adjacent lowlands (the *Qa*). To what degree these dams also operated as storage dams is debated, but their height and strength indicate that they were used for at least short-term storage of water.

The dam of Sedd al-Ajmar, some 90 m long, 13–15 m wide, and c. 14 m high, together with a second dam, intercepted water from two wadis and directed it downstream to irrigate some 60 to 90 ha of valley-floor land (Figures 3.7 and 3.8; Gibson and Wilkinson 1995:172–174). Despite its magnificence, this dam was incapable of withstanding peak floods, and now it remains (like many high, strong dams in the highlands) severed by the floods (Figure 3.9). The surrounding area was well settled during the Himyarite period, with a major hilltop town (Masna'at Mariyah) overlooking the irrigated lands from the west, together with several smaller sites in the adjacent lowlands. Present-day agricultural production relies on terraced fields scattered over both hill slopes and valley floors, and the meager evidence concerning the history of terraced cultivation in the neighborhood suggests that such cultivation systems can be

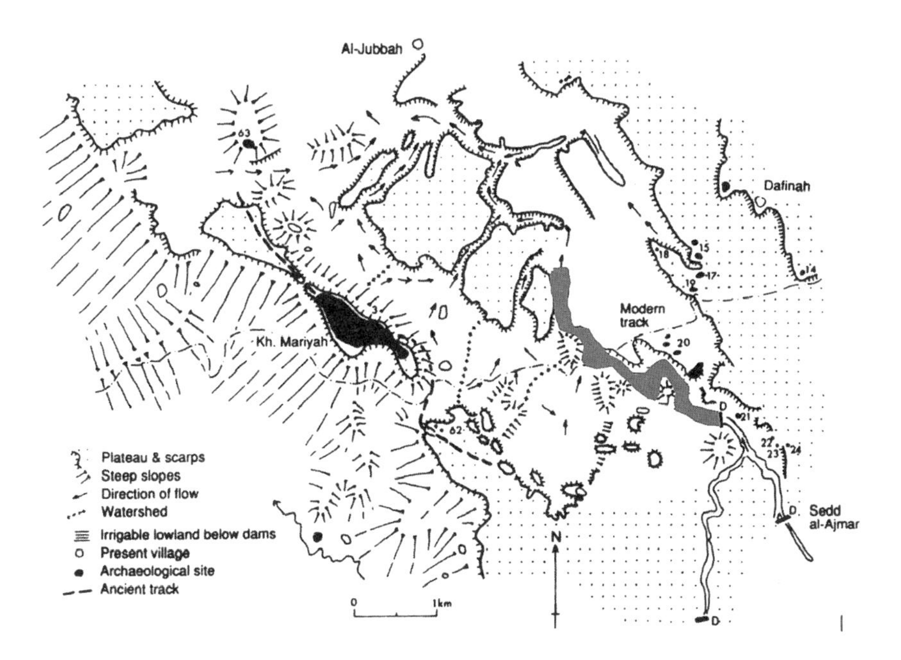

Figure 3.8. Area of Sedd Ajmar and its associated irrigated area (shaded). Note the massive bulk of the site of Masna'at Mariyah looming over the west-facing escarpment, but with oversight over the dams (D) and irrigated terrain (from Gibson and Wilkinson 1995).

Figure 3.9. Sedd al-Ajmar severed by high-energy floods. Photo by T. J. Wilkinson.

traced back to at least the Iron Age if not earlier (Gibson and Wilkinson 1995: Site DS 63). In summary this Himyarite agricultural system appears to have relied primarily on terraced cultivation, with supplementary production coming from irrigated agriculture supplied by the two dams at al-Ajmar.

Elsewhere in the highlands numerous monumental dams constructed in Himyarite-style masonry demonstrate that such dam-irrigated modules were frequent throughout the core of the Himyarite kingdom (Gibson and Wilkinson 1995). Dams include those at Sedd Adh-Dhra'ah, to the east of Dhamar (Gibson and Wilkinson 1995), as well as a complex of smaller dams near the Himyarite capital of Zafar (Barceló et al. 2000). Somewhat counterintuitively, many of these irrigated modules were not constructed in the dry or marginal fringes of the highlands but were in operation in the moister highland fringes, where rainfall today exceeds 400 mm per annum. Other modules occurred in areas that were marginal for rain-fed cultivation where it was necessary to irrigate to make up for deficiencies in the annual rainfall. One such enclave, comprising two hydraulic dams, at least one monumental terrace wall (*harrah*), and a pocket of alluvial soils around the Himyarite town

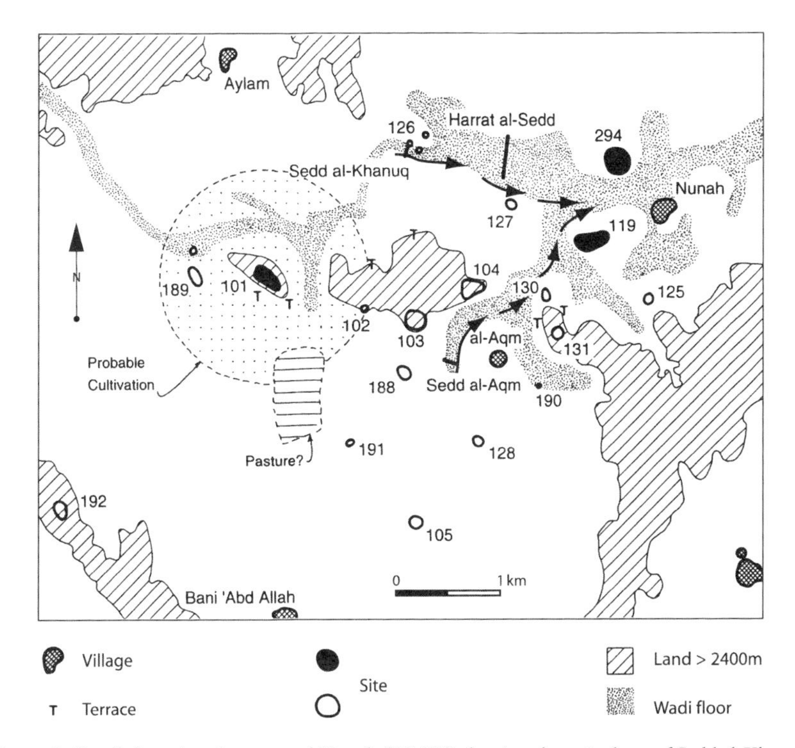

Figure 3.10. Irrigated enclave around Nunah (DS 294) showing the twin dams of Sedd al-Khanuq and Sedd al-Aqm, as well as the direction of flow of irrigation waters toward the Himyarite site of Nunah. Note the earlier sites, which included Iron Age DS 119 and 3rd and 2nd millennium BC Hammat al-Qa (DS 101), both perched on hilltops. Drawing by Peggy Sanders.

of Nunah (DS Site 294; see Figure 3.10), gives an idea of the scale and layout of an irrigation module in the vicinity of the limit of rain-fed cultivation.

Overall, such dam-irrigated modules provided water for significant areas of lowland but were incapable of supplying sufficient irrigation water to feed the entire population of the region in question. In many cases irrigation could have supplied either small towns (such as Nunah on Figure 3.10) or estates, the latter being distinguished by occasional formally constructed fields defined by monumental terrace walls. Although the precise date of such systems, as well as the details of their administration, remains uncertain, rock-cut commemorative inscriptions in epigraphic Old South Arabian (typically in the Himyarite language), when present, suggest that most irrigation structures were in use during the last centuries BC and first centuries AD. In one case an inscription on a rock-cut sluice by a relict dam hints that the dam may have been administered from the Himyarite capital of Zafar, some 45 km away (Norbet Nebes, personal communication). Extreme floods destroyed many of the larger dams, but when such failures occurred, that part of agricultural production vested in terraced agriculture appears to have continued. Although vulnerable to high floods, therefore, this system had a built-in safety mechanism in the form of the large area of terraced cultivation.

Within the wettest part of the central highlands near the Himyarite capital of Zafar, monumental walls or dams (*asdad*, singular *sedd*) illustrate the considerable long-term continuity of these landscapes. Many features have names that are also referred to in the writings of al-Hamdani, who compiled a series of geographical texts in the tenth century AD. Moreover, al-Hamdani observed when such place-names referred to features that were considered (in the tenth century) to be of Himyarite construction. By combining landscape survey, traditional toponyms, and historical sources, it is therefore possible to trace key landscape features back some 1,500 to 2,000 years.

Such qualitative research has been supplemented by detailed field surveys of water systems, termed "hydraulic archaeology" by Barceló and colleagues (2000). Because the flood (*sayl*) water still follows seemingly ancient pathways between demonstrably ancient field systems, the ancient systems of water distribution can be reconstructed, thereby allowing the pattern of Himyarite irrigation to be inferred with some degree of sensitivity. In general, an articulated and linked system of small dams and monumental cross-valley walls forms the basic skeleton of the present landscape around Zafar. This highly controlled landscape consists of valley-floor irrigated fields, which are partly terraced as a result of the accumulation of soil, overlooked by staircases of rain-fed fields on adjacent hill slopes. Not only do the hill-slope terraces trap the eroding soil and hold it in place, but the valley-floor dams (*asdad*) and related cross-valley walls have, over the years, intercepted large quantities of soil so that both valley floor and adjacent slopes are now highly controlled and geomorphologically stable, that is as long as the system of soil-retaining walls is maintained.

Barceló and colleagues, although recognizing the possibility that there was a general design behind each hydraulic system, demonstrate that these valley-floor irrigated landscapes can be subdivided into a remarkable system of interlinked irrigation modules based around now-silted-up dams and basins. Three coherent systems defined in one area near Zafar cover a total irrigated area of some 25.8 ha, and in general each irrigated module, with its constituent dam or dams, appears to have covered some 10 ha. Such an area is only slightly larger than is necessary to feed a single Yemeni family under modern conditions of *sayl* agriculture. For example, in the Wadi Rima (Tihama plain [Figure 3.1]) during the 1970s irrigated holdings for single families were mainly in the range of 2.5 to 7.5 ha (Makin 1977:60). This implies that a 10-ha module could have been the holding of a single large extended family (*bayt*) or perhaps a minimal lineage (a *dayma*; Varisco 1983:379) consisting of several closely related families sharing a house or living in close economic cooperation.

The Yemen highlands around and to the northeast of Dhamar are climatically marginal for rain-fed cropping (mainly wheat and barley in winter and sorghum in summer). Nevertheless, it is noteworthy that many Himyarite dams are situated not in these drier areas but where rainfall exceeds 400 mm. Similarly, during the 1970s irrigation appears to have been more common in the moister western mountains (rainfall of 350–500 mm) than either the montane plains (250–350 mm) or the driest eastern volcanics (rainfall around 250 mm) (see Table 3.2). Investment in irrigation may have been greater in these moister areas because they provide a greater surplus for investment in the necessary pumps or irrigation technology. Moreover, families who practiced irrigation in addition to rain-fed cultivation were not only somewhat larger than those that practiced rain-fed cultivation alone but also had larger holdings of land and were more prosperous in terms of holdings of animals and hired labor. In such cases irrigation appears to have been employed as a means of intensifying cropping rather than simply to offset a deficit in rainfall or soil moisture.

The location of many Himyarite irrigation systems within the moister parts of the highland suggests that they too were intended for the intensification of production rather than simply to offset soil moisture deficits. Overall, the positive feedback mechanisms set in motion by highland intensification (both terraced agriculture and irrigation) arguably promoted growth in agricultural production and population. Precisely how such feedback mechanisms developed is not entirely clear, but in the Himyarite period the capital, Zafar, was established in the wetter highlands to the south of Dhamar, where it formed a locus of irrigated production centered, according to al-Hamdani, around some 80 dams (Gibson and Wilkinson 1995). In other words the initial choice of a verdant area for the capital and its estates resulted in royal investments in agriculture in the immediate vicinity, which in turn resulted in further increases in production. These two examples of irrigation for

Table 3.2. Percentage of irrigated land in the Yemen highlands around Dhamar.

Zone	Region	Rainfall mm	% irrigated 1977	% rain-fed 1977	% rangeland	% villages
A	Western mountains	350–500	3.5	11.0	84.5	1.0
B	Western montane plains	300–350	2.5	55.0	41.0	1.5
C	Eastern montane plains	250–300	1.5	28.0	69.0	1.5
D	Eastern volcanics	250	0	42.5	56.5	1.0

Source: Kessler 1987:28–29 (data for 1977).

agricultural intensification underscore the fact that irrigation can develop along divergent pathways, one that seeks to address moisture deficits in arid areas and an alternative that grows from the processes of positive feedback that develop out of growth in investment.

Flood Runoff Systems in the Highlands

In the semiarid margins of the Yemen highlands rainfall is frequently insufficient for successful crops, but by gathering excess runoff from neighboring parcels of land, surplus water can be led onto the field plot to provide a significant bonus of soil moisture. In regions that lie between the moister highlands and the desert, runoff agriculture therefore forms the main means of addressing shortfalls in rainfall. As a result of this practice cultivation can extend into areas that receive as little as 150 mm per year (Eger 1987). In theory, runoff agriculture is intermediate in scale between the rain-fed systems of the moist highlands and the flood irrigation systems of the desert fringe, although in reality there is a considerable interdigitation of different types of systems, depending on the vagaries of local topography and farming practice.

The most thorough study of traditional runoff systems in southwest Arabia (around ʻAmran in northern Yemen: Eger 1987) distinguishes between two classes of runoff irrigation: *sayl* irrigation represents the use of wadi floodwater for irrigation, whereas *sawaqi* agriculture refers to runoff agriculture in which overland flow is guided from adjacent parcels of land to irrigate neighboring fields (Eger 1987:47). Helmut Eger (1987:94–151) subdivides the *sawaqi* systems of the ʻAmran area into nine different rainwater harvesting systems, which themselves can be grouped into five runoff agricultural systems (Figures 3.11 and 3.12). In general, the larger the ratio of catchment area to field area, the lower is the hydraulic efficiency of the irrigation system. This is because with larger ratios there is a tendency for more water to be lost to infiltration, transmission losses, and other factors. A noteworthy feature of these landscapes is the way that each runoff landscape is specific to a certain topographical/geological land unit. For example, the limestone hill runoff systems (ratios 1:1 to 3:1) form the most efficient systems, whereas those of the valley floor that utilize runoff from neighboring fields or from valley-floor soils (ratios of 1:1 to 20:1) result in considerable inefficiencies in the collection and transmission of runoff water to the recipient fields.

Few archaeological examples of runoff agriculture have been analyzed in the highlands, but the technology can be traced back some 6,000 years according to preliminary records of a valley-floor terraced wall (or check dam) encapsulated within a valley fill near Sedd Adh-Dhraʻa. This feature, which is buried by more than 5 m of overlying sediment, has been dated by radiocarbon to c. 4000 BC

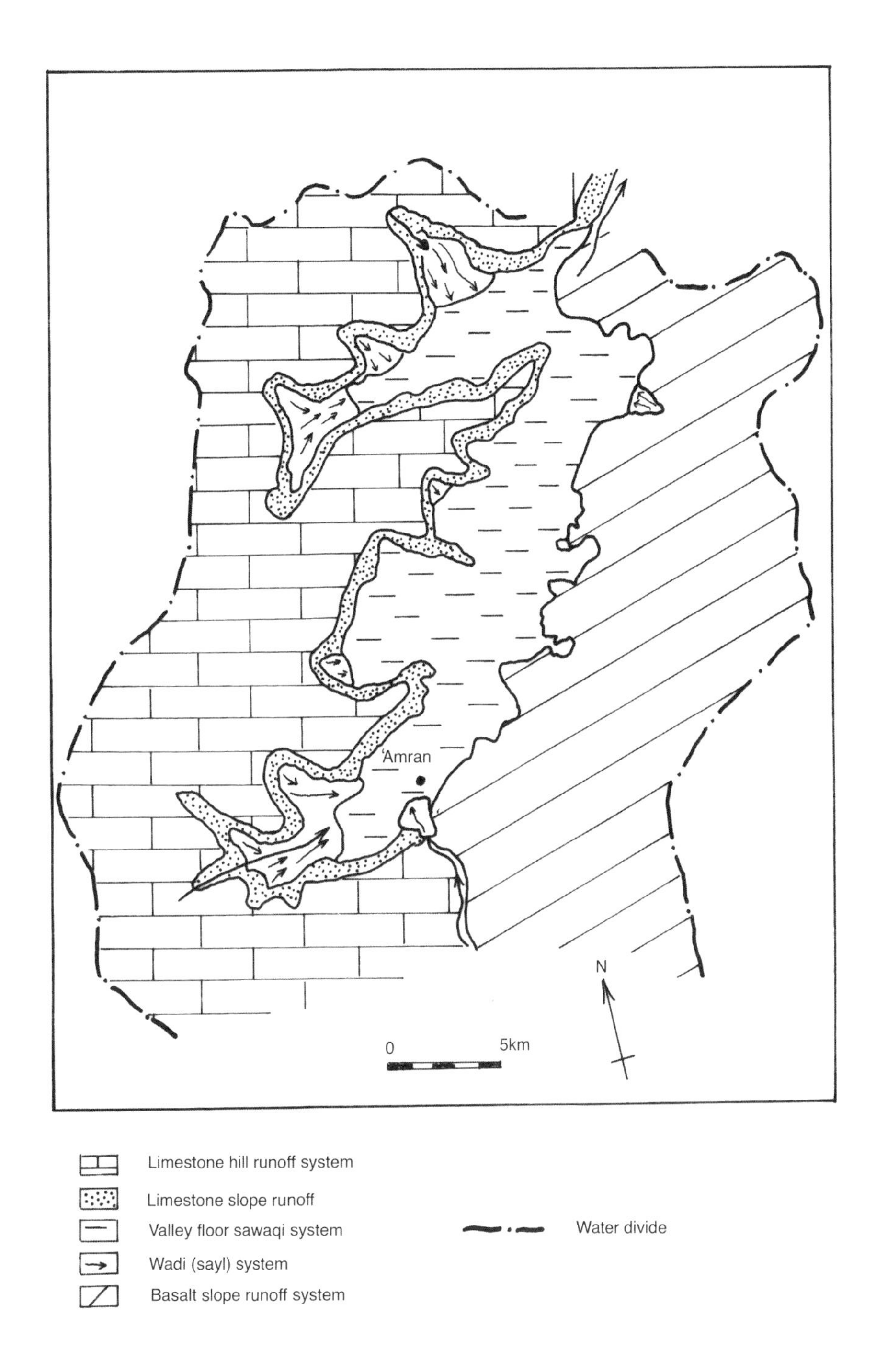

Figure 3.11. Distribution of runoff zones in the area of 'Amran (after Eger 1987:Appendix 3).

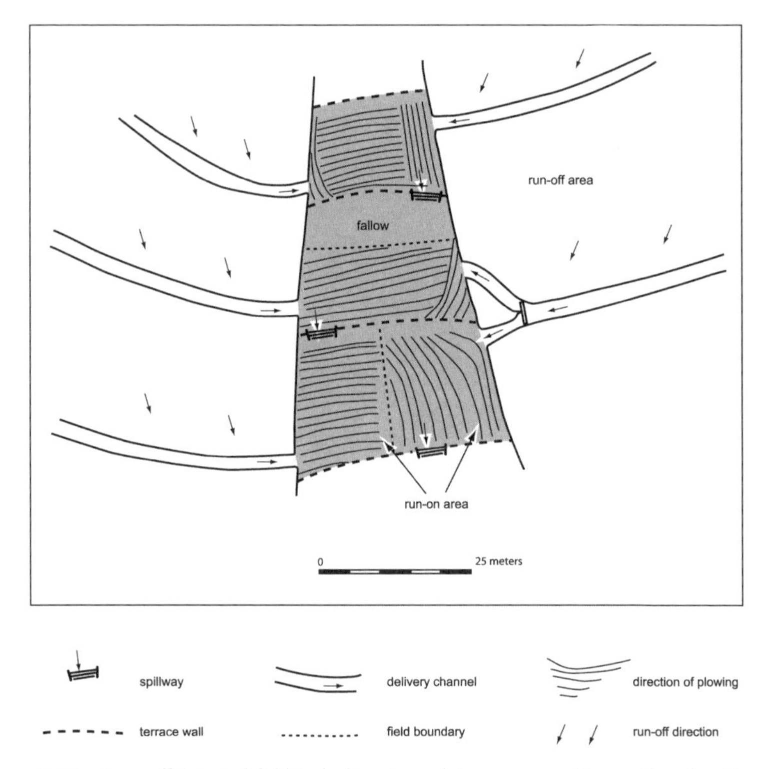

Figure 3.12. A runoff-irrigated field in the limestone plateau area near ʿAmran (based on Eger 1987: Figure 22). Drawing by Peggy Sanders.

(Wilkinson 1999; Wilkinson and Edens 1999), but because of its deep burial it cannot be placed in any spatial context (Figure 3.13).

Less well dated, but providing a more coherent spatial context, are traces of long-continued runoff agriculture recorded near the village of al-Hajir, northeast of Dhamar (profile G of Gibson and Wilkinson 1995:Figures 1 and 2). Here, a small valley containing a valley fill associated with terrace walls contains the sedimentary accumulations of a long sequence of runoff farming. The c. 4.5-m accumulation of pale brown sandy silts overlies a humic paleosol horizon, broadly dated to the range 6000–2500 BC (Wilkinson 1997). The proximity of runoff silts and their associated terraces to a Bronze Age site of third or second millennium BC date provides further associational dating (Gibson and Wilkinson 1995:Figure 1:DS 45). Although these relationships are hardly sufficient to provide an unambiguous date for the silts, their considerable depth does suggest a long period of accumulation. Taken together with the Sedd Adh-Dhraʿa terrace or check dam, runoff agriculture appears to have a long history, one of sufficient duration to demonstrate that such systems fill in the continuum of water-gathering technologies between rain-fed cultivation and

Figure 3.13. Buried terraced wall or check dam (above and left of figure) at Sedd adh-Dhra'ah and dated to c. 4000 BC. Photo by T. J. Wilkinson.

flood irrigation. This early history for flood-runoff agriculture is supported by the discovery of third to fourth millennium BC features in the arid eastern parts of Yemen (McCorriston et al. 2002:77–79)[1] and canals of second millennium BC date in the southern coastal plain around Ma 'layba (Vogt et al. 2002).

The Marib System

During the first millennium BC and AD the basin of the Wadi Dhanah housed the main settlements of the frankincense-trading states of the kingdom of Saba', including its capital of Marib (Sabaean Maryab; Mariaba). The ca. 100-ha town was surrounded by between 8,000 and 9,600 ha of irrigated fields (Brunner and Haefner 1986; Hehmeyer 1989), which received their irrigation water from the main Wadi Dhanah via the 680-m-long Marib Dam (Figure 3.14). The end-member of the sequence under discussion, namely the flood runoff irrigation system, was not only the biggest system physically, but if the textual records can be relied on, it was also the largest in terms of its organizational scale. Inscriptions in Sabaean demonstrate that organization or coordination at the scale of the community or larger groups operated since the early first millennium BC. A major phase of dam construction appears to have been commissioned in the sixth century BC by two rulers of Saba': Sumh 'aliy Yanuf and his son (Schmidt 1988:59, 61). Conventionally, the organization of these irrigation systems could be seen as "inextricably part of the theocratic administration and governmental apparatus" of the Sabaean kingdom (Schmidt

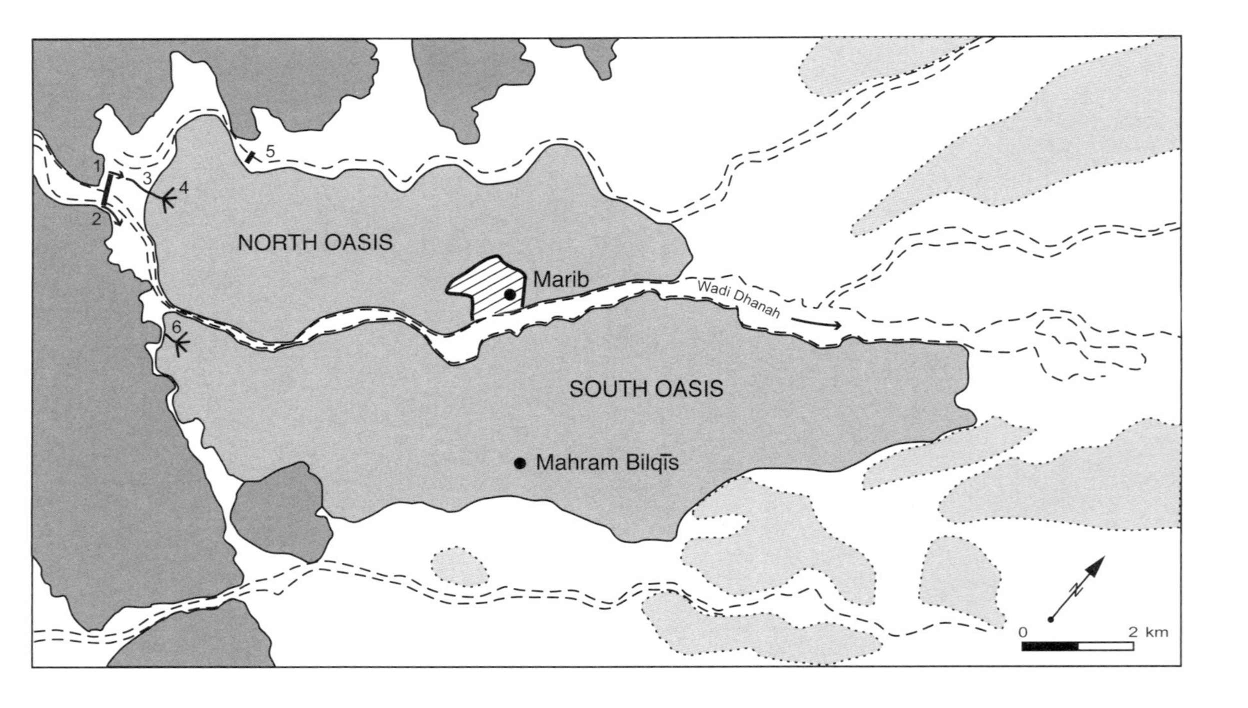

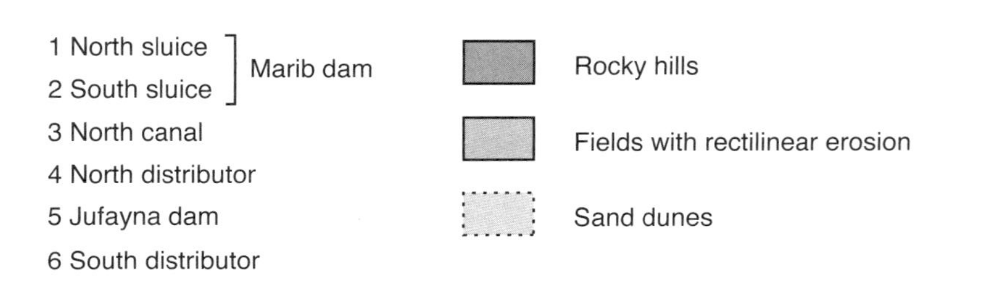

Figure 3.14. The irrigated areas of the Marib oases, which surround the Sabaean town of Marib (hatched). Redrawn by Peggy Sanders from data in Brunner (1983:Figure 24); Hehmeyer and Schmidt (1991); and Wagner (1993).

1988:61). Schmidt emphasizes this point by stating that the entire water-supply system belonged to the state and that every type of work was understood to be an act of worship dedicated to the gods.

Despite such confident assertions, the terminology employed in the inscriptions makes the role of the state in such administrations less clear. During the earlier first millennium BC (perhaps from around the eighth century), the ruler was referred to as a *mukarrib*, which can be translated as "maker of covenants" (Audouin et al. 1988:74), "federator," "person who rallies the people together," or "unifier" (Breton 1998:33, based on Robin 1996). The *mukarrib* did not therefore possess absolute authority but rather played a consultative role in dealing with "councils" or "assemblies" (Breton 1998:33). Like the traditional tribal shaikh he was presumably expected to have the charisma and authority to be able to mobilize, when necessary, large numbers of tribesmen to perform major feats of collective action. Neither is a state-oriented model accepted by Andrey Korotayev (1994), who asserts that Saba' was neither a state nor a country but rather a collectivity of people (a *sha'b*) who at various times were led by either a *mukarrib* or a *malik* (literally a king). These were the political leaders of the Sayhad communities, who at the beginning of the first millennium BC created a large commonwealth of such *sha'bs* that occupied most of South Arabia (Korotayev 1994:69). Such communities, which can be inferred as loosely resembling the traditional Arabian tribe, forged various alliances, and it was probably the mobilization of such allied communities that was the key to the success of large-scale public works such as dam construction or reconstruction.

Although massive in scale, the Marib Dam system may simply be viewed as the end-member of a system of flood/spate irrigation that has operated in parts of Yemen back to at least the third millennium BC (Brunner 1999). In such systems the discharge of major wadis receiving their flow from extensive mountain catchments are intercepted by water diversion structures of soil, rock, and brushwood placed either part way across the flooding river or across the entire channel width (*aqm*). Such structures raise and divert the water into major earthen canals that guide the flow to upper fields, which, when they have received their due of water, release the flow to the next field downstream (Pratt 1977:58; Serjeant 1988). Although such systems can operate under state control, more frequently they are operated on behalf of landed proprietors, the basic administration being in the hands of administrators variously referred to as *wakīl al 'aqm*, *shaykh al-sharij*, and other related terms (Serjeant 1988:148). The landowners as well as the water master and his assistants receive a cut of the harvest according to a predetermined set of local rules. For example, the water master (*wakīl al 'aqm*) traditionally would receive between 1:10 and 1:20 of any grain crop; the watchman (*qayyim*) around 1:20 (or less), plus various additions assessed in sorghum; the brushwood supplier and ox master (*khadim*) receives two sheaves of sorghum, various sweepings and so on (Serjeant 1988:149). Overall, this political economy forms some sort of collective action rather than resulting from the state or the authoritarian rule of a king. Typically, those in charge need to be able to negotiate truces in existing feuds so that crucial repairs can be conducted, and in certain cases a major figure (such as a *mansab* or lord) will declare a corvée for essential construction work to take place (Serjeant 1988:151). Needless to say the ability to mobilize large numbers of tribesmen requires a combination of brute strength, charisma, political acumen, and personal authority, as well as the existence of self-interest and some financial or personal incentives on the part of the labor that is mobilized. On the one hand, self-interest among the irrigation community can bring cohesion to often disparate tribal groups; on the other hand, problems stemming from upstream allocation of irrigation water at the expense of downstream users can result in competition or conflict (Varisco 1996:241–242).

Although massive dams required greater organizational clout than traditional *sayl* irrigation management, there is little evidence to show that these were necessarily state-organized systems. Rather, as Jean-François Breton has alluded, systems of irrigation such as that at Marib required a collective approach to management (1998:95), albeit one that was subsumed under the overall responsibility of the sovereign (*mukarrib* in the earlier first millennium BC, *malik* in the late first millennium BC).

Because the Marib Dam and similar structures straddled large wadi systems that generated very great variations in flow, they were vulnerable to occasional devastating floods. If a breach occurred in the main dam the entire crop supply system could be threatened; consequently, the management and mobilization of large quantities of labor were crucial to both the construction and the repair of these systems. Evidence of the scale of these enterprises can be gleaned from textual references to the labor that was employed to rectify any damage incurred by peak floods. According to a text dated to c. AD 449, efforts to strengthen the dam entailed some 14,000 camels, 200,000 sheep (presumably to feed the workers employed), 217,000 pounds of flour, and 630 camel loads of beverages (Brunner and Haefner 1986:83). In addition, some 2,460 camels were sacrificed to the gods. In the next year (AD 450), when the great dam collapsed, the king succeeded in rounding up some 20,000 workers, with the result that the dam was rebuilt (Brunner and Haefner 1986:83; Glaser 1897). The role of the king in effectively mobilizing the tribes was crucial to such rebuilding programs, as is well illustrated by the response in AD 542 when parts of the dam were again damaged, and in this case neighboring tribes were mobilized to complete the repair in some 58 days (Brunner and Haefner 1986:84).

HIGHLAND AND SAYHAD SYSTEMS IN CONTEXT

During the last 5,000 years the desert fringe (Sayhad) and highlands of Yemen experienced rather different settlement trajectories. In the Sayhad there is relatively little evidence for settlement during the Bronze Age, although this is in part because many such sites may be buried below the deep accumulations of irrigated silt (Orchard 1982). Settlement and irrigation then grew rapidly during the first millennium BC to attain a high point in terms of both population and the scale of irrigated area, during the main phase of the incense trade.

In the highlands, however, there is good evidence for the existence of complex societies from the late fourth millennium BC (Edens and Wilkinson 1998; Edens et al. 2000; Ekstrom and Edens 2003). Small towns located mainly on hilltops became fairly common during the third and second millennia BC, after which settlement increased somewhat during the first millennium BC and first millennium AD. Unlike in the Sayhad the peak settlement during the first millennia BC and AD did not significantly exceed the settlement of earlier periods. In terms of landscape development, Bronze Age cultivation of near-site fields, perhaps associated with the exploitation of basin-floor pastoral resources, appears to have given way during the Iron Age to the gradual encroachment of cultivation onto basin floors. By the Himyarite period it is possible to infer, in many areas, the existence of a "dual economy" consisting of a traditional long-term sector of rain-fed terraced cultivation supplemented by dam-irrigated systems. In some localities the major irrigation systems appear ultimately to have been unsustainable because many dams eventually succumbed to damage by high floods. As a result, high-investment dam-irrigated systems were frequently replaced by simple flood-cultivation systems (*sayl* cultivation) or by a reemphasis on terraced agriculture. The cultivation of rain-fed crops on terraced fields, as well as some of the smaller-scale irrigation systems, appears, however, to have continued in use to the present day.

In the narrow upland valleys and neighboring intermontane basins of the Yemen highlands it is likely that demographic growth took place in the context of a positive feedback cycle in which dual strategies of moderate to high levels of intensification (terracing and at a later stage irrigation) could have been associated with larger families and possibly the accumulation of greater wealth. Such growth fueled further population growth and increased the potential for investment in monumental architecture (dams, irrigation systems, and monumental terrace walls). Although seemingly administered by the Himyarite "state," these relatively modest-sized irrigation systems could have been built and managed by the mobilization of local communities in the manner described by Robert Fernea for irrigation systems in southern Iraq (Fernea 1970). There was no compelling need for a formalized bureaucratic state apparatus in the modern sense of the word, and such systems may have been more akin to hydroagriculture according to the terminology of Wittfogel (Price 1994).

The flood agriculture of the Sayhad was not only large in scale, but it was also brittle and inflexible. Furthermore, it required more management and administration than the dual systems of the highlands. Collapse of the polities of the desert fringe during the early first millennium AD was partly caused by specific dam bursts but also probably resulted from a complex downward spiral resulting from shifts and eventually a decline in incense trade, lack of investment in irrigation systems, siltation in the field areas, decline in local population (and therefore labor), sociopolitical disruptions, and major flood events. Because the incremental accumulation of silts necessitated the gradual raising of the level of the dam and its sluices, it was necessary for dam builders to construct higher structures that, in turn, became more vulnerable to major floods. Fortunately, the more flexible agricultural strategies of the highlands enabled most communities to absorb such catastrophes, because the reserve of terraced agriculture provided a convenient insurance that enabled production to continue if the dam-irrigation systems failed. As a result, it seems that in post-Sabaean times it was possible for the main political and population centers to persist and continue in the highlands.

Although many Yemeni irrigation systems might be perceived as having been planned and maintained by a large state-level organization, most show evidence of some form of collective action in which the appropriate labor force was mobilized by political leaders in consultation with other notables, more in keeping with a tribal mode of social organization. Nevertheless, state control has occurred at times, as during the Rasulid period of the thirteenth and fourteenth centuries AD (Varisco 1983:368). The administration of Yemeni flood irrigation is therefore comparable to other irrigation communities, where the major duty of the irrigation leader was the mobilization of labor and materials for the annual repair of diversion structures (Coward 1980:205). In the case of Sabaean irrigation systems, administration may simply have been affected by scaled-up versions of the traditional irrigation communities such as those described by Robert Serjeant (1988). In such traditional irrigation communities a water master settled most disputes and, with his assistants, effected most day-to-day administration such as the mobilization of labor for the maintenance of the water diversion structures (*aqm*). Of course, whether such functionaries exerted the political clout to mobilize the huge quantities of labor required to repair systems such as that at Marib is unclear, and presumably such tasks would have been the function of the *mukarrib* or *malik*.

Overall, in southwest Arabia one can discern a continuum of land-use systems from rain-fed terraced agriculture maintained largely by family groups in the highlands, through small-scale runoff systems in the semiarid margins (again managed by families or small lineages) to the maximum-scale systems operated and managed by the leaders of large communities in excess of 10,000 people. This situation is sketched in Figure 3.15, although it must be appreciated that this does not simply represent a continuum from wet to dry areas because, as noted above, the moister parts of the highlands also exhibited dam- or *sayl*-irrigated systems that were used to intensify agricultural production. In terms of the social organization of production, the degree of household control over the process of production increases to the left in Figure 3.15, that is as the ratio of catchment to field area approaches unity.

The data from Yemen demonstrate the significance of household levels of production but also show that flood- or dam-irrigated systems clearly formed part of a political economy. For example, in the recent past, various levels of functionary performed allocated tasks for a share of the product. Such "officials," who may have been part-time, appear to have formed part of a kin-based social system rather than being embedded within a state level system. Even in the rain-fed highlands the rulers and/or elite families played a role in the redistribution of food products, specifically during times of food stress. Hence, when a drought struck Yemen in 1723–1724, and large numbers of people in the region of San'a were dying from hunger, the imam Al-Mutawakkil Qāsim bin al-Husayn redistributed grain from his storehouses (Dresch 1993:208).

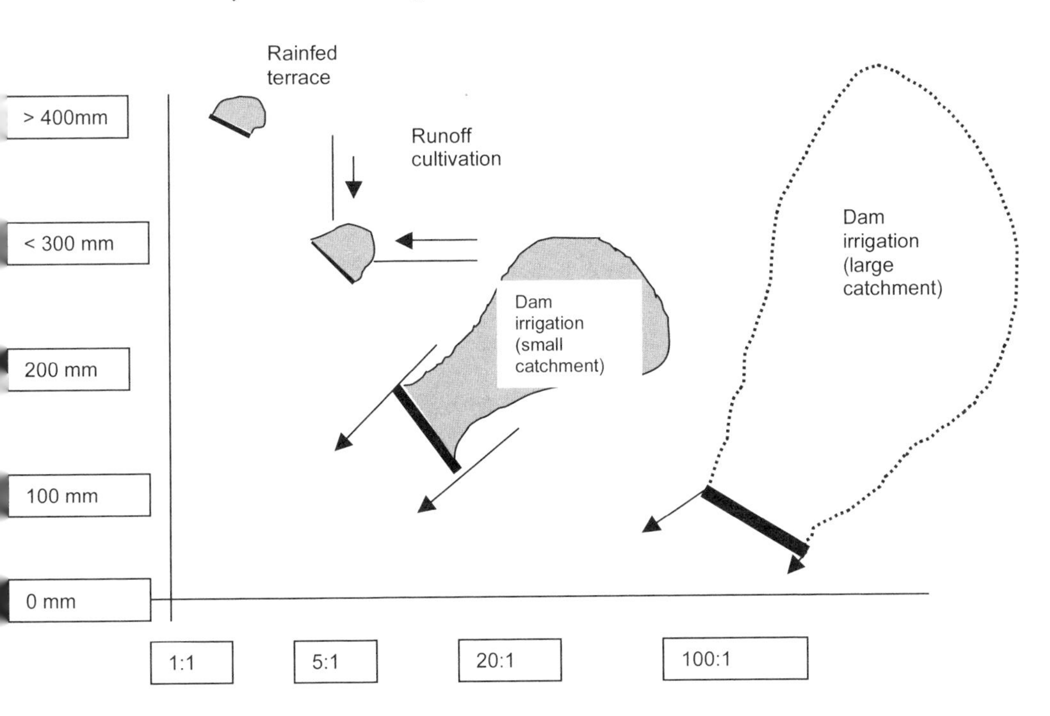

Figure 3.15. Sketch showing the theoretical trend as runoff catchment: field ratios increase toward progressively drier areas. Arguably, the scale of social groups involved in water management also increases to the right.

Fundamental to the resilience of the highland systems of rain-fed terraced agriculture (as well as some of the smaller irrigation systems) is that they were built and maintained by families or small social groups. Consequently, if they suffered a devastating flood, the fields could be reconstructed with a relatively small labor force. If, however, the larger dams in the highlands collapsed, reconstruction required the mobilization of larger labor forces. In such cases total demographic collapse was prevented because the systems of terraced fields ensured continuity of agricultural production, thereby supplying a safety net. From what we see today, such mobilization of labor for dam reconstruction did not always materialize; instead, downstream of broken dams (such as Sedd Adh-Dhra'a), simple forms of flood (*sayl*) irrigation, organized and administered by the local village population, have replaced the dam-irrigated systems and have continued in use to the present day.

In contrast, the essential but brittle systems of the Sayhad were vulnerable to the powerful and capricious wadi floods and became steadily more vulnerable through time, until they were eventually unsustainable, even with the collective action of the communities, whether it was a state or tribal confederation. The consequence of dam failures was therefore total system collapse because there was no supplementary production system to act as insurance.

Overall, what is fascinating about the agricultural systems of ancient Yemen is not their relationship with the state or other polities; that dilemma is still to be fully resolved. Rather, it is the way that scale of social organization and labor mobilization maps onto the ecology of water-gathering systems. As a result of this relationship each system is vulnerable in different ways, with the Sayhad dam-irrigation system being subject to catastrophic failure, whereas the dual system of the highlands has built-in positive feedback mechanisms and resiliencies that contributed to the overall sustainability of the system.

Acknowledgments

I am extremely grateful to the following organizations that contributed funding for fieldwork: the National Geographic Society, the American Institute for Yemeni Studies, the National Science Foundation, and the Oriental Institute. The following members of the General Organization of Antiquities and Museums (Yemen) contributed immensely to the fieldwork: Ali Sanabani, Ahmed Haidari, and Khaled al-Ansi. And gratitude must especially go to officials of the General Organization of Antiquities and Museums, San'a, especially Dr. Yusuf Abdullah, and to Ahmed Shemsan, for help and advice before and during the season. Special thanks go to my colleague Dr. Christopher Edens, who participated in fieldwork and provided superb administrative services as director of the American Institute of Yemeni Studies in San'a; also to Professor

McGuire Gibson of the Oriental Institute, who participated in the early seasons of fieldwork. I also wish to thank team members Krista Lewis, Eleanor Barbanes Wilkinson, Jerry Lyon, Alan McCune, Charley French, and Caroline Davies for their help during fieldwork; thanks also go to Peggy Sanders for providing some of the illustrations. I am particularly grateful to Charles Stanish, who organized the workshop at which an earlier rather vaguely defined version of this chapter was presented. The essay also benefited from the comments of two anonymous reviewers, to whom I am very grateful.

NOTE

1. This dating is now supported by radiocarbon dates of the fourth millennium BC (M. Harrower, personal communication 2004).

REFERENCES

Abdulfattah, K.

1981 *Mountain Farmer and Fellah in 'Asir, Southwest Saudi Arabia: The Conditions of Agriculture in a Traditional Society*. Erlanger Geographische Arbeiten, Erlangen.

Audouin, R., J.-F. Breton, and C. Robin

1988 Towns and temples: The emergence of South Arabian civilization. In *Yemen: 3000 Years of Art and Civilization in Arabia Felix*, edited by W. Daum, 55–62. Pinguin-Verlag, Innsbruck.

Barceló, M., H. Kirchner, and J. Torró

2000 Going around Zafar (Yemen): The Banu Ru 'ayn field survey: Hydraulic archaeology and peasant work. *Proceedings of the Seminar for Arabian Studies* 30:27–39.

Beck, R. (editor)

1990 *Environmental Profile: Dhamar Governorate*. DHV Consultants, Amersfoort, Netherlands.

Bowen, R. L.

1958 Irrigation in ancient Qataban. In *Archaeological Discoveries in South Arabia*, edited by R. L. Bowen and F. P. Albright, 43–131. Johns Hopkins University Press, Baltimore.

Breton, J.-F.

1998 *Arabia Felix: From the Time of the Queen of Sheba*. University of Notre Dame Press, South Bend, Indiana.

Breton, J.-F., J.-C. Arramond, B. Cique-Delhuille, and P. Gentelle

1998 *Une vallée aride du Yémen antique. Le wadi Bayhân*. Ministère des Affaires Etrangères. Éditions Recherche sur les Civilisations, Paris.

Brunner, U.

1983 *Die Erforschung der antiken Oase von Marib mit Hilfe geomorlogischer Untersuchungsmethoden*. Deutsches Archäologisches Institut, San'a. Verlag Phillip von Zabern, Mainz am Rhein.

1997 Geography and human settlements in ancient south Arabia. *Arabian Archaeology and Epigraphy* 8:190–202.

1999 *Jemen: Vom Weirauch zum Erdöl*. Böhlau Verlag, Wien.

Brunner, U., and H. Haefner

1986 The successful floodwater farming system of the Sabaens, Yemen Arab Republic. *Applied Geography* 6:77–86.

Caton-Thompson, G., and E. W. Gardner

1939 Climate, irrigation, and early man in the Hadramaut. *Geographical Journal* 93:18–38.

Coward, E. W.

1980 Management themes in community irrigation systems. In *Irrigation and Agricultural Development in Asia: Perspectives from the Social Sciences*, edited by E. W. Coward, 203–218. Cornell University Press, Ithaca, New York.

Dresch, P.

1990 Imams and tribes: The writing and acting of history in Upper Yemen. In *Tribes and State Formation in the Middle East*, edited by P. S. Khoury and J. Kostiner, 252–287. University of California Press, Berkeley.

1993 *Tribes, Governments, and History in Yemen*. Clarendon Press, Oxford.

Edens, C., and T. J. Wilkinson

1998 Southwest Arabia during the Holocene: Recent archaeological developments. *Journal of World Prehistory* 12(1):55–119.

Edens, C., T. J. Wilkinson, and G. Barratt

2000 Hammat al-Qa: An early town in southern Arabia. *Antiquity* 74:854–862.

Eger, H.

1987 *Runoff Agriculture: A Case Study about the Yemeni Highlands*. Ludwig Reichert, Wiesbaden.

Ekstrom, H., and C. M. Edens

2003 Prehistoric agriculture in highland Yemen: New results from Dhamar. *Yemen Update: Bulletin of the American Institute for Yemeni Studies* 45:23–35.

Evenari, M., L. Shanan, and N. Tadmor

1982 *The Negev. The Challenge of a Desert*. Harvard University Press, Cambridge, Massachusetts.

Fernea, R. A.

1970 *Shaykh and Effendi: Changing Patterns of Authority among the El Shabana of Southern Iraq*. Harvard University Press, Cambridge, Massachusetts.

Gentelle, P.

1991 Les irrigations antique à Shabwa. *Syria* 68:5–54.

Gibson, M., and T. J. Wilkinson
1995 The Dhamar plain, Yemen: A preliminary study of the archaeological landscape. *Proceedings of the Seminar for Arabian Studies* 25:159–183.

Glaser, E.
1897 Zwei inschriften über den Dammbruch von Marib. *Mitteilungen der Vorderasiatischen Gesellschaft* 6:405–410.

Hehmeyer, I.
1989 Irrigation farming in the ancient oasis of Marib. *Proceedings of the Seminar for Arabian Studies* 19:33–44.

Hehmeyer, I., and J. Schmidt
1991 *Antike Technologie—Die Sabäische Wasserwirtschaft von Marib, 1.* Deutsches Archäologisches Institut, San'a. Verlag Phillip von Zabern, Mainz am Rhein.

Hopkins, D. C.
1985 *The Highlands of Canaan: Agricultural Life in the Early Iron Age.* Almond Press/ASOR, Decatur, Georgia.

Hunt, R. C.
1988 Size and the structure of authority in canal irrigation systems. *Journal of Anthropological Archaeology* 44(4):335–355.

Korotayev, A.
1994 The Sabaean community in the political structure of the Middle Sabaean cultural area. *Orientalia* 63(2):68–83.

Makin, M. J. (editor)
1977 *Montane Plains and Wadi Rima Project: A Land and Water Resources Survey.* Land Resources Division. Project Report 16. Ministry of Overseas Development, Surbiton.

McCorriston, J., E. A. Oches, D. E. Walter, and K. L. Cole
2002 Holocene paleoecology and prehistory in highland southern Arabia. *Paléorient* 28(1):61–88.

Orchard, J.
1982 Finding the ancient sites in southern Yemen. *Journal of Near Eastern Studies* 41:1–21.

Pacey, A., and A. Cullis
1986 *Rainwater Harvesting. The Collection of Rainfall and Runoff in Rural Areas.* Intermediate Technology Publications, London.

Pratt, D. J. (editor)
1977 *Montane Plains and Wadi Rima Project: A Land and Water Resources Survey.* Land Resources Division. Project Report 17. Ministry of Overseas Development, Surbiton.

Price, D. H.
1994 Wittfogel's neglected hydraulic/hydroagricultural distinction. *Journal of Anthropological Research* 50:187–204.

Robin, C.

1996 Sheba dans les inscriptions d'Arabie du Sud. *Dictionnaire de la Bible*, supplement. Letouzey, Paris.

Schmidt, J.

1988 The Sabaean irrigation economy of Marib. In *Yemen: 3000 Years of Art and Civilization in Arabia Felix*, edited by Werner Daum, 55–62. Pinguin-Verlag, Innsbruck.

Serjeant, R. B.

1988 Observations on irrigation in southwest Arabia. *Proceedings of the Seminar for Arabian Studies* 21:145–153.

Sidky, H.

1996 *Irrigation and State Formation in Hunza: The Anthropology of a Hydraulic Kingdom*. University Press of America, New York.

Varisco, D. M.

1983 *Sayl* and *ghayl*: The ecology of water allocation in Yemen. *Human Ecology* 11:365–383.

1991 The future of terrace farming in Yemen: A development dilemma. *Agriculture and Human Values* 8:166–172.

1996 Water sources and traditional irrigation in Yemen. *New Arabian Studies* 3:238–257.

Vogt, B., V. Buffa, and U. Brunner

2002 Ma ʻlayba and the Bronze Age irrigation in coastal Yemen. *Archäologische Berichte aus dem Yemen* 9:15–26. Verlag Philipp von Zabern, Mainz.

Wagner, W.

1993 *Bodenkundliche Untersuchungen in der Oase Marib*. Deutsches Archäologisches Institut, Sanʻa. Verlag Philipp von Zabern, Mainz am Rhein.

Wilkinson, T. J.

1997 Holocene environments of the high plateau, Yemen: Recent geoarchaeological investigations. *Geoarchaeology* 12(8):833–864.

1999 Settlement, soil erosion, and terraced agriculture in highland Yemen: A preliminary statement. *Proceedings of the Seminar for Arabian Studies* 29:183–191.

2003 *Archaeological Landscapes of the Near East*. University of Arizona Press, Tucson.

Wilkinson, T. J., and C. Edens

1999 Surveys and excavations in the central highlands of Yemen: Results of the Dhamar Survey Project, 1996 and 1998. *Arabian Archaeology and Epigraphy* 10:1–33.

Asia

Intensification as Situated Process

Landscape History and Collapse

Kathleen Morrison

> So far as I have observed in *Mysore*, ground, once brought into cultivation for rice, is universally considered as arrived at the highest possible degree of improvement; and all attempts to render it more productive by a succession of crops, or by fallow, would be looked on as proofs of insanity.
>
> (Francis Buchanan 1807)

In nearly all discussions of the intensification of production, and especially of agricultural intensification, questions of cause have dominated. Even in my own previous work (Morrison 1994a, 1995, in press), which focused on process, and in particular on what I referred to as the *courses* (or trajectories) of change, critical discussion (e.g., Stone and Downum 1999) concentrated largely on the very aspect of intensification—cause—that I was *not* discussing. Such critiques missed my primary point that intensification needs to be studied as a process that is itself complex and variable, consisting of numerous coexisting strategies, processual complexity that seems to mirror the organizational complexity of real productive systems. Perhaps because of the initial dependence of the field on the Boserup model, which presented a very limited range of intensification strategies, changes in production are too often viewed in terms of only a few kinds of strategies (fallow reduction, timing of cropping cycles, and construction of irrigation facilities, for example), even though these important changes represent just a few of the many possibilities open to agriculturalists.[1] Among the strategies of intensification little-discussed in the literature are those that perhaps seem unrelated to agriculture (e.g., wage labor, craft production; cf.

Morrison 2001a) or that apparently violate the normative Boserupian sequence of change. As an example of the latter, my own work in southern India revealed sequences involving the later addition of land-extensive forms of production to an existing repertoire involving more land-intensive forms, as well as the adoption of a range of intensification strategies other than fallow reduction (Morrison 1995).

In this chapter, although I do consider some of the causes of long-term agricultural change in southern India, I want to focus instead on how we might use the strengths of the archaeological record to examine both the *courses* and *consequences* of changes in production.[2] Historically, the intensification of both agricultural and nonagricultural production has resulted in radical transformations of human life. High-yield intensive agriculture, for example, undergirds urban life; both the earliest cities and the contemporary megalopolis would be unthinkable without the long history of agrarian change that led to intensive, commercial farming. Agricultural intensification, too, is necessary for the elaboration of occupational specialization. Long histories of intensification have, in some parts of the world, radically transformed vegetation, landforms, soils, and even climate. Past consequences become future possibilities, so that understanding intensification at one point in time often requires analysis of longer-term change. In spite of this, archaeologists rarely discuss intensification in terms of its consequences or potential consequences, something, as I discuss below, our data are uniquely suited to address.

THE CONTEXTS OF PROCESS: STUDYING LANDSCAPE HISTORIES

We would do well to think of intensification as an example of a *situated process*. Here I stress both *process*, as noted above, as well as *context*. Because I have already written extensively about intensification as an internally complex process (Morrison 1994a, 2001a, in press a), I will leave this aspect of the equation aside for now. In terms of context I emphasize here three related but distinct issues: (1) the embeddedness of intensification strategies in total socioeconomic systems; (2) the historically accretional nature of landscapes on and within which agriculture takes place; and (3) the specific (but temporally shifting and never fully shared) cultural contexts in which decisions about production are made.

The first of these three issues relates to overall contexts of production—intensification, like other changes in the organization of production, must be understood in terms of total economies.[3] Here I stress what is no doubt noncontroversial to most anthropologists: that changes in agricultural production are always linked to and cannot be understood apart from the larger productive picture. The latter might include, for example, collecting and hunting, wage labor, military service, or craft production, and its overall contours will depend critically on issues such as the availability of markets, transportation routes, structures of tribute or taxation, and many other factors. Such concerns enter into decisions about agricultural production no less than more commonly discussed factors such as the availability of labor and irrigation or the nature of soils, climate, and rainfall. Analysis of the intensification of production entails analysis of production, conceived broadly.

The second way in which context enters into the process of intensification is the location of human action within long-term landscape histories (Morrison in press b). I use the term *landscape* to mean a particular spatial and temporal field, both material and ideal, a kind of *constructed context* constituted both by the structure and dynamic of the natural environment and by human action. Landscapes are culturally constructed—both materially and conceptually *built*—variable in their perceived structures and meanings, even at a single point in time, phenomenal and ideational variability that is balanced by a less-shifting materiality. The landscapes I consider here are largely anthropogenic, highly restructured by generations of human action. By *landscape history* I mean the specific historical trajectories of landscapes (although I am certainly not making a claim of total historical uniqueness or lack of general process). Study of specific landscape histories show us that history matters—producers do not begin anew each time they make a decision but are instead constrained by features of the landscape itself, that accretional product of the past.[4] Without an understanding of landscape history, individual instances of intensification (or its reverse) might seem inexplicable, as I discuss below.

The third way in which the process of intensification must be viewed as situated is its operation within specific cultural contexts. I have suggested above that landscapes be seen as historically produced outcomes created by both human and nonhuman actions; obviously landscapes are always viewed by people and acted on by people with specific cultural understandings. At the same time, however, these cultural understandings are not nearly sufficient to understand historical process—humans cannot make the world any way they choose. Still, the motivations that drive production, while significantly constrained by both basic human needs and conditions of plant and animal growth, are to a large extent, flexible, as I show in the following brief case study. Agricultural production, in particular, is significantly structured by cultural notions about food—what is appropriate to eat, for whom, and in what contexts. Food restrictions and desires may engender technological change, labor increases, or forms of production that a strict cost-benefit logic could never predict. In the case discussed here, one critical factor in long-term agricultural histories has been the high cultural value placed on rice, a crop whose production entails both high costs and high rewards.

In this chapter I summarize some aspects of long-term agricultural change in southern India, a history marked by a sustained but irregular trajectory of agricultural intensification, punctuated by brief periods of "disintensification" or "collapse." In the context of a more general discussion of the long-term productive history of this region between the Neolithic (3000 BCE) and the Late Middle Period (1600 CE), I focus on one episode of "collapse," the rapid transformation in land use, settlement, and political organization following the abandonment of the city of Vijayanagara in the late 1500s. This case illustrates the three points made above about the contextualization of the process of intensification. First, changes in agricultural strategies cannot be separated from larger political-economic structures; following the fall of the city, both the loss of urban markets and changes in taxation radically restructured the organization of production. Second, contingent trajectories of landscape history matter—the consequences of change in one period resonate on the landscape, shaping the possibilities for later change. In this case we can see that the course of intensification did not operate in reverse; instead, understanding this specific instance of collapse requires attention to long-term landscape history. Finally, both the shorter history of the Vijayanagara urban collapse and the longer history of agrarian change in dry interior South India make clear the importance of specific cultural regimes of value in shaping production choices. Here I focus specifically on the great importance attached to rice, but it would be possible to identify many other such examples.

LANDSCAPE HISTORIES IN SOUTH INDIA: LONGER-TERM CONTEXTS OF AGRICULTURE

The low and annually variable rainfall of the dry interior of southern India is temporally concentrated, falling primarily during the southwest monsoon of the late summer/early autumn. Unlike the tropical vegetation of India's southwestern coast, the semiarid interior contains only minor concentrations of thorny deciduous forests amid extensive field systems and savannah-like expanses of grass and shrubs. Today, the high granitic hills of northern Karnataka, in the heart of the rain shadow that covers all of the peninsular interior, consists largely of exposed rock with minimal soil development. Agricultural fields are concentrated in (but not restricted to) the colluvial valleys between outcropping ridges and along the limited alluvium of the Tungabhadra River. In spite of its modest natural endowments, however, this region has managed to support large, permanent villages and towns for at least the last 3,000 years and has been an agricultural landscape for the last 6,000.

The earliest agriculture in southern India dates only to about 3000 BCE, the beginning of a period known as the Southern Neolithic (ca. 3000–1000 BCE), when extensive dry farming of millets and pulses supplemented an economy based largely on cattle pastoralism (Korisettar et al. 2001; Possehl and Rissman 1992). The Southern Neolithic was marked by a considerable amount of experimentation with new crops (and animals). According to Fuller (2003) the staple crops of the earlier part (2800–2200 BCE) of the Southern Neolithic—*Vigna radiata* (Mung bean), *Macrotyloma uniflorum* (Horse gram), *Bracharia ramosa* (Browntop millet), *Setaria verticillata* (Bristly foxtail millet), and *Dioscorea* sp. (yams)—were supplemented during the middle years of the period (2200–1800 BCE) by *Lablab purpureus* (Hyacinth bean), *Pennisetum glaucum* (Pearl millet), *Triticum* sp. (wheat), *Hordeum vulgare* (barley), and perhaps *Lathryus sativus* (Grass pea). By the latter Neolithic *Cajanus cajan* (Pigeon pea), *Gossypium* (cotton), *Linum usitatissimum* (flax/linseed), and perhaps *Eleusine coracana* (Finger millet, an African import) were added to the mix. During the Southern Neolithic, too, sheep, goats, pigs, and domestic cattle were kept for the first time, animals that would be part of faunal assemblages of all subsequent time periods.

The earlier part of the Southern Neolithic appears to have been a period of relative aridity in this already-dry region, making dry farming even more problematic than it is at present. Because Neolithic people were certainly at least occasionally mobile, moving out from more tethered locations for grazing and raw material procurement as well as visits to other settlements and ashmound locations, and because few of the settlements in this area have been excavated or studied in any detail, it is impossible at this point to know the scale of regional population, though clearly it was larger than that of preceding periods.[5] Many excavated sites, including Tekkalakota (Nagaraja Rao and Malhotra 1965) and Brahmagiri (Wheeler 1947), have very deep and substantial Neolithic deposits, so if these represent periodic reoccupations by mobile people, reoccupation must have been very regular and may have involved large groups of people over a substantial period of time. Not all Neolithic sites are large and deeply stratified, however, and we have as yet only a very superficial understanding of how these smaller locations fit into the overall pattern of land use and the use of space on a regional and supraregional scale.

We may, however, be more certain that the Neolithic was the beginning of the long trajectory of agricultural history in this area, and, although we know little at present about what went before it, we do know that the Neolithic represents both rupture and continuity—continuity in that some excavated sites show stratified deposits of the Paleolithic and/or Mesolithic under Neolithic levels, rupture in that the herding and farming carried out in this period stand in stark contrast to the economic strategies of those preceding it and in the way in which places were marked and modified on the landscape.

Although the landscape of the peninsular interior, especially during the increased aridity of the Southern Neolithic, but today as well, was inhospitable to rice (*Oryza sativa*), rice is nevertheless found in some botanical assemblages of the Southern Neolithic. Never amounting to more than a trace amount in this region, rice was better known elsewhere on the subcontinent, where its pedigree is also longer. By sometime in the Iron Age (1000–300 BCE) and certainly by the Early Historic period (300 BCE–500 CE), rice cultivation became well-established in northern Karnataka, irrigated by a range of strategies from annual river inundation to wells and runoff-fed reservoirs. The locally intensive production of rice, as well as more extensive farming of pulses, millets and other grains, may have, in some important respects, underwritten the trend toward settlement aggregation and allowed the establishment, during the Iron Age, of very large towns or small cities such as Kadebakele, which sprawled more than 40 hectares across a high granitic outcrop.[6] Iron Age and Early Historic settlements in our study area cluster densely along the banks of the Tungabhadra, the only perennial water source in the region and the only river reliable enough to support the production of water-demanding crops such as rice. How did this grain, so poorly suited to local environmental realities, come to dominate not only the historical record of southern India but also local imaginations?[7] Why was so much effort expended in creating the conditions under which rice could grow?

Although we are not yet in a position to answer these questions, our ongoing research on the Iron Age and Early Historic in this region may eventually help us better understand the long-term courses of change. Already, we can suggest that the transition to locally intensive production of rice and other crops, necessary to support the dense population aggregations and increased social differentiation of the Iron Age and all subsequent periods, led to significant changes in regional environments, including localized deforestation and perhaps large-scale erosion off the ubiquitous granitic hills. Certainly by the early fourteenth century CE, the Early Vijayanagara period, pollen analysis (Morrison 1995) has shown that the regional vegetation was already largely anthropogenic, with both cultigens and field weeds constituting a significant part of the regional flora. In the Iron Age and Early Historic, too, some of the basic technologies of production, including a wide range of facilities designed to capture monsoon runoff, were developed, technologies that by the British colonial period would be seen as iconic of South Indian agriculture (Morrison in press b). Thus, the nearly 4,000-year history of agriculture that preceded the Vijayanagara period created conditions of possibility for Vijayanagara-era producers and consumers, conditions that would affect choices about intensification and, later, about how to respond to the fall of the city of Vijayanagara and the subsequent demographic collapse of the region.

MIDDLE PERIOD SOUTH INDIA: COURSES AND CONSEQUENCES OF CHANGE

If the long period between the beginnings of agriculture and the establishment of the great imperial city of Vijayanagara in the early fourteenth century set many of the conditions of possibility for agriculturalists, investors, pastoralists, and others, it is also the case that human action over the 300-year Vijayanagara period (ca. 1300–1600 CE) transformed local landscapes to an unprecedented extent, reworking local vegetation, slopes, and soils and creating new configurations of settlement, transportation, irrigation, and sacred spaces. The history of Late Middle period (Vijayanagara) agriculture is discussed in more detail in Morrison (1995, 1997, in press b), Kotraiah (1995), and Stein (1980); here I simply summarize some trends. The information below derives from several sources, including archaeological survey and excavation (Sinopoli and Morrison 1995, in press a), analysis of texts (Kotraiah 1995; Morrison and Lycett 1994, 1997; Morrison 1995; Patil 1991a, 1991b, 1992, 1996), and paleoenvironmental analyses of pollen, sediment, and microscopic charcoal (Morrison 1994b, 1995).

Causes

Much of the impetus for agricultural expansion and intensification in our study area came from the expanded population created by the fourteenth century establishment of the city of Vijayanagara atop an older sacred place, in a location that had not supported a large settlement since the Early Historic period and, later, by the massive expansion of the urban population in the early sixteenth century (Morrison 1995).[8] Urban population growth, mostly fueled by in-migration, was to a large extent also responsible for rural population expansion, especially in the sixteenth century, when agriculture was pushed to the margins (Morrison 1997), both in terms of distance from the city and of movement into areas (and forms of production) that entailed greater risks of crop loss.

Urban markets and the changing structure of tax demands also transformed local agriculture in a number of ways. Production for markets was attractive to some producers (chiefly those with access to irrigation who could grow the products in demand by local elites, temples, and others dependent on purchased foodstuffs). For others, market participation was a necessary outcome of taxation practices that extracted in-kind payments from rice growers and in-cash payments from all others (Morrison 2001b, in press b). Taxation and other forms of revenue extraction had direct effects on productive strategies, including decisions about intensification. Within this context both regional environmental variability (soils, slopes, water sources, availability of forage) and transportation possibilities (distance from the city, connection to road networks) played important roles in the eventual structure of agricultural production in that place. Finally, culturally specific food preferences and restrictions were also salient. Muslims and foreigners such as the Portuguese, who came from the coastal colonies to trade, preferred grains such as wheat and were not averse to kinds of meat eschewed by most local residents. At the same time, local elites and deities had strong preferences for rice, as well as for other products of irrigated lands (coconuts, for example), and these players exercised their power to obtain such produce. Although we have less information about the preferences of the nonliterate, some contemporary residents of dry farmed regions express a preference for millets and other grains that they perceive as more filling and strength-giving and whose production is less risky. Decisions about agricultural practice took in all of these factors.

Courses, Part One: The Founding to the Fall of the City

Briefly, the founding and rapid growth of a large city in a place never politically central and in between existing settlement clusters required rapid and profound transformations in production. Vijayanagara city was located in a bend in the Tungabhadra River, one of the few places in this region where it was possible to build long, river-fed canals that (mostly) ran across rich alluvial soils. Canals were supplied by diversion dams or *anicuts* built across the braided channels of the Tungabhadra. In flood season river water ran right over the tops of the anicuts, whereas at other times the weir would divert all the flow into a canal. Although one small canal system had been built in the area prior to the establishment of the city, there was a rapid investment in canal construction starting in the early fourteenth century. At least two reservoirs were supplied by canals; one of these (the Kamalapuram reservoir) was constructed in these early years of the city's history. All the early canals and canal-fed reservoirs (as well as most Early Vijayanagara settlements) of necessity lay close to the city, in the limited alluvial zone along the river. Several factors seem critical in understanding the course of this fourteenth-century intensification. First, irrigated production was geared toward supplying the city and was financed in large part by elites. Second, the early focus on irrigated agriculture, in spite of the massive labor requirements not only of facility construction but also of the day-to-day production of paddy rice (Morrison 2001b), made possible the production of rice, as well as other desired crops such as coconuts, areca nuts, flowers, fruits, and vegetables.

At the same time, regional agricultural strategies included extensive rain-fed dry farming of millets, pulses, and sorghum, some of this aided by small-scale facilities such as terraces, gravel-mulched fields, and check dams. Larger runoff-harvesting facilities such as reservoirs (Morrison 1993, in press b) were also important, but, again, these were clustered in areas nearer the river. In general, however, the extent of nonperennially irrigated lands was limited, and there are only a moderate number of Early Vijayanagara villages in zones beyond the reach of canals.[9] Thus, we see a significant early investment in facilities designed to support irrigated, wet agriculture. Both the construction and maintenance of these facilities, as well as the forms of production they supported, required significant inputs of capital and labor. The initial intensification of agricultural production, in this case, then, did not run from more extensive to more intensive forms of production, nor did different strategies follow one another in succession; instead, a diversity of forms of production was maintained throughout (Morrison 1996).

During the Late Vijayanagara period, especially the early sixteenth century, there was a dramatic spatial expansion of agriculture in the region, with entire valleys becoming home to long chains of interconnected reservoirs supporting small patches of more intensive cultivation within a landscape of long-fallowed dry-farmed fields. These outlying areas also saw the establishment of many new villages. It is clear that, at a regional scale (Morrison in press a), one strategy of intensification in the sixteenth century was the expansion of both grazing and dry farming, extensive forms of land use.

Simultaneous with this outward expansion was an in situ intensification of areas closer to the city. North of the river, several runoff-fed reservoirs fell out of use as canal irrigation was extended. Canal water was even brought to a large island in the river near the city, which was supplied, at considerable cost, by the construction of an aqueduct. Perhaps the most interesting feature of sixteenth-century agricultural expansion and intensification was the focus on runoff-fed reservoirs (Morrison 1994b, in press b). Reservoirs could sometimes support limited cultivation of rice or other wet crops, but more commonly, it seems, they watered traditionally dry crops such as millets. During the later Middle Period widespread textual discussion of both the religious merit and political value of investment in reservoirs is unmatched by a similar focus on canals, perhaps because the latter were rarer. The construction of canals, a more costly undertaking, was more exclusively a royal activity in the area immediately around the city (Morrison and Lycett 1997), although canals in the northern part of the empire as a whole were endowed by a variety of kinds of elites.

Both canals and reservoirs require large investments in labor, time, and resources. Canals, despite their significantly larger costs, seem, however, to be ultimately less risky enterprises, producing larger--and more highly valued--harvests. Analysis of actual use-lives makes it clear that canals are more stable in the long term than reservoirs, though this stability needs to be contextualized in light of the cultural value placed on rice and other wet crops. There is no doubt that canal construction and extension required significant resources and specialist labor; the services of specialist engineers, as well as laborers, were required. It is not, however, clear that engineers were routinely engaged for reservoir construction, although specialist reservoir "diggers" were certainly in existence. Furthermore, canal construction is locationally constrained in powerful ways, and canal zones are inevitably small in this environment.

In many ways reservoirs are, in contrast to canals, a dubious economic investment. Although they may allow for limited wet cultivation or make dry farming more secure and productive, reservoirs silt in, breach—sometimes with disastrous consequences, as when the entire village of Daroji was swept away by a catastrophic dam breach in the nineteenth century—and in this semiarid environment, fail in some years to collect any water at all. In at least one historically recorded famine (Morrison 2000), reservoir-irrigated areas actually fared worse than dry-farmed fields because of the rapid evaporation of water needing to flow some distance before reaching fields below—in this instance rain-fed fields actually failed less disastrously. Reservoirs create only isolated patches of irrigated ground, unlike the continuous distribution of canal-irrigated lands, and the very dispersal of such patches also increases transportation costs.

On the other hand, reservoirs encode considerable cultural value of their own, particularly through their close association with Hindu temples and religion (Morrison in press b). Constructing a reservoir was a highly valorized act—more so even than canal construction, if the inscriptional record is any guide. Reservoirs, like temples, sons, treasures, and literary works, were also seen to endure. They constituted one of the "sevenfold progeny," links between the present and the future. If modernist architecture can be seen as a "defense against the terror of time" (Harvey 1989:206), a striving for eternity against a universe of flux, South Indian progeny might be seen in an analogous but somewhat different light. Here there is a focus on extension and longevity but one that adopts a procreative rather than creative stance. To "give birth" to a reservoir, temple, son, or other form of progeny is an ironic, if apt, metaphor in this male-dominated society where the actual work of reproduction is often elided. The donor, as the text asserts, "causes the structure to be built," but in texts the actual labor of building, plowing, growing, and so on takes place invisibly—transparent work that nevertheless leaves enduring remains.[10] Further, as the archaeological data make clear, reservoirs vary tremendously in size so that the ambition of building or rebuilding a reservoir was much more possible for most people than the excavation of a canal. Inasmuch as patronage of irrigation was linked to political legitimacy and as political ambition was widespread and far-reaching in this decentralized empire, the plurality of power holders alone makes the proliferation of reservoirs somewhat less surprising.

Here, too, landscape histories matter. Reservoirs can be placed in areas where canals could never reach and thus are suited to the rocky, dissected landscape of northern Karnataka, with its seasonal rainfall. Places with long histories of cultivation, too, may be hemmed in with preexisting rights and restrictions on the use of land. The construction of a new facility inevitably disrupts these social arrangements as well as existing physical patterns of water flow and transport. Canals must of necessity follow a carefully laid path to maintain an appropriate gradient, and such a path may well burst through existing features and lands already under cultivation. Reservoirs, on the other hand, are more modular, and new facilities, within limits, can be placed in the interstices of older ones. Political clout, as well as resources, then, necessarily paves the way for major changes in irrigation regimes.

By the early sixteenth century, then, agricultural practice on a regional scale became both more intensive and diverse. Areas where more canals could be built or older canals could be extended came under irrigation, and north of the river this meant that land under older runoff-fed reservoirs was converted to permanent irrigation. To the south of the river a rapid process of filling-in proceeded so that small spaces within the canal zone where canals could not reach were served by new runoff-fed reservoirs and wells. Farther out, well past the limited alluvial strip of the Tungabhadra River, entire areas such as the Daroji Valley and the Dhanayakanakere area were converted to cultivation by the residents of a series of relatively closely spaced villages, the landscape transformed into a patchwork of reservoir-watered and dry-farmed fields. Pollen and microscopic charcoal (Morrison 1994b, 1995) confirm this archaeological picture, showing that regional vegetation was already significantly transformed by centuries of human action well before the founding of the city.

In the complex history of agrarian change through the 300-year Vijayanagara period we can see a multifaceted process of intensification involving, among other things, the construction of new facilities, changing fallow regimes, multicropping, adoption of new crops, extension of cultivation to new areas, changing market systems, changing tax structures, and significant mobility of cultivators (Morrison in press b). Even this brief example from a single region points to the importance, within an overall trajectory of intensification, of the maintenance of diversity and of the demonstrated flexibility of farmers who were willing to adopt new crops, shift crop assemblages, reorganize production toward markets (and away from them, too), leave their homes and land when necessary, and even engage in armed rebellion when access to land and oppressive taxation threatened (Morrison 2001b). Under adverse environmental and social conditions farmers showed an impressive technological virtuosity, constructing (and maintaining) a wide range of facilities and employing a range of strategies such as manuring, mulching, weeding, and intercropping, keeping themselves and others fed.

Courses, Part Two: The Collapse and Aftermath

The fall of the city of Vijayanagara was both precipitous and dramatic. Situated near the northern boundary of the empire, it had been besieged several times but had never fallen. While many of the early military expeditions to the south had succeeded in (sometimes loosely) incorporating territory, near-constant warfare with polities to the north and northeast did not bring those areas into the Vijayanagara state (Stein 1980, 1989). In 1565, near the village of Talikota, in one of the contested regions north of the city, the Vijayanagara armies fell to a combined force created by an alliance of the five Sultanates who divided this territory. This defeat occurred at a time of dynastic uncertainty, and there has been much discussion of why, specifically, this defeat occurred and why the city was so quickly abandoned. In any case the city was burned, sacked, and looted, actions that were not only documented by both sides but that also left a clear trace on many of the monuments. The king and court, as well as most of the elite stratum of the city and the service personnel they relied on, fled, relocating to one, then another, of the empire's southern strongholds. After the abandonment of the city and as the zone of Vijayanagara control moved south, the capital of the empire was also shifted southward in a series of moves as the power and spatial scale of the empire slowly contracted.

The study area came under the control of the Sultanate of Bijapur and, soon thereafter, under various local leaders in what is usually called the Nayaka period. Many of these leaders claimed a mandate from the Vijayanagara state (e.g., Ota 1999), and indeed some of them were local governors or other leaders (*nayaka*) during the earlier period of greater centralization. The Nayaka period was one with a patchwork of small rulers, among whom we must count the Vijayanagara king, who now oversaw a much smaller territory south of the erstwhile capital. This was also a period of some unrest in the countryside, a complex political field entered decisively by the British in 1799, when they defeated the indigenous ruler, Tipu Sultan, whose reach extended almost as far north as our region. Although the territory south of the Tungabhadra was briefly claimed by another indigenous ruler, the Nizam of Hyderabad, it became one of the districts ceded by him to the British in 1800. Thus, the area south of the river, where the abandoned city of Vijayanagara lay, became part of British India, while the region north of the river remained part of the princely state of Hyderabad until Indian Independence in 1947.

Although the city of Vijayanagara, whose population has been estimated at between 250,000 and 500,000, was quickly and completely abandoned after 1565, nearly all of the small agricultural villages surrounding it remained occupied, even those, such as Kamalapuram, that had been engulfed by urban sprawl during the sixteenth-century expansion. Demographic loss was not restricted to the urban elite but also included many lower-ranking craftspeople, servants, merchants, temple personnel, and others.[11] Even though most rural villages were not abandoned, the loss of urban markets (and shifts in the structure of tax demands) meant that production was radically restructured. The only area where we have documented significant settlement abandonment and contraction is in the outlying dry-farmed areas first used intensively in the early sixteenth century. Even here, however, farmers, herders, and others in many small villages continue to eke out an existence.

The complex sixteenth-century agricultural landscape did not (and perhaps could not) last very long. With the defeat of the imperial armies and the plunder and abandonment of the capital city, the majority of consumers disappeared from the local scene; and, no doubt, many of the preexisting arrangements around the control of land, water, labor, and produce were suddenly disrupted or called into question. Within a relatively short period the majority of these outlying reservoirs, many already choked with silt, were either abandoned or allowed to grow smaller and less effective each season (Morrison in press b).

At the same time that many (but not all) dry-farmed areas fell out of use, the extensive canal network in and around the city continued to be used and maintained. Although the city itself lay in ruins, changes to the landscape, both material and conceptual, continued to shape the ways local people could and did interact with it. Should we not be aware of local landscape histories, we might assume that canal irrigation was beyond the reach of local farmers and the decentralized patchwork political system they lived in, since they had neither the resources nor the structure to create such a system. However, the prior existence of the canals made such distinctions moot. The Vijayanagara channels are enduring landscape features, "facts on the ground," and the post-Talikota history of the region makes little sense unless this is taken into account. Although canals are costly to maintain, they do allow both high and consistent yields. More than this, however, they also supported culturally valued crops. Rice continued, certainly, to be valorized, but farmers working in the irrigated zones along the river shifted production to commercial crops such as sugarcane as soon as the conditions for its sale became sufficiently appealing under the British.[12]

Significantly, as the Vijayanagara canals continued in use long after the fall of the city, they also continued to attract "royal" attention. British colonial irrigation records show extensive work on repair and maintenance of canals and their headworks; the British even extended a few Vijayanagara-area canals, cutting through solid rock to extend their reach. Thus, there was a sporadic but sustained government effort to preserve and even extend valuable canal systems.

The history of collapse in this region, even in the sketchy outline given here, can thus be seen to follow a course dependent on a large number of local variables, including political organization, cultural values, and environmental change. Just as the course of intensification at Vijayanagara did not move simply from more extensive to more intensive forms of cultivation, this particular trajectory of collapse was not "intensification in reverse." Many of the (already failing) runoff-fed reservoirs were abandoned for both solid environmental reasons (they were full of silt, unreliable, even dangerous) and explicable political reasons (their patrons were gone, making repair very difficult). At the same time, land- and labor-extensive forms of production such as dry farming, grazing, and collection continued, as did labor-intensive irrigated agriculture under canals. Even in this later period, then, there was a simultaneous diversity of production strategies on a regional scale; subsistence production existed side by side with commercial cropping, a pattern that continues to this day. This course of change, while not illogical from the standpoint of local farmers, astute calculators of benefit and loss, was critically dependent on local landscape histories and on the enduring changes created by earlier generations.

DISCUSSION: CONSEQUENCES OF CHANGE

The long-term history of agricultural intensification (and collapse) in southern India may be seen as, in a real sense, underwriting this region's political, economic, and cultural history; without significant localized concentrations of agricultural produce, the highly stratified urban society that began to form in the Iron Age and that has continued to the present would not have been possible. At the same time, however, this changing sociopolitical context itself provides an important context for understanding long-term agrarian change, a set of contingent outcomes that became part of the field of possibility for those who come later. It would of course be impossible to dissect all of the consequences of agricultural change in this region, even in a longer format than this chapter; here I simply outline a few of these.

The environmental consequences of local land-use histories have been particularly profound. The adoption and independent domestication during the Neolithic of a range of cultigens and domestic animals had a profound effect on regional flora, fauna, and soil regimes, as well as on cuisine, labor regimes, mobility, and a host of other cultural and economic factors. African millets, introduced in the late Neolithic or early Iron Age, were followed later by New World crops and weeds, some of which now pose major ecological challenges. It is probable that the region had already experienced major episodes of erosion and deforestation by the first millennium CE, and it is quite clear that the scale of erosion and loss of woody vegetation cover during the sixteenth century remains (as yet) unsurpassed. The construction of canals, reservoirs, terraces, and other features has modified both slopes and water flows, affecting not only sediment and vegetation but also aquatic animals and even climate. Where wet rice has been grown, what are known as paddy soils have formed, a permanent change in soil type.

The cultural consequences of intensification and collapse are also profound, though, as noted, neither cultural logics nor environmental possibilities can be seen as determining. Neolithic peoples enjoyed a range of animal foods eschewed by many later inhabitants of the region. Conversely, new tastes such as the elite obsession with rice, emerged historically and became factors in restructuring agriculture, even to the point of creating the extreme landscape modifications required to support this water-loving crop in a semiarid environment. The demographic and spatial expansion of the sixteenth century resulted in the founding of many new villages, some of which continue to be occupied, as well as the construction of roads, temples, wells, reservoirs, and other features that still structure transportation, pilgrimage and worship, residence, and, of course, agricultural strategies.

In what might be seen as an unintended consequence of the great sixteenth-century expansion of runoff-fed reservoirs, the landscape of many areas outside the city is now dotted with contemporary shrines made of converted reservoir sluice gates. These gates often contain religious sculpture, but even where they do not, both reservoirs and sluices are symbolically (and institutionally) associated with temples and with potent religious images (lotuses, the eternal ocean, mythological animals, and so forth; Morrison in press b). Here, then, the specific course of agricultural intensification—the extension of runoff-fed reservoirs into dry areas—has also restructured a religious landscape, one whose longevity has proven to be significantly greater.

Politically and economically, intensive agriculture underwrote both urbanism and specialization, including the specialization that constitutes the development of a food-consuming elite class. The mobilization of resources by that class, in turn, allowed the development (through the labor of others, to be sure) of the infrastructure (canals, aqueducts, anicuts) that allowed the production of wet rice, coconuts, flowers, areca nuts, mangos, and other desired produce in an area in which they would not normally thrive. Temples, too, were both consumers of (primarily irrigated) produce and investors in and owners of land. The specific institutional forms that prompted both urbanism and agricultural intensification—kings, nayakas, and large temple complexes, for example—also left architectural markers on the landscape—palaces, pavilions, shrines, and waterworks—that attract tourists and archaeologists to this region today.

I do not deny that it is possible to develop general understandings of the process of intensification; indeed, I would argue that it is imperative we do so. However, I would also suggest that this understanding will require close comparison of many well-documented and long-term trajectories of change. Intensification is a process, not a quantity or an essential quality, a process that can indeed be understood in broad terms outside consideration of any particular case. Such broad understandings are, however, limited by their very generality, a quality at once their strength and their weakness. They gain analytical teeth only when applied to new cases where they may prompt new ways of looking at old information or compel researchers to gather new information as a consequence. Out of this foray into the specific we both derive our ideas and evaluate them. As such, the problem of contextualization is a critical one. I have argued that the process of intensification is itself complex, multiple, and may take many courses; here I have tried to consider a little more closely how we might situate our understanding of this complex process in terms of actual courses and consequences of change, in a world where human landscapes, both material and conceptual, with their scars of the past and resources for the present, are the places where production happens. Understanding production, then, requires attention to these places.

NOTES

1. Archaeological research on the intensification of production has primarily been oriented toward the development of more general causal models of change, an orientation that has led to what, for the social sciences, seems to be the amazing staying power of the Boserup (1965, 1981) model of intensification (see discussion by Morrison 1994a), a relatively simple scheme that proposed a straightforward causal relationship between population pressure and (one form of) agricultural intensification, elaborating a stepwise model of cultural evolution (Morrison 1996) in which different stages of intensification were also marked by changes in land tenure and agricultural tools, among other things. Despite a flurry of criticism and abandonment of the model by some of its early advocates (e.g., Brookfield 2001), various iterations of this scheme are still being proposed (e.g., Johnston 2003; Stone 2001), even though some of these modify its basic assumptions so fundamentally that it is difficult to credit arguments that such changes are simply friendly amendments. The purpose of this chapter is, however, *not* to discuss Boserup's view of intensification, an exercise that hardly needs repeating. At this point it is possible to say a great deal more about actual historical courses of change (as opposed to a quasi-historical sequence assembled out of contemporaneous variation that Boserup had recourse to) and thus to the range of possibilities for change than was possible some 40 years ago.

2. A careful reading of my previous work will make it clear that I am in substantial sympathy with and acknowledge the value of Boserup's (1965) original contribution to this question. Clearly, people intensify (within which I would include failed attempts to increase production or productivity as well as successful ones, meaning that the

definition of *intensification* cannot be operationalized in terms of either productivity or output) production because of a perceived need to do so. Population-resource imbalances (leading to actual shortages or to the awareness of the potential for shortages) are one critical example of conditions leading to such perception. However, they are not the only such condition (Morrison 1994); it is possible to document examples of intensification where production far outstrips some (difficult-to-define) biological baseline of need, pointing to the problematic nature of distinctions between production for "subsistence" as opposed to "exchange" or "surplus" (cf. Gardner 2001). Furthermore, decisions about production strategies are always made within the context of regional power relations (Morrison in press a), so discussion of the causes of intensification also need to address issues of power in ways they have not to date done so.

3. And, indeed, in terms of "total socio-cultural systems" (Minnigal and Dwyer 2001:269), an approach I think is also exemplified in my own empirical work.

4. To a certain extent we can include as parts of these landscapes features sometimes viewed as entirely "natural," such as rainfall, crop pests, slopes, and soils given that generations of human action have helped construct these as well.

5. Ashmounds, a distinctive feature of the Southern Neolithic in this region, are actually large mounds of fired cattle dung, material that has vitrified into relatively impervious slaggy concretions. For more discussion of ashmounds see Morrison (1995), Johansen (2004), and Paddayya (1993).

6. Critical, too, is the history of *Sorghum bicolor*, locally known as *jowar*, which became one of the primary dry-farmed crops of the dry regions of South Asia by the Middle periods. Introduced from Africa around the early Iron Age, the expansion of jowar production, like that of rice, seems to have accompanied and supported urbanization in the south.

7. As I have discussed elsewhere (Morrison 2001b), the historical record of the Middle period might lead one to believe that the production of rice and other irrigated crops was the primary form of agricultural production in the region. In fact, our archaeological work makes it clear that millets and other rain-fed crops predominated throughout the period and that efforts toward intensification often included the expansion of extensive dryland cultivation (Morrison 1995, 1997).

8. There is no evidence of an Iron Age or Early Historic site directly underneath the area around Hampi that constituted the early core of the city, nor indeed is there for most of the area subsequently covered by this huge city. However, there is very good evidence for several large Iron Age and Early Historic sites alongside the Tungabhadra River in locations that would become the near suburbs of the city. Thus, the concentration of population in this area was not completely novel (see Morrison in press b for an extended discussion).

9. Here I am discussing primarily the large area immediately around the city, what we have called the greater metropolitan region (GMR). However, an important exception to this pattern is provided by the level plains that lie to the east and west of the

GMR. In these areas there are significant clusters of tenth-century to thirteenth-century (Early Middle period) settlements, most of which lie along seasonal streams that would have supported inundation agriculture of dry crops and perhaps some rice. Many of these places continued to be occupied into the Vijayanagara period (though some may have experienced population loss), so that while it is the case that *new* Vijayanagara settlements in the fourteenth century were primarily oriented to the production of wet crops such as rice, older places continued to be occupied and older forms of production continued alongside innovations such as the expansion of runoff-fed reservoirs.

10. It may go without saying that it is not the donor who supplies this labor. The work of lower-status men and women is the foundation for the merit (*punya*) accrued to the donor, just as the reproductive labor of the wife, invisible labor in these texts, is the foundation for the son.

11. Significantly, many of the large temple complexes were also abandoned, either at this time or shortly afterward. These complexes, designated as cities in their own right within the urban landscape, supported large cadres of specialists, held extensive rights in land and produce, and often had considerable stores of wealth. Still, they depended on elite patronage and protection. Elsewhere (Morrison in press b) I discuss the differential patterns of temple abandonment and reuse and their connections with local landscape histories.

12. Without discussing commercial agriculture during the colonial period in depth I would point out that tax demands also sometimes forced producers into cropping choices that they might not otherwise have followed. The extension of railroads in the region in the nineteenth century had some effect on cropping choices, as did the expansion, under colonial rule, of the towns of Hospet and Bellary.

REFERENCES

Boserup, E.

1965 *The Conditions of Agricultural Growth: The Economics of Agrarian Change under Population Pressure*. Aldine, New York.

1981 *Population and Technological Change: A Study of Long-Term Trends*. University of Chicago Press, Chicago.

Brookfield, H.

2001 Intensification and alternative approaches to agricultural change. *Asia Pacific Viewpoint* 43(2–3):181–192.

Buchanan, F.

1807 *A Journey from Madras through the Countries of Mysore, Canara, and Malabar, Performed under the Orders of the Most Noble the Marquis Wellesley, Governor General of India, for the Express Purpose of Investigating the State of Agriculture, Arts, and Commerce; The Religion, Manners, and Customs; The History Natural and Civil, and Antiquities, in the Dominions of the Rajah of Mysore, and the*

Countries Acquired by the Honourable East India Company. T. Cadell and W. Davies, London.

Fuller, D. Q.

2003 An agricultural perspective on Dravidian historical linguistics: Archaeological crop packages, livestock, and Dravidian crop vocabulary. In *Assessing the Language/Farming Dispersal Hypothesis*, edited by P. Bellwood and C. Renfrew. McDonald Institute for Archaeological Research Monographs, Cambridge, UK.

Gardner, D.

2001 Intensification, social production, and the inscrutable ways of culture. *Asia Pacific Viewpoint* 42(1–2):193–207.

Harvey, D.

1989 *The Condition of Postmodernity: An Enquiry into the Origins of Cultural Change*. Blackwell, New York.

Johansen, P. G.

2004 Landscapes, Monumental Architecture and Ritual: A Reconsideration of the South Deccan Ashmounds. *Journal of Anthropological Anthropology* 23 (3): 309–330

Johnston, K. J.

2003 The intensification of pre-industrial cereal agriculture in the tropics: Boserup, cultivation lengthening, and the Classic Maya. *Journal of Anthropological Archaeology* 22 (2): 126-161.

Korisettar, R., P. C. Venkatasubbaiah, and D. Q. Fuller

2001 Brahmagiri and beyond: The archaeology of the southern Neolithic. In *Indian Archaeology in Retrospect, Vol. 1, Prehistory: Archaeology of South Asia*, edited by S. Settar and R. Korisettar, 151–239. Manohar and Indian Council of Historical Research, Delhi.

Kotraiah, C. T. M.

1995 *Irrigation Systems under Vijayanagara Empire*. Directorate of Archaeology and Museums, Mysore.

Minnigal, M., and C. P. Dwyer

2001 Intensification, complexity, and evolution: Insights from the Strickland-Bosavi region. *Asia Pacific Viewpoint* 42(1–2):269–85.

Morrison, K. D.

1993 Supplying the city: The role of reservoir irrigation in an Indian urban landscape. *Asian Perspectives* 32:133–151.

1994a Intensification of production: Archaeological approaches. *Journal of Archaeological Method and Theory* 1(2):111–159.

1994b Monitoring regional fire history through size-specific analysis of microscopic charcoal: The last 600 years in South India. *Journal of Archaeological Science* 21:675–685.

1995 *Fields of Victory: Vijayanagara and the Course of Intensification*. Contributions of the University of California Archaeological Research Facility, No. 53. University of California, Berkeley.

1996 Typological schemes and agricultural change: Beyond Boserup in South India. *Current Anthropology* 37(4):583–608.

1997 Agriculture at the edges: Archaeology and history in the Vijayanagara hinterland. In *South Asian Archaeology 1995*, edited by B. Allchin, 783–791. Oxford and IBH, New Delhi.

2000 Naturalizing disaster: From drought to famine in South India. In *Environmental Disaster and the Archaeology of Human Response*, edited by G. Bawden and R. Reycraft, 21-34. Maxwell Museum of Anthropology, Albuquerque, New Mexico.

2001a Intensification and specialization, archaeology of. In *International Encyclopedia of the Social and Behavioral Sciences*, Vol. 11, 7678–7681.

2001b Coercion, resistance, and hierarchy: Local processes and imperial strategies in the Vijayanagara Empire. In *Empires*, edited by S. Alcock, T. D'Altroy, K. Morrison, and C. Sinopoli, 253–278. Cambridge University Press, Cambridge, UK.

in press a Rethinking intensification: Power relations and scales of analysis in precolonial South India. In *Rethinking Intensification*, edited by T. Thurston and C. Fisher. Plenum, New York.

in press b The Daroji Valley: Landscape history, place, and the making of a dryland reservoir system. Vijayanagara Research Monograph Series. Manohar, New Delhi.

Morrison, K. D., and M. T. Lycett

1994 Centralized power, centralized authority? Ideological claims and archaeological patterns. *Asian Perspectives* 33(2):312–353.

1997 Inscriptions as artifacts: Precolonial South India and the analysis of texts. *Journal of Archaeological Method and Theory* 3(3–4):215–237.

Nagaraja Rao, M. S., and K. C. Malhotra

1965 *The Stone Age Hill Dwellers of Tekkalakota: Preliminary Report of the Excavations at Tekkalakota*, Deccan College Postgraduate and Research Institute, Poona.

Ota, N.

1999 Bëda Nayakas and their historical narratives in Karnataka during the Post-Vijayanagara period. In *Kingship in Indian History*, edited by N. Karashima, 163–194. Manohar, New Delhi.

Paddayya, K.

1993 Ashmound investigations at Budihal, Gulbarga District, Karnataka. *Man and Environment* 18(1):57–88.

Patil, C. S.

1991a Mummadi Singa, Kampila, and Kumara Rama. In *Vijayanagara Progress of Research 1987–88*, edited by D. V. Devaraj and C. S. Patil, 179–198. Directorate of Archaeology and Museums, Mysore.

1991b Kummata. In *Vijayanagara Progress of Research 1987–88*, edited by D. V. Devaraj and C. S. Patil, 199–216. Directorate of Archaeology and Museums, Mysore.

1992 *Temples of Raichur and Bellary Districts, Karnataka, 1000–1325 AD*. Directorate of Archaeology and Museums, Mysore.

1996 Doravati. In *Vijayanagara: Progress of Research 1988–91*, edited by D. V. Devaraj and C. S. Patil, 243–247. Directorate of Archaeology and Museums, Mysore.

Possehl, G. L., and P. C. Rissman

1992 The chronology of prehistoric India: From earliest times to the Iron Age. In *Chronologies in Old World Archaeology*, 1:465–490, 2:447–474. University of Chicago Press, Chicago.

Sinopoli, C. M., and K. D. Morrison

1995 Dimensions of imperial control: The Vijayanagara capital. *American Anthropologist* 97:83–96.

in press The regional landscapes of the imperial city of Vijayanagara: Report on the Vijayanagara Metropolitan Survey Project. In *South Asian Archaeology 1999*, edited by K. R. van Kooij and E. Raven. Egbert Forsten, Groningen.

Stein, B.

1980 *Peasant State and Society in Medieval South India*. Oxford University Press, Delhi.

1989 Vijayanagara. *The New Cambridge History of India*, vol. 1, part 2. Cambridge University Press, Cambridge, UK.

Stone, G. D.

2001 Theory of the square chicken: Advances in agricultural intensification theory. *Asia Pacific Viewpoint* 42(1–2):163–180.

Stone, G. D., and C. E. Downum

1999 Non-Boserupian ecology and agricultural risk: Ethnic politics and land control in the arid Southwest. *American Anthropologist* 101:113–128.

Wheeler R. E. M.

1947 Brahmagiri and Chandravalli 1947: Megalithic and other cultures in the Chitaldurg district, Mysore State. *Ancient India* 4:181–309.

Water Supply, Labor Requirements, and Land Ownership in Indus Floodplain Agricultural Systems

HEATHER M.-L. MILLER

The development of ancient floodplain civilizations like the Indus is often associated with the need for managers to direct irrigation systems, leading to the rise of an elite class through their control of the water supply necessary for surplus agricultural production. Such ideas are most strongly associated with Karl Wittfogel's (1976 [1957]) classic statement of the role of hydraulic agriculture in the rise of oriental despotism. As several of the chapters in this volume note, the recognition of the efficient creation and management of small-scale irrigation systems without hierarchically ranked managers resulted in general repudiation of Wittfogel's hypothesis. Nonetheless, many texts continue to cite irrigation agriculture as one of the factors in the development of state-level political units in the Indus Valley during the third millennium BCE, in spite of the lack of any evidence for large-scale irrigation systems. A recent exception is Vernon Scarborough's (2003:145–146) conclusion that water management was unlikely to have been a method of promoting state power and elite control.[1] In this chapter I will go even farther, and argue that Indus Valley Tradition systems for the management of agriculture may in some ways have restricted the hierarchical power of state-level political units, particularly with regard to control of wealth from elite land ownership, yet encouraged the existence of state-level corporate mechanisms for arbitrating disputes about land ownership.

This is not to say that agricultural technology and management were not important to the Harappan Phase state. On the contrary, a stable agricultural supply would have been essential. And that is the point of this volume, which addresses topics that arise from two well-known but heavily modified or rejected theories: Wittfogel's original statement about the role of hydraulic agriculture in state formation and Ester Boserup's original strong linking of population increase with agricultural intensification. Both of these authors addressed the more general idea that forms the basis for this volume, that agricultural intensification in some form is related to the rise of states. It is the exact way in which agricultural intensification is related to state power that is the source of debate, and which varies so much from case to case, even in similar environments.

For example, for ancient Indus Valley farmers the supply of water would have been the limiting factor in agricultural production, as in any classic Wittfogelian state. Their choice of methods for increasing water supply would have been affected primarily by the type of land available. For some water-supply systems the available labor was also important, particularly for the most labor-intensive water-supply system discussed here, that of well and lift irrigation. But for the most part, rather than seeing the need for large-scale irrigation, I conclude that a multiplicity of water-supply systems were likely employed by Indus farmers, with particular choices about intensification and/or extensification reflecting the type of crops in use, the type of land obtainable, the labor availability, and the demand for increased production.

Whether or how political leaders were actively involved in agricultural management is another question, however. As with later periods and other regions, leaders may have benefited from and even encouraged agricultural intensification, but focused their energies on extracting agricultural surplus as tribute or taxes in exchange for religious or military protection, rather than trying to control agricultural production directly via land ownership or infrastructure development. Therefore, I am not merely interested in management of water supply, but have more broadly examined the connections between agricultural production and the Indus state in terms of wealth production.

Food production is usually seen as one of the two major sources of wealth in ancient economic systems, the other being manufactured goods. Wealth from craft production and trade has seen considerable attention for the Indus Valley Tradition, especially for the urban integration period of the Harappan Phase, c. 2600–1900 BCE (chronology after Shaffer 1992). In contrast, the accumulation of wealth and prestige from food production, and particularly from land ownership, has not yet been a focus of models of Indus Valley Tradition social and political organization, even though much recent research has been devoted to the investigation of Harappan Phase food production systems more generally. Although an early debate on the nature of the Harappan Phase political system included a discussion of cattle herds as a source of wealth and political power (Fairservis 1986; well summarized in Possehl 1998), agriculturally based wealth and prestige has generally been taken as a given (Leshnik 1973 is a crucial exception). The lack of appropriate data has been a major reason for this state of affairs, especially the lack of textual information, although the great advances in South Asian plant analysis over the past decade are starting to inspire more complex interpretations. Nevertheless, it is useful to explicitly examine the implications of the different water supply and land ownership systems possible for this region because they have very different implications for sources of wealth and power, and thus contribute substantially to broader interests in the sources of Indus political power and social structure (Kenoyer 1998; Possehl 1998 for recent summaries).

In the spirit of Timothy Earle's (2000) call for optimism, creativity, and critical methods in the archaeological investigation of land ownership, I present here several possible models of the ways water supply, labor requirements, and land ownership might have affected Harappan Phase economic, social, and political systems. There are a limited number of possible agricultural systems for this region, as is true of most areas. This is particularly true with respect to potential water-supply systems, and given the semiarid nature of this region, water supply has been a key part of all agricultural systems. From the perspective of water-supply requirements, there are four basic agricultural water-supply systems possible for the Indus Valley floodplains during the Harappan Phase: (1) riverine inundation, (2) rain-fed, (3) small-scale canal irrigation to extend the inundation, and (4) well or lift irrigation. Given our very incomplete data for any kind of Harappan Phase agricultural water-management system, my strategy here will be to construct a range of possible agricultural models for the regions of the Punjab (northern floodplains) and Sindh (southern floodplains) (Figure 5.1) during the Harappan Phase and analyze their socio-economic implications.[2]

The various water-supply systems have a crucial effect on land ownership, especially as the rivers shift often and make previously valuable land worthless. Therefore, for the cities and large-scale cultural system of the Indus civilization to have been sustained, agricultural systems must have been organized to allow some degree of long-term stability of the agricultural base. To evaluate strategies ensuring stability, I indicate how each of the four water-supply systems possible in the Indus floodplains during the time of the Harappan Phase would affect and be affected by labor requirements and land ownership, and assess the varying levels of risk of crop failure. Throughout, I provide any available archaeological evidence for these hypothetical models. I conclude with the case of agricultural change toward the end of the Harappan Phase and offer a discussion of how land ownership practices might have affected Harappan Phase socio-political systems. To provide the necessary background for assessing likely agricultural choices, I first briefly outline the two main cropping regimes and the agricultural processing stages employed in the Indus Valley floodplains in the Punjab and Sindh.

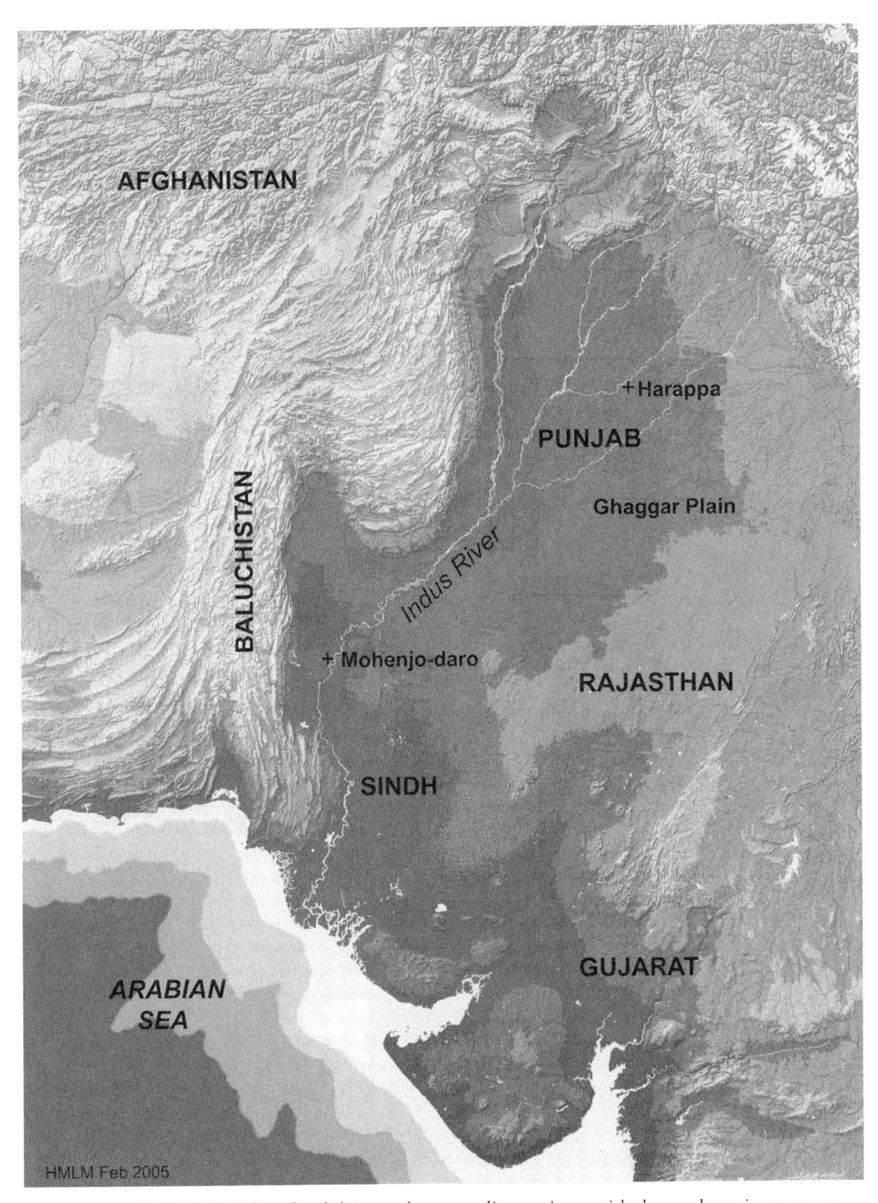

Figure 5.1. The Indus Valley floodplains and surrounding regions, with the modern river system. Base map courtesy of Randall Law.

CROPPING REGIMES AND AGRICULTURAL PROCESSING STAGES IN THE INDUS FLOODPLAINS

My reconstructions of Indus floodplain cropping regimes and agricultural processing stages, like others in the literature, are based on the few archaeological finds to date, on generalized modern and historic agricultural accounts, and on casual ethnographic reports. There has been little ethnoarchaeological research on crop management and processing for the Indus floodplains as yet, in contrast to Seetha Reddy's (1994, 1997, 2003) work from very different environments in Gujarat. Detailed overviews of these cropping systems and the archaeological finds to date as they pertain to the Harappan Phase situation are given in Richard Meadow (1996) and Dorian Fuller and Marco Madella (2000), updated by Steven Weber (2003) and Madella (2003). Ifran Habib (1999) is also a particularly useful source, providing a painstakingly researched account of agricultural systems in the Indus and Gangetic valleys during the Mughal period (1556–1707 CE).

Cropping Regimes and Types of Crops

The two main cropping regimes for the Indus floodplains are divided by season of planting: the *rabi* (winter-planted) crops and the *kharif* (summer-planted) crops. The third cropping regime, perennials, includes cotton as well as tree crops and vegetable gardens (Figure 5.2). At the moment, researchers are concluding that past environmental conditions were roughly the same as the present, as is discussed in the section on rain-fed agriculture below. These cropping regimes make use of rainfall and especially river water, both of which are affected by the two rainfall systems that serve the Indus Valley: the winter cyclonic system of Western Asia, which reaches as far east as the Indus Valley, and the South Asian summer monsoon system, which reaches as far west as the Indus Valley then moves back across the northern mountains (Himalayas), increasing the summer water flow of the Indus River system.

As the Indus River system falls exactly at the meeting point of the two rainfall systems, it is the point at which both systems are weakest and most irregular. Today the average annual rainfall for the Indus Valley is less than 25 cm, making it the most arid portion of the South Asian subcontinent (Oxford University Press 1955). The northern Punjab is slightly wetter, with 25–50 cm mean annual precipitation, and is exactly bisected by the two weather systems so that the winter rains are the stronger system to the west, and the summer rains are stronger to the east (Figure 5.1). Given the lower temperatures, the most reliable rain-fed agriculture is in the winter in the northern Punjab, especially to the northwest. For the entire Indus Valley region, however, river water is far more important and reliable for agricultural water supply than rainfall.

<u>***Rabi* (Winter-planted) crops**</u>
- Plant in Oct-Dec
- Harvest in March-May
- Water: winter rains, receding inundation

examples:
wheat, barley, peas, lentils

<u>***Kharif* (Summer-planted) crops**</u>
- Plant in April-August
- Harvest in Sept-Dec
- Water: summer monsoon, active inundation

examples:
millets, rice, various pulses

<u>**Perennials**</u>: year-round use of land; example: cotton, harvested with *kharif* crops

Figure 5.2. Cropping regimes in the Indus Valley floodplains.

The riverine inundations are thus paramount in the agricultural regime, with flood peaks in the late summer, allowing planting of winter crops in the early fall on the newly deposited wet silts.

Lawrence Leshnik (1973:Table 6.2) gives the cycle of the rivers and cropping regimes for the Punjab during the nineteenth and twentieth centuries:

Low water mid-January to mid-April:	*rabi* crops harvested mid-March to mid-May
Rising water mid-April to mid-August:	*rabi* crops harvested mid-March to mid-May
	kharif crops planted mid-April to August or later
High water mid-August to mid-September	
Subsidence mid-September to mid-January:	*rabi* crops planted October to mid-December
	kharif crops harvested mid-September to December

This cycle might be a few weeks different for Sindh, with the delay in the inundation downstream as well as the greater heat and lesser rainfall.

These cropping regimes are still used today, although they are often heavily modified by the modern water-supply system and the introduction of new crops. The crops discussed here are those known to have been used during the Harappan Phase, and this list is almost completely dependent on work done at the ancient city of Harappa itself, the only Indus floodplain site where large quantities of plant remains have been systematically recovered. My comments here about water supply refer to the system of management prior to the large-scale building of massive riverine irrigation canals by the British in the early 1900s and prior to the modern tube-well irrigation systems; primarily small local ditch or canal systems supplied by well, oxbow, or riverine water.

Rabi crops are planted in November or December and are harvested in April or May. They primarily depend on the wet soils of the receding summer riverine inundations and on the winter rain system in the north, although some crops are also watered by various artificial water-supply systems. The ancient crops typical of the rabi system from Harappa and other sites are the winter wheats (*Triticum* spp.), the barleys (*Hordeum vulgare* spp.), common peas (*Pisum sativum*), chickpeas/gram (*Cicer arietinum*), grass peas (*Lathryus sativus*), lentils (*Lens culinaris/esculenta*), linseed/flax (*Linum usitatissimum*), and probably mustard (*Brassica* sp.). Almost all of these are also the main crops in Western Asia.

Some of these species may have been indigenously domesticated here as well as in Western Asia, such as the barleys. Some species must have been domesticated in Western Asia and traded or brought east, such as the einkorn or emmer wheats (present at the site of Mehrgarh, albeit sparsely documented). Bread wheat (*T. aestivum*) could have been indigenously domesticated here as well, although only if tetraploid wheats had been brought from the west (Zohary and Hopf 1988). Certainly some species were modified in this eastern region to form new varieties or morphological types, such as the various sphaeorococcoid barleys and wheats. (As a side note, Percival 1921 is completely unreliable for his classification of these varieties, as he relied on a few specimens sent to him from India, out of the great range that existed.) The full tale of distant versus indigenous domestication and modification must await further work on the earliest periods of the Indus Tradition, as we currently only have early plant remains from the Baluchi site of Mehrgarh, which has some domesticates even in the earliest levels (Costantini 1984; Costantini and Costantini Biasini 1985; Meadow 1996). However, we do not have to wait to determine that from the beginning the people of the Indus Tradition were active experimenters and selectors of new varieties of plants and animals. This should not surprise us, given the records of the early British agricultural specialists in this region, who recorded dozens of varieties of wheats in the Indus Valley and especially Baluchistan, varieties selected for characteristics from relative drought resistance to cooking properties (Howard and Howard 1909a, 1909b; Howard et al. 1910). (See also Thomas 2003 for modern and ancient risk- and food-related strategies of farming in the Bannu region to the northwest of the Indus floodplains.)

Kharif crops are typically planted in May or June and harvested in October or November. They may make use of low-level river inundation from snowmelt in the early summer planting period, or monsoon rainfall in the northeast of the Indus Valley. Standing kharif crops might also be watered by river inundation in July to September, with floods from the northwestern edge of the summer monsoon, although this runs the risk of drowning the crops around harvest time. However, more often they depend on artificial water-supply systems of various types during the growing period, although it should not be assumed that these water-supply methods were necessarily complex or labor-intensive, as described below. Crops typical of this system from the site of Harappa and elsewhere in the plains include some of the millets (*Panicum* spp., perhaps *Setaria*, but not *Sorghum* or *Eleucine*), rice (*Oryza sativa*, based on both charred grains and phytoliths from glumes), various gram beans (*Vigna* sp., *Medicago* sp.), sesame (*Sesamum indicum*), and fruits such as melon (*Cucumis* sp.). No information is yet available as to whether this may have been wet or dry cultivated rice.

As Weber (1998, 1999, 2001, 2003) has elaborated in several articles, at Harappa most of these crops are present in the Harappan Phase, but their ubiquity increases at the end of this period and in the following Punjab Phase ("Late Harappan" period). The inhabitants of Harappa during the late third and early second millennia BC appear to have extended their agricultural regime to increasingly employ two cropping regimes, which may have been done by multi-cropping on a single plot and/or seasonal use of different lands, as appropriate for water supply and environmental conditions. However, the rabi (winter-planted) crops are clearly the most important staple crops at Harappa throughout the Harappan Phase. Note that double-cropping has not been suggested for the beginnings of the Indus civilization but rather for its end.

Perennials form a third class of crops. The potentially most important case is that of cotton (*Gossypium* sp.), as it may well have been a major crop during the Harappan Phase. There is some question as to the type of Old World cotton used by the Harappan Phase people, as both species of Old World cotton are known in both perennial and annual forms today, although all are ready for harvesting in the late autumn with the kharif crops (Watson 1983; Meadow 1996). It seems likely, however, that the annual forms of Old World cotton were only developed in the past two thousand years, so varieties grown by the Harappan Phase people were perennial plants (Watson 1983:31–41 and footnotes). Because cotton is a perennial, land would have to be dedicated to its growth year-round and could not be multi-cropped with winter crops, such as barley and wheat. Cotton cloth (in the form of metal pseudomorphs) is known from Harappan sites, but little is known about its prevalence as a fabric or as a crop. It may have been a major trade item for the Harappan Phase people, or it may not have been widely grown till a later date; we simply do not know, although the former seems more likely. The dye plant indigo, so important in this part of the world historically, is still completely unknown for the Harappan

Phase but might have been another crop grown perennially to allow multiple cuttings, as was done in this region historically (Habib 1999:47).

Other perennials are tree, shrub, and vine crops. Particularly well-attested in the archaeobotanical record is the fruit of the jujube shrub (*Ziziphus mauritania/jujuba*), found in both domesticated and wild forms and harvested with the rabi crops in the spring. Finds of dates *(Phoenix dactilifera)* at several Indus sites indicate that they were traded in from elsewhere or cultivated, as they do not grow naturally in the floodplain itself. Cultivation seems more likely than trade alone, particularly in the south, especially given recent *Phoenix* phytolith finds from the Sindhi site of Kot Diji (Madella 2003). Finds of domesticated grape seeds (*Vitis vinifera*) are also reported in low quantities from several Indus sites, including Harappa, although more research is needed to help determine its importance as a crop. Madella (2003) also reports finds of phytoliths from the banana family from Harappan Phase deposits at Kot Diji but indicates that these cannot yet be identified to species, whether wild species of various types or the domestic varieties of banana and plantain. Andrew Watson (1983:51) notes that Greek accounts of Alexander's march through the Indus Valley reported bananas being cultivated in 325 BCE, but this is the earliest firm report for the region as yet. Other possible shrub and tree crops have yet to be positively identified; mango in particular will be an interesting case, as the earliest reports in Sindh presently come from Arab writers (Watson 1983). Certainly in the historic and modern period, small orchards were a habitual part of the agricultural regime in the Indus floodplains.

Finally, many vegetable crops likely were grown together or in rotation in garden plots requiring a "perennial" dedication of land use. Unfortunately, vegetables and greens are always elusive in archaeobotanical research because of a variety of problems affecting preservation of identifiable remains. Although we have no evidence for the nature or the very existence of Harappan Phase mixed gardens for growing vegetables and small-scale crops, such gardens were very important in the Punjab region historically, with farmers growing multiple crops per year for exchange with local towns and cities, sometimes even in mono-cropped fields (Habib 1999:52 and elsewhere). Unlike many early cities, the Harappan Phase cities appear much too densely packed to have allowed much garden space within the city walls. While city families may have had garden space just beyond the city walls, exchange for garden foods may have been necessary for some city families. During the Harappan Phase, such crops likely included peas and beans, melons, greens, other vegetables, and even small patches of grains, particularly special varieties for special purposes. Northwestern South Asia is the suspected origin location for eggplant (*Solanum melongena*) and spinach (*Spinacia oleracea*), to name just two of the many vegetables and greens grown here historically (Habib 1999; Watson 1983).

The *Brassica* spp. reported as mustard (above, as a rabi crop for seeds) could also have been grown as a green in other seasons. Semi-wild (weedy) and wild plants were also collected for food and other uses, including jujube (*Ziziphus*) and no doubt many greens and herbs, as well as grasses for mats and basketry (Madella 2003; Miller 1991).

AGRICULTURAL PROCESSING STAGES AND LABOR REQUIREMENTS

All of the agricultural systems discussed here required various stages of processing, including plowing and other soil preparation stages, planting, possibly weeding, supply of water, and harvesting and other crop processing techniques. Each of these stages is briefly described, any available archaeological evidence given, and the potential labor requirements assessed.

Soil preparation for this region in the recent past consisted of both large-scale field preparation techniques, including plowing, and smaller-scale "garden" preparation methods, such as the use of digging sticks and hoes. We know that fields were plowed in at least some parts of the Indus civilization, based on the find of a plowed field at the edge of the site of Kalibangan (Lal 1979; Thapar 1973). The evidence for clay models of plows, as well as yokes, numerous types of carts, and other agricultural and traction artifacts, is well summarized by Laura Miller (2003). She is also documenting the use of zebu cattle for traction, either for carts or for plows, through analysis of faunal remains from the site of Harappa (Miller 2003). Nothing is directly known about the existence of other methods of field preparation, such as the use of digging sticks or hoes, as no digging implements of either stone or metal have been identified, although such large fragments of metal would certainly have been recycled. It is also most likely that wood would have been used to make digging implements, plows, and rakes, as in ethnographic examples from the region. Other stages of soil preparation, such as the breaking up of clods and possible fertilization techniques, are also unknown. In the recent past, however, inundation lands along the river banks were often planted with no prior preparation of the ground (Leshnik 1973), so that water-supply systems employing inundation would have had little labor associated with soil-preparation stages, allowing farmers to plant a larger area.

After the ground was prepared, *planting* was probably done by hand using broadcast sowing of seed, barring future evidence for seed drills. Although we do not have evidence either way for Indus agricultural fences restricting access by animals, modern gardens and orchards in the Punjab are often enclosed with a mud-lump wall, but fields are left open (personal observation). Whether or not Indus Valley fields and gardens were regularly *weeded* is a question for which future analyses of plant remains may provide crucial evidence. Present-day agriculturalists frequently weed their crops by hand and use the weeds as fodder for their few farm animals, including cattle and water buffalo used for dairy and traction (K. Thomas, personal observation in the Northwest Frontier Provinces; author's personal observation in the Punjab). Some of these crop weeds, the *Medicago* species, are collected, sold in market, and eaten as greens in the northwest today (Thomas 2003:418). Weeding practices of this sort, for fodder or greens collection, should be visible in archaeobotanical assemblages if practiced in the past; for example, such a suggestion was made for plant remains from the third millennium site of Tarakai Qila, northwest of the Indus floodplains (Miller 1988). Archaeobotanical studies of other sites for information on ancient weeding practices would contribute a great deal to our reconstructions of past labor requirements, as such intensive weeding requires significant additional labor input throughout the growing season. Given that this fodder can be especially suited for animals kept for milking, intensive weeding could also provide supporting evidence for a well-developed mixed-farming economy, in combination with other types of bioarchaeological data.

Water supply is the agricultural processing stage central to this chapter and will be described below. Work has been done elsewhere on the identification of water supply through irrigation via the weed seeds found in ethnoarchaeological assemblages (Charles et al. 1997, 2003; Jones et al. 1997), and it is possible that such an approach might be useful for the Indus Valley. However, it will be important to do ethnoarchaeological work in the region to see what the local weed assemblages would look like for irrigated and unirrigated fields, and the possibility of intensive hand weeding, as discussed above, complicates such an approach.

Finally, evidence for *harvesting methods* for the Harappan Phase in the Indus Valley itself also awaits future ethnoarchaeological studies (see Reddy 1994, 1997, 2003 for such studies for Harappan Phase Gujarat). No root crops are known to have been used by the Harappan Phase people, so that harvesting would have been by hand-picking or cutting with a knife or sickle. Chert blades are plentiful, but those with "sickle polish" are not so prevalent, surprisingly (Hoffman and Cleland 1977:17), and there are no sickle handles. Harvesting knives of metal may have been widely used, since the lack of metal blades of all sorts recovered from Indus sites is probably due to poor preservation conditions and a high level of recycling of metal (Kenoyer and Miller 1999). This question remains open until more work is done on both lithic and metal assemblages. No evidence exists for later stages such as winnowing and separating of grain and some pulse crops. Ethnoarchaeological and archaeobotanical work done elsewhere in Eurasia has been extremely helpful in the identification of crop processing stages, including harvesting techniques, so that this method may be of use in the Indus Valley region (Hillman 1984; Jones 1984, 1987; Miller 1991; Reddy 1994, 1997, 2003; Fuller and Madella 2000). But once again, if intensive weeding was a part of the Harappa Phase agricultural regime, as suggested above, the use of weed seeds as diagnostics for water supply and harvesting methods will be seriously compromised, and Indus specialists will need to find alternative methods of investigating these processing stages.

This brief background gives some idea of the complex agricultural systems and labor scheduling decisions faced by Indus Valley Tradition farmers. The choices they made about water-supply systems had to harmonize with these decisions, as well as other economic and social situations and goals.

POSSIBLE WATER SUPPLY SYSTEMS AND AGRICULTURAL REGIMES

There are four types of water-supply systems possible in the Punjab and Sindh, in the regions surrounding the Indus River and its tributaries (Figure 5.3). Particularly useful sources for water-supply systems employed in the Indus floodplains are Habib's (1999) detailed historical research for the Mughal period, and the portions in Walter Fairservis (1967), Lawrence Leshnik (1973), and H. T. Lambrick (1964) that summarize British and other accounts from the late nineteenth and early twentieth centuries. (The archaeological data employed by Fairservis, Leshnik, and Lambrick are out of date and largely incorrect, but their historical agricultural data are still valid.)

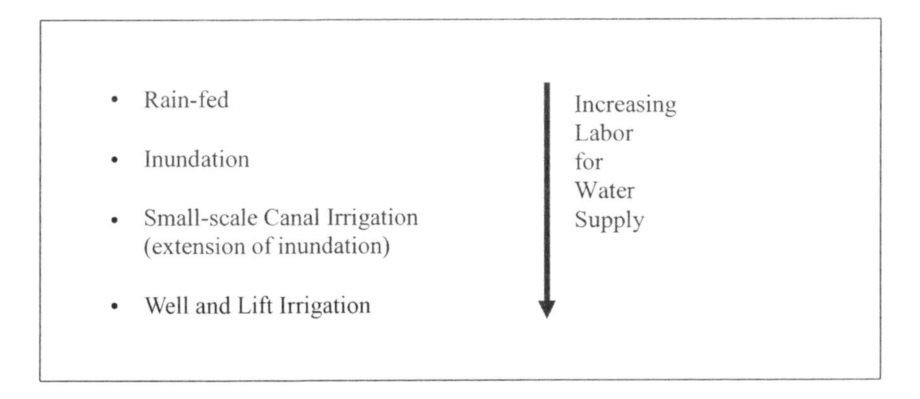

Figure 5.3. Possible water-supply systems in the Indus Valley floodplains.

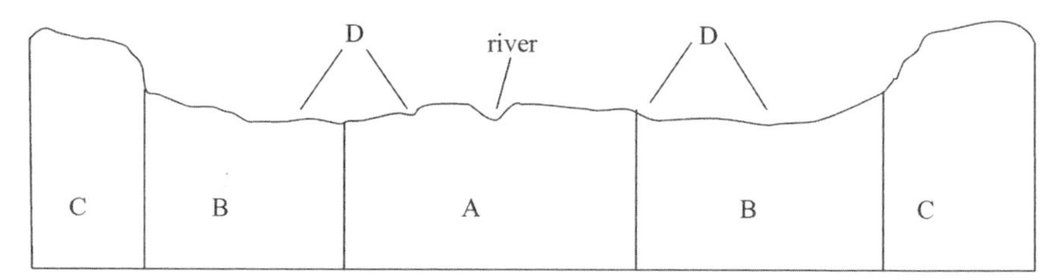

Figure 5.4. Land-use zones of the Indus Valley floodplains (after Leshnik 1973; vertical scale greatly exaggerated). A. Cultivated areas, parallel and close to the river (heavily inundated). B. Cultivated areas farther from rivers (inundated, extensions of inundation made via natural and artificial channels, and use of wells). C. Uplands (for grazing and rain-fed crops). D. Depressions, oxbow lakes, and marshes (oxbows most common in Punjab, marshes most common in Sindh).

These four water-supply types are primarily distinguished on the basis of the *land type* where they are in use (Figure 5.4). However, these water-supply types also have varying labor requirements. For all of the Indus agricultural systems, labor was needed for plowing and other soil-preparation stages, for planting, for weeding, for supply of water, and for harvesting, as discussed above. In general, the labor requirements other than water supply would be the same for all of these systems, with the exception of decreased need for soil preparation in inundation areas. The primary difference in labor relates to water supply; there is a roughly increasing hierarchy in labor requirements for their water-supply aspects between these four types, as shown in Figure 5.3.

However, these agricultural systems should not be seen as exclusive alternatives. It is likely that most or all of these types of water supply were practiced during the Indus period. The particular choice of type(s) would have depended on local environmental conditions, climatic conditions (including annual variations), labor availability, and land availability. The Harappan Phase people in the Indus floodplains almost certainly used multiple agricultural systems in varying proportions rather than one type versus another, depending on socio-economic goals and micro-environmental conditions.

Inundation

The Indus tributary rivers of the Punjab and especially the Indus River in Sindh are very aggrading, with the river perched above the floodplain around it, so that it easily overspills its banks with any increase in water supply. Such floods can cover the land for several kilometers, or even result in shifting the river to a completely new bed (Fairservis 1967; Flam 1981, 1986; Habib 1999:31–33; Lambrick 1964; Leshnik 1973). This phenomenon was the norm for the river systems of the Indus Valley during the third millennium BCE as well. Seasonal inundation was probably the single most important source of dependable water for crops in the Harappan Phase of the Indus Valley Tradition, particularly in the areas of lower rainfall to the south in Sindh. Winter crops would have been planted in areas inundated by the annual summer floods after the water had receded, leaving behind newly deposited sediments (Figure 5.4, zone A and portions of B). The soaked sediments contained extra water and nutrients for the young plants. Only a little winter rain would then be necessary to produce many types of crops. This was the primary agricultural system used for most of the region until the early 1900s, prior to the British-built large-scale canals and the deep tube wells of the last half-century.

Variability and risk. The annual summer inundation of the Indus River system is primarily dependent on snowmelt from the Hindu Kush and Himalayan mountains (which is dependent on winter precipitation), and secondarily on the summer monsoons. If the past climate was similar to the present, the inundation system was likely about the same during the Harappan Phase. Being dependent on rain and snow, the extent and to some degree the timing of the inundation is variable within the annual cycle.

More important, the frequent river avulsion (movement of the river to a new bed) common in these aggrading rivers made previously valuable inundation land into much less valuable land useful only for risky rain-fed agriculture. This would have had a major impact on stability of agricultural produce, unless farmers were able to quickly make use of the new inundation lands near the new bed of the river.

Labor requirements. Inundation farming required the least labor input of these four agricultural systems. As with rain-fed farming, there were no labor costs associated with water supply in this system. In addition, labor costs associated with soil preparation might also have been avoided, with direct planting of seed in the newly deposited riverine sediment.

Likely importance. River inundation would have been the primary method of agricultural production for winter crops, as long as enough land was available. This method could only have been cautiously used for summer crops, which ran the risk of being washed away by the summer inundation. Therefore, the proposed additional focus on summer crops at the end of the Harappan Phase most likely involved extension onto non-inundation land, employing other water-supply systems.

Land ownership implications. The prime areas for agriculture would be those inundated by riverine flooding in the summer and available for planting of winter crops. The use of inundation land would supply both fresh silt (highly fertile land) and moisture for the germination period. There would thus be competition for this prime land. However, frequent river shifting in the Indus Valley region would complicate ownership of it. It may have been advantageous to hold land in common, by kin group or clan for example. A very large region could have been owned by the group, allowing common planting and harvesting of land plots within this region as the river shifted within it. Any other systems of land ownership would have required some sort of structure for the reassignment of land after the river shifted, to maintain economic and social stability. All of these systems would have required an authority with the accepted power to reapportion land within the group on a regular basis, and also an authority to settle disputes between groups about the reapportioning of the larger tracts of land. These two authorities might be the same, or they might be different authorities; for example, group leaders might do the reapportioning within groups, while a larger corporate or single authority might settle disputes between groups.

Rain-Fed

Rain-fed agriculture could theoretically be practiced on any type of land within the Indus Valley floodplains. In this section I will discuss rain-fed agriculture as practiced on non-inundated (Figure 5.4, zone C and portions of B) land. However, in this semiarid landscape with often irregular rains, the land inundated by the summer floods was by far the most regularly used for "rain-fed" agriculture during the recent past. Thus, although rain-fed and inundation water-supply systems are treated separately here, as they potentially relate to different categories of land, they were preferentially used in conjunction in most historically known cases, except for drought-resistant crops.

In many ways the Harappan Phase use of rain-fed agriculture is the most difficult to reconstruct, as relatively minor changes in the climatic system might have a large effect on the amount, timing, and duration of the two rain systems affecting the Indus Valley: the Indian summer monsoon from the east and the winter Mediterranean or cyclonic rains from the west. The Indus is the border between the two systems, and thus the extreme range of both, so it is sensitive to any climatic change. Unfortunately, our knowledge of the climatic regime for this region during the Harappa Phase is sketchy and controversial, in large part because of the lack of suitable locations for pollen cores. The data and conclusions to date are well summarized in Fuller and Madella (2000), with additional notes in Fuller (2003; especially see footnotes) and Marie-Agnes Courty (1995). Courty's (1995) use of micromorphological as well as geomorphological research has provided much-needed new environmental evidence for this region, and has led her to conclude that conditions were not wetter during the Harappan Phase, as seems to be the general consensus given the current data.

Variability and risk. If we assume that the Harappan Phase climate was not so different from that of today, we can assume that this was a semiarid to arid region where rainfall could be quite variable from year to year. Such conditions would make strictly rain-fed agriculture risky, especially to the south in Sindh where rainfall would be low in the best of years. Furthermore, the existence of two different rainfall regimes has to be taken into account, as the relative strength of these two regimes is reversed between the eastern and western parts of the Indus Valley. At least at present, rainfall is usually greater in the winter to the west and in the summer to the east, but the decreased temperatures and transpiration rates of the winter period can make this season of greater importance for rain-fed agriculture, even in the eastern Punjab.

Labor requirements. The primary labor costs for rain-fed agricultural regimes are plowing and other ground-preparation methods, planting, perhaps weeding, and harvesting. Water supply involved no labor cost at all.

Likely importance. Strictly rain-fed agriculture would most likely have been used for rabi (winter-planted) crops, but likely only in the north for more water-demanding crops. It was perhaps most substantially used for perennial crops, with additional watering in poor rainfall years (see Well and Lift Irrigation below), since perennial crops could not be planted in the inundation floodplain. Strictly rain-fed agriculture in non-inundation areas could possibly have been used with risk for kharif (summer-planted) crops, particularly drought-resistant crops.

Land ownership implications. The prime areas for rain-fed agriculture, those inundated by riverine flooding in the summer and available for planting of winter crops, are discussed in the inundation section above. Land not inundated by the river could also be used for rain-fed agriculture, especially for drought-tolerant crops, and there would be considerably less competition for this land. It is possible that the increase in drought-tolerant crops like millets at the end of the Harappan Phase was encouraged by the desire to make use of this additional land. At least for the region around Harappa, populations became more concentrated at the urban center at the end of the Harappan Phase, abandoning other settlements (Wright 2000), so that land in the immediate vicinity would have become a matter of competition, either for group or individual ownership.

Small-Scale Canal Irrigation from Rivers (Extension of Inundation)

The Harappan Phase people certainly had great technical knowledge of water movement in the form of well-engineered water-disposal systems within their cities (Jansen 1993). This movement of water and sewage within the city is extremely impressive as a coherent, complex system that must have required extensive planning, technical ability in construction, and dedication to maintenance to keep the entire system flowing.

Such technical knowledge and managerial ability could have easily been applied to small-scale irrigation canals, which would have been on a similar scale of size and complexity of drainage. Given these abilities, the archaeological data we have, and the advantages of extended water supply, such small-scale irrigation canals were no doubt regularly employed by Harappan Phase farmers to extend the reach of the annual inundation or perhaps the regular riverine water flow. Such canals, described in eighteenth-century British accounts for this region (Lambrick 1964; Fairservis 1967; Leshnik 1973; Habib 1999: 33), simply breached the ridges along the sides of the active river channels, allowing annual summer inundation to overflow into canals, which were often natural channels that had been widened or straightened. These canals used gravity flow to extend the reach of the annual inundation, allowing lands beyond the reach of the natural inundation floodplain to be watered (Figure 5.4, zone B).

However, large regional-scale canals are considerably less likely. Construction and especially maintenance of such large canals would have been difficult because of the aggrading, shifting nature of these rivers, and their benefits might have been ephemeral. Not only would periodic investments of large quantities of labor be needed to clean out and realign large canals, but within a decade or less the river at the head of a canal might well shift its course some distance away. Certainly there is no evidence from the distribution of Harappan Phase settlements that large-scale or long irrigation canals were employed. Even four thousand years later, during the great Mughal Empire of the seventeenth and eighteenth centuries CE, the major irrigation canals in this region were constructed by simply digging out and extending old abandoned river channels, with a few notable exceptions of very large-scale works (Habib 1999:33–39).

The most thorough investigations of possible Harappan Phase irrigation canals have been done by Courty and Francfort in the eastern Punjab region in the Ghaggar plain, around the relict beds of the seasonal Ghaggar-Hakra or Sarasvati hydrological system (Figure 5.1). As noted above, Courty (1995) has done considerable work on the ancient environment of this region, but she has also focused on the possible evidence for irrigation by Harappan Phase farmers. The only irrigation canal found so far dated to 2000 BP, two thousand years after the Harappan Phase (2600–1900 BCE). However, soil crusts found in farmed, buried soils of the Harappan Phase period are "typical of irrigated soils regularly flooded by water with a high suspended load" (Courty 1995:120). This could represent the inundation of ancient fields by the small-scale canal extension of seasonal inundations in natural channels, but as Courty says, more field data are needed to support such a conclusion. In an earlier paper, Francfort (1992) refers to (undated) 1 m deep by 300–500 m wide "canals" that he describes as "the courses of ancient natural waterways which were used and perhaps, in some places, rerouted by man." The lack of a similar description by Courty suggests that her subsequent work either dated these features to a later period, or attributed them to shrewd use of areas naturally irrigated in rainy years by natural extension channels.

It is also noteworthy that Courty concludes that the Ghaggar-Hakra was never a perennially flowing river during any part of the Harappan Phase, but only seasonal meandering channels primarily based on monsoon rainfall, with occasional years of stronger flooding. This finding contradicts earlier conclusions based on the tracing of old riverbeds and associated archaeological sites, assuming that the river flowed perennially in the past because sites were located there. Clearly there is a substantial need for sophisticated geoarchaeological work of the sort Courty carried out to resolve this critical issue.

Variability and risk. Temporal variability in the water supply from small-scale canals extending the seasonal inundation into additional lands would match the variability and risk of inundation agriculture. Years with low water levels would be more risky, when inundation floods might not reach to the end of the extended system. The only other risk would be that of unusually strong floods bursting through canal walls and flooding areas planted with summer (kharif) or perennial crops.

Labor requirements. Labor associated with short canals to extend inundation areas would be of two types: initial construction and periodic maintenance. Initial construction could range greatly in labor costs, from canals that were simply short breaches in river ridges, to those constructed by digging out and straightening natural overspill channels, to canals dug for some distance without benefit of previous natural channels. For the first two types of canals, periodic maintenance might well require as much labor as the original construction, given the very high silt load of the Indus hydrological system and the tendency for floods to wear away canal walls.

These construction and periodic maintenance labor costs would be all that was required for small-scale canals used to extend the seasonal inundation. We do have evidence for sluices in the Indus city water systems (Jansen 1993; Kenoyer 1998), so if gravity flow was possible for a given field irrigation canal, the Harappan Phase people certainly had the technical skill to create a fairly sophisticated gravity-flow system, if desired. More likely, however, would have been the simple opening or blocking of access to a given canal by opening a gap or filling it with sediments, as desired. Other types of small-scale canals, those used for water supply beyond the inundation season, required more labor as gravity flow alone would not be sufficient to transport the water to the canals. These types of canals are discussed in the next section, Well and Lift Irrigation.

Likely importance. Although we have very little if any solid evidence for small-scale irrigation canals, I think they must have been widely used in the floodplains during the Harappan Phase, particularly toward the end of the period. Short canals would have been invaluable for opening up new areas for agriculture. They might also have been extremely important in evening out agricultural risk, allowing additional sources of water to otherwise rain-fed crops during periods of poor rainfall, although at the expense of some labor. Leshnik (1973:74) notes that because of the nature of the land, canals used to extend the inundation were especially prevalent in Sindh, where rainfall is also the lowest.

Information about whether late Harappan Phase rice was wet-cultivated or dry-cultivated will be particularly interesting in relation to the use of small-scale canal irrigation combined with summer inundation for rice and other kharif (summer-planted) crops. During the summer, when the rivers were full or flooding, a canal could be extended to a field just beyond the inundation, or fields within the inundation area could be surrounded by low walls and water allowed in as desired. Water supply would be plentiful, and the only risk would be of too much water breaking in on the field (Lambrick 1964; Fuller and Madella 2000:349).

Land ownership implications. This type of water-supply system has some of the same implications for land ownership as inundation. Given the movement of rivers, holding land in common might be the most desirable option for long-term security. This would also create a labor force of sufficient size to create and maintain the small-scale canal systems envisioned under a kinship-based managerial system.

Well and Lift Irrigation

Unlike the case in peninsular India (Morrison 2000, this volume), there was little use made of artificial tanks for water storage in the Indus Valley floodplains during any time period, perhaps because natural water storage bodies in the form of oxbow lakes and depressions were available (Figure 5.4, zone D). Instead, the floodplains of the Indus River systems were famous during historic and early modern times for the very widespread use of wells for agricultural water supply, especially in the Punjab (Habib 1999:28–30; Leshnik 1973), and particularly in lands adjacent to the active floodplain (Figure 5.4, zones B and A).

The same lift-irrigation techniques used in well irrigation could have been used to obtain water from other sources, including oxbow lakes and inundation-filled depressions, as well as the active river channels themselves (when not in flood). Such techniques involved the use of lifting devices to raise the water and then pour it into small-scale field canals, canals identical in construction to those described in the previous section. Alternatively, water could have been lifted into a small holding device or directly into containers, which would then be carried to individual plants. This last is such a labor-intensive method that it could only have been used for a small number of plants, perhaps household trees or small orchards.

We know that the Harappan Phase people were capable of building deep wells because we find wells lined with specially made wedge-shaped baked bricks within the sites, especially Mohenjo-daro (Jansen 1993). We know they were capable of building and maintaining fairly sophisticated gravity-flow drains, as discussed in the previous section. But what about lifting devices? Historic accounts mention three types of lifting devices for this region: the "chain-of-pots" or "Persian wheel"; a rope-and-pulley system with a container; and *shaduf*-type devices (Habib 1999:28–30; Lambrick 1964; Leshnik 1973; Watson 1983:105, 191).

Lambrick references a Harappan Phrase representation of a figure operating a *shaduf*-type lifting device but provides no provenance for this representation, as Fuller and Madella (2003:349) point out. While we have no other evidence as yet for shaduf-type lifting devices (a container on a weighted pole supported by a pivot) during the Harappan Phrase, such a device is certainly a likely possibility. There is no direct evidence for a rope-and-pulley system with a container either, but wells of the period at Mohenjo-daro show characteristic rope marks in the bricks at the top of the well, indicating the use of ropes to draw up water. Historically, these were often pulled by animals. However, the more usual animal-run lifting device, the chain-of-pots or so-called Persian wheel (most likely developed in India), is almost certainly a much later development, although the period when these water-lifting devices are first developed is uncertain (Watson 1983:105, 191 footnote 15; Habib 1999: 28 footnote 23). Thus, water-lifting devices available to the Harappan Phase peoples had relatively high labor requirements, as compared to historic populations in this region, and may have required primarily human rather than animal labor.

Variability and risk. Risks were very low for these methods; droughts would have to be very severe to dry up these sources of water. The danger of summer flooding of fields might be an issue for kharif or perennial crops planted near the river. The chief risk for this water-supply system would be insufficient labor to raise the water to the irrigation canals.

Labor requirements. This water-supply system is by far the most labor intensive of the systems presented here, no matter what variation of the system was employed. First, labor would have to be invested to create and maintain irrigation canals, as with the system described previously. Second, lifting the water into the canals would have required either seasonal or year-round labor, on a more or less frequent basis, depending on the type of land being irrigated and the season of the year. Crops planted beyond inundation areas would have to be watered periodically whenever rainfall was insufficient. Even within the inundation-extended areas, crops might need to be watered during the winter if the rains failed, or during the early summer or late fall if the inundation was low in a given year.

Likely importance. This water-supply system is likely to have been very important for perennial crops and those requiring year-round care in smaller plots, such as orchards, house gardens, and extensive vegetable plots, providing a large part of the daily food and also exchange foods. This system would also have been necessary for double-cropping on the same plot of land, except in years of high rainfall. A major question, when using the historic analogy of widespread use of well irrigation, is whether the great savings in energy provided by the chain-of-pots/Persian wheel greatly increased reliance on lift irrigation, so that the Harappan Phase people are unlikely to have used well irrigation in such abundance, given the higher labor costs. I think a substantial use of well irrigation is still likely for the Harappan Phase, since the other two lifting techniques could certainly have been used, although human labor demands probably greatly limited the extent of land utilizing this system. Therefore, it is less likely that well and lift irrigation would have been widely used for extensive field crops, as is recorded in historic and early modern accounts for the Punjab.

Land ownership implications. With sufficient labor, the use of well and lift irrigation could theoretically extend the available agricultural land to any part of the floodplain, although in actuality the higher water levels closer to the rivers would make distant areas less attractive. However, labor requirements might result in smaller parcels of land being farmed with this system, given the much higher daily and seasonal labor costs for water supply. The crops grown were likely to be those of high value or types needing year-round water. Taking all of these factors into consideration, lands watered by well and lift irrigation are the most likely to have been owned by individuals or close family groups.

SUMMARY POINTS: WATER-SUPPLY SYSTEMS AND AGRICULTURAL REGIMES

There is obviously a gradient of labor inputs from rain-fed and inundation agriculture (which require no labor for water supply), to canal irrigation extending the inundation, to well and lift irrigation (Figure 5.3). For the last two, periodic labor crews would be needed to build and maintain the canals, but this would be easily handled by a small group of people employing kin-group or village-level managerial organization (Leshnik 1973), as we see for most parts of the world. The major question from a labor perspective is how water was supplied to small-scale irrigation canals. The differing labor requirements of extending the annual inundation versus lifting water by the bucketful from a well or lake are immense. There are major distinctions in labor as well as season of use between these two types of irrigation that need to be made clear when referring to "canal irrigation" as a source of water. The easiest archaeological method to differentiate between these two types is by looking for the heads of the canals—are they from the inundation floodplain channels, or from lakes or even wells? Determining this is not as simple as it sounds for the Indus Valley floodplains because of the great alterations to the floodplains since the Harappan Phase. Literally meters of alluvium have been deposited across the old land surface, from a few meters in the Punjab to as much as 12 m in Sindh.

Based on recent and historic analogies, I expect that Harappan Phase agricultural systems in the Punjab depended on all four of the water-supply types described here: inundation, rain-fed, canal irrigation to extend inundation, and well and lift irrigation. Rainfall is higher here and more predictable than Sindh, and the lower temperatures reduce transpiration of moisture from the soil. The regions of the Punjab flooded by inundation and capable of extended inundation by canal irrigation are less extensive than in Sindh but still impressive. Finally, the consistent references throughout the last half-millennium to high amounts of well irrigation for the Punjab indicate that we must seriously explore this method of water supply, especially for intensive agriculture, even though it is by far the most labor intensive. For Sindh there was probably a much greater dependence on natural inundation and canals extending the inundation, given the natural landscape. Habib (1999:32–33, footnote 49) cites two authorities, Lambrick and Hamilton, to the effect that the vast majority of crops in Sindh prior to the nineteenth century were rabi crops produced using inundation agriculture, and Leshnik (1973) supports this. However, even in Sindh, well irrigation likely played an important role for selected crops.

The employment of each of these water-supply systems has different implications for labor requirements, as well as for land ownership systems—such as the amount of land one family could farm, the long-term profitability of a particular portion of land, or the way labor for improvements was provided (Fuller 2001:412; Morrison and Sinopoli 1992; Pollock 1999; Earle 2000; Stanish, this volume). But the multiplicity of systems available to Harappan Phase farmers was in itself a force for stability for the overall agricultural system. Farmers could employ a variety of systems in a given year, depending on the land type and labor available to them. This diversity allowed for security against the loss of one particular plot from either drought or flood or from the unexpected shifting of the rivers.

ECONOMIC, SOCIAL, AND POLITICAL IMPLICATIONS OF AGRICULTURAL SYSTEM MODELS

There are two more topics to which this modeling exercise can contribute. One is an ongoing debate about the nature of agricultural change at the end of the Harappan Phase, in relation to labor and land use. The other is the unexplored topic of land ownership systems during the Harappan Phase, as a basis for group or individual wealth and power. Both of these topics provide examples of the connections between agricultural production and evolving social or political systems.

Agricultural Change at the End of the Harappan Phase

Working at the urban site of Harappa, Weber (2001, 2003) has written about the increased ubiquity of kharif crops at the end of the Harappan Phase and the beginning of the following Punjab Phase ("Late Harappan" period), although rabi crops are still in the majority for both periods. In other words, the inhabitants of Harappa during the late third and early second millennia BC appear to have increasingly used two cropping regimes. While it would be helpful to have other sites in the Indus Valley floodplain for comparison, continued work on the Harappa material appears to be verifying this change. As specialists working in the region have indicated, the issue at present is whether this change consisted of an intensification or extensification of the agricultural system in the region (Weber 2001; Fuller 2001; Fuller and Madella 2000:353–355). For purposes of assessing water supply and labor requirements, however, we require more than a simple dichotomy of intensive versus extensive, as I will show. Instead, more detailed analyses of the combination of particular agricultural systems employed and added need to be made in order to assess the likely effects on economic, social, and political systems.

Fuller (2001:412) notes and Weber (2001:414) agrees that the term *intensification* should be reserved for the original Boserupian sense of "strategies used to increase yields from a given unit of land." (See Morrison [2000] for more extensive discussion of the term *intensification*). I also use *intensification* in this sense, which in this case would refer to multi-cropping on a single plot, primarily alternating rabi and kharif crops, but also possibly a continuous cycle of fast-growing vegetable crops on one plot. I use *extensification* to refer to the addition of new lands to an agricultural system. This includes the use of different lands during different seasons, which Fuller (2001) seems to include as part of his category of "diversification."

Intensification and extensification can both come about through the addition of new crops, new technologies, increasing input of labor, and, for this region, increasing water supply. The addition of new crops is a fact, although their relative importance is still somewhat debated. Technological innovations could potentially have had an impact on these agricultural systems. For the Harappan Phase case discussed here, however, long-term technical *experience* (where to plant each crop type, where to drill for wells, where to locate and how to shape field canals for maximum efficiency) was probably more important than new technical *inventions* (plows, digging implements, sluice gates, channel construction for gradient flow). Most of the relevant technical inventions were simple and were already known to the Harappan Phase people, and the new cropping regimes seem to have been gradually adopted, allowing time to develop experience. Except for the possibility of fertilization, still undetermined, increasing input of labor to increase yields for this region would relate directly to increasing the water supply. So how does the water-supply system relate to this discussion?

Extensification. Planting distant natural inundation areas that might or might not receive water in a given year could have been used as a somewhat risky extensification strategy, as documented ethnographically for Gujarat to the southeast (Reddy 1994, 1997). Such a strategy would involve minimal labor input, with the primary gamble being the potential loss of seed. Reddy's work on food versus fodder production shows that the use intended for each crop would be important in the effort put into their production. Rain-fed agriculture is likely to have been similarly used as a risky extensification strategy, with Harappan and Punjab Phase farmers regularly planting distant fields with drought-resistant crops in the hope that some, if not all, would produce a small crop, as is attested ethnographically to the northwest of the Indus Valley (Thomas 2003). Requiring more labor but less risk, small-scale canal irrigation extending the inundation would have been the primary system of extensification used during the Harappan Phase, in my opinion. It would have expanded the land that could be used for cultivation by extending the inundation area farther away from the river, particularly for the planting of rabi crops in wet soils (Figure 5.4, zone B). Extensification could also be accomplished by using well and lift irrigation beyond the active inundation floodplain, but the high labor requirements make this water-supply system a better candidate for intensive agriculture, unless land was at such a premium that even the distant uplands were needed for production. Given the relatively low percentage of available land under cultivation in this region during the much later Mughal period (Habib 1999), when populations were considerably higher, it is extremely unlikely that the Harappan Phase farmers were pushed to such a measure.

Intensification. Regular inundation areas are unlikely to have been used for intensification, as standing kharif crops would be flooded by the inundation. The same is probably true of areas irrigated through canal extension of the flood, as fields carefully protected from inundation waters, to keep from drowning young kharif, perennial, or vegetable crops, would not receive the inundation sediments and soaking needed to grow the most successful rabi crops. Nor would rain-fed agriculture alone have allowed reliable intensification in the form of multi-cropping in any region of the Indus. Depending on two separate and rather unpredictable rain regimes would not be a reliable agricultural strategy, although it might have been successful on occasion in the Punjab, especially for drought-resistant crops. However, winter rabi crops watered by rain or distant canal extensions could have been grown on the same ground as well- or lift-irrigated kharif crops. Canals far enough beyond the inundation to carry only slower-moving water, and not silt, could be controlled during the summer high waters to prevent flooding the kharif or perennial crops. The addition of well- or lift-irrigation systems to rain-fed or distant canal-extension lands is thus by far the most likely candidate for intensification systems in the Indus floodplains.

As in most classic intensification systems, the main limiting factor to intensification becomes labor supply. For rain-fed lands farmers would need summer labor to lift water from wells or other sources of water such as oxbows or marshes. For areas where canal extensions of the inundation provided reliable water (but not sediments) for the winter (rabi) crop, labor requirements would be especially intense: building and maintaining canals for the winter crop, making sure the summer crops were protected from flooding, and lifting water from the canals or other sources for the summer (kharif) crop. With labor, however, the previously less-desirable non-inundation lands (Figure 5.4, zone B farther from the river and lower areas of zone C) become the areas of richest continuous production. Some care must be taken by farmers to fallow or wash these lands periodically, though, as these non-inundated lands are also the most vulnerable to salinization with long-term use, especially in periods of low rainfall when the salts are not washed down below the surface.

Overall, my first three systems—inundation, rain-fed, and canal-extension water supply—offer the best agricultural systems for extensification projects and growing large amounts of staple crops. The fourth system, well and lift irrigation, is the most likely water-supply system for creating dependable intensification through multi-cropping on a single plot of land. This system is particularly suitable for small-scale farming, including vegetable gardens and orchards, for personal consumption as well as exchange. It could also have been used on a larger-scale level if enough animal or human labor was available to supply sufficient water by operating the lifting devices. This combination of water-supply and land-use systems is well suited to the sort of mixed farming system reconstructed for the Indus Valley Tradition, with cattle available for traction and possibly fertilizer (Miller 2003; Meadow 1996).

Thus, I stress again that the Harappan Phase people in the Indus basin had several options for water supply, using multiple agricultural systems in varying proportions rather than one type versus another. Therefore, we are ultimately interested in examining the relative proportions of the agricultural systems employed, and in *how the relative dependence on particular systems may have changed over time*, in relation to both extensification and intensification. Such proportional rather than absolute changes in the agricultural regime with regard to water supply are similar to Weber's (1999, 2001, 2003) discussions of changes to the crops used toward the end of the Harappan Phase at the city of Harappa, changes involving addition of crops and increased use of certain cropping regimes rather than replacement. A summary of proportional use of water-supply systems during the Harappan Phase is not possible with current data, but it needs to be a component of future research.

There are some existing examples of what we might expect to find. Courty's (1995) research in the Ghaggar plain (eastern Punjab region) indicates that the Harappan Phase farmers were favoring micro-depressions where rainfall was retained, as well as lowlands where seasonal inundation dropped fresh silt as well as water. She speculates, based on soil evidence, that canals extending the seasonal inundation waters may have been constructed. This one example shows Harappan Phase farmers using two, possibly three, of my four water-supply systems in one relatively small region. There is no other archaeological data of a similar nature, but Leshnik's account of early modern agricultural systems in Sindh might be a good analogy to the Harappan Phase case, and certainly Leshnik uses it as one. Leshnik (1973:80) notes that agricultural practices were both extensive and intensive, with extensive use made of large amounts of land covered by inundation for rabi crops, but intensive use of areas around wells. In the lands around wells kharif crops were grown closest to the wells, to reduce the distance of water transport, and the best rabi crops grown on this same land, but additional rabi crops grown at a farther distance, in an "infield/outfield system" (Leshnik 1973:80).

It is noteworthy that double-cropping has not been suggested for the beginnings of the Indus civilization as an aspect of state development but rather for its end. Weber discusses this apparent contradiction of increasing agricultural complexity at a time of political and urban dissolution; as usual, the Indus seems to be different from other early states. Not the origin but the end of the Indus state may be the period when we begin to have major changes in agricultural management, changes that may or may not be linked to the dissolution of the Indus cultural and political entity. It will be particularly important to discover if the double-cropping systems involved extensification or intensification. For example, if extensification was the primary strategy, the agricultural changes may be related to such diverse factors as increasing population density requiring use of additional lands; increasing rainfall allowing use of rain-fed agricultural lands; or decreasing river shifting allowing increased canal building. If intensification was the primary strategy, increased use of well and lift irrigation would require more labor, either human or animal.[3] However, farmers would then be less tied to inundation as a reliable basis for agriculture, and there might be less need to manage the redistribution of large tracts of land when rivers shifted if farmers are now more densely packed around the city. All of these scenarios have very different implications for the dissolution of the Indus political structure, as is clear when examining issues of land ownership more closely.

Agricultural Demand and Land Ownership

Leshnik (1973:80) notes that the limitations on the area cultivated during the Harappan Phase was likely to depend on the availability of water, the availability of labor (animal and human), and the size of the cooperating social unit. All of these factors affect the long-term stability and productivity of the agricultural systems used and, ultimately, the wealth and power that can be derived from them. Within these limitations, expansion of the system would have been related to demand for agricultural products.

To date, the main discussion of the role of agricultural production in the social and political organization of the Harappan Phase relates to its place in a safety net of food-exchange systems, proposed as one of the reasons for the Harappan Phase cultural and economic integration (Kenoyer 1998). Beyond acting as a cross-regional hedge for bad years (which implies a truly large-scale trade system in staple foods), such widespread exchange in foods, if it occurred, would encourage the production of crops of all kinds for exchange, for local demand as well as distant. Alternatively or additionally, rising populations of non-producers during the urbanizing Harappan Phase would require additional food production for exchange purposes. Whatever the impetus, farmers prior to and during the Harappan Phase would have had incentives to increase their crop production, whether intensively or extensively. Especially for areas near towns and cities, we would expect to see increasing competition for prime land, increasing investment into new lands, and/or increasing use of methods to produce more from the range of land available. At the same time, the high value placed on inundated land would have exacerbated disputes about land ownership that were sure to arise whenever the rivers shifted.

Given this scenario, two strategies may have been pursued, depending on the crops involved, the land and labor available, and, as Leshnik put it, the cooperating social unit. First, the best solution to river shifts was likely common ownership of huge amounts of inundated and canal-extended inundated land within a kin group (Miller 1991:122). Such an ownership system would still require a structure for rapid reallocation of fields to farmers within the group if the river moved, preventing loss of production while land squabbles were solved. This structure could easily be kin-based, with elders allotting land to families and settling disputes; an external managerial elite would not be required. However, there would likely be a need for a structure beyond the kin level to settle disputes between kin groups, once population became more densely packed around growing towns and eventually cities. These might be institutionally based structures, with representatives (whether temple priests, elected members, or a high-ranking hereditary elite) settling land disputes between the kin groups. Given the much higher rate of river movement for the Indus hydrological system than for other early states, rapid reallocation of inundation land would have been a priority for the long-term stability of the agricultural system. A form of allocation based on general consensus within a corporate group organization, rather than imposition by an elite, is much more likely given the ability of Indus farmers to move elsewhere in the vast alluvial plains if too much pressure was applied. Such a more or less consensual structure and communal land ownership could be an essential social aspect of the peculiar invisibility of Harappan Phase elite. Any power based on land allotment would depend on at least appearing to appeal to consensus within communal ownership, rather than imposing one's will, so that excessive display of power or wealth would be counterproductive. In line with Jonathan Kenoyer's (1998) suggestion for the continued role of kin groups during the urban Harappan Phase, Indus kin groups with power bases set firmly in the control of land ownership may have functioned as leveling structures, cross-cutting other non-kin-based hierarchies.

The second strategy for dealing with increased demand relates to increased improvements to and labor investment in non-inundation land, in order to increase production for exchange. This includes extending the inundation to new areas by building canals, although these new areas would be even more vulnerable to river shifts than natural inundation land and subject to the same needs for reallocation of land. It also includes building canals, wells, and lifting devices to allow the use of well and lift irrigation in a variety of land types. On a small scale this could be done by individual farm families, allowing the accumulation of family wealth through increased production. Such increased production could be by either intensive or extensive methods, adding new crops grown in other seasons or allowing production of perennial crops; the possibilities are extensive. But such improvements could also be sponsored by corporate groups of different kinds in exchange for a proportion of the returns, as Kathleen Morrison and Carla Sinopoli (1992) document for much later periods in South India. In either case, agricultural production could become a source of wealth which could be reinvested in more land, more labor, or other venues. It may be that this second strategy became more important at the end of the Harappan Phase, with increasing population densities and decreasing land availability around the cities, at least in the area around Harappa (Wright 2000). If the basis for agricultural complexity in the form of double-cropping was being managed at the family or kin-group level, it is not surprising to see this complexity remain or even increase at a time when more centrally controlled systems (such as record keeping and long-distance exchange) were tied to political dissolution.

CONCLUSION

My picture of Harappan Phase agriculture is very different from Wittfogelian hypotheses about the origins of the state stemming from the need for state-level institutions to manage and supply labor for large-scale irrigation projects designed to provide needed agricultural surpluses. In contrast, I propose that there was no need for the major extension of agriculture into new areas via large-scale irrigation projects, as smaller-scale methods of water supply were adequate to supply the needs of the early Indus state. What is more, the aggrading, shifting nature of the Indus River and its Punjab tributaries do not favor large-scale irrigation projects, particularly in the southern regions of Sindh.

I also have suggested that the majority of agricultural land may have been commonly owned and redistributed by extended kin groups, rather than by individuals or institutions. Only relatively small plots of intensively watered land, requiring high labor input, were likely to have been owned by individuals or small family groups. Such a land ownership system could help explain the unusually "flat" (truncated) hierarchy characteristic of the Indus civilization, as one major focus of wealth and power would continue to be controlled by kin-based mechanisms. In contrast, shifts from communal land ownership to some individual ownership have been postulated for Mesopotamia during the Akkadian period, a time of regional state formation (Pollock 1999:121–122). As noted, Indus kin groups with power bases set firmly in the control of land ownership may have functioned as leveling structures, cross-cutting other non-kin-based hierarchies within the social and political system. Such a strong kin-based power base within a state-level society would function to limit one of the key elements of a state—hierarchies based on power over non-kin. The political leaders of the Indus state(s) may instead have based their power on religious authority, control of trade or manufacturing, or other methods of legitimation. The lack of state involvement in the management of agricultural systems also explains why agricultural systems remained complex, or even increased in complexity, at the end of the Harappan Phase when other complex economic, social, and political structures were dissolving.

This is an entirely hypothetical reconstruction of power bases—but other conclusions about Indus power bases are currently grounded in equivalently hypothetical data. This combination of cross-cutting groups controlling different aspects of the political economy does help to explain some of the puzzling features of the Indus evidence for social status that have resulted in debates about its status as a state—the lack of differences in burial goods, the lack of representations of rulers, the meager evidence for extreme socio-economic hierarchies (Possehl 1998). Some later South Asian states have similarly invisible rulers, with regard to burial goods and representations, even though they wielded considerable power, but such power was seldom absolute and seldom extended to all aspects of the economy. Overall, the modeling of potential water-supply systems for the Indus Valley floodplains during the Harappan Phase is a surprisingly productive exercise, given the relatively small amount of archaeological data available. Whether correct or incorrect, these suggestions about Harappan Phase agriculture and its potential impact on social and political systems provide, I hope, a point of reference for future debates about the nature of Harappan Phase agriculture and the Harappans themselves.

Acknowledgments
Many thanks to Chip Stanish for kindly including me in this seminar, allowing me to dredge up memories of topics I have missed discussing for the past pyrotechnologically focused decade. I am eternally grateful to the 1987 core faculty of the M.Sc. in Bioarchaeology (Archaeobotany) at the Institute of Archaeology in London: David Harris, Gordon Hillman, and Ken Thomas. They cannot be blamed for my errors after all these years, but they must shoulder the responsibility for teaching me to think about plant use as procurement and production systems that were integral parts of economic and social systems. My gratitude goes to Seetha Reddy, Richard Meadow, and Bill Belcher for many past discussions about Harappan agricultural systems and to Steven Weber for his recent dedicated work on the topic. Finally, I much appreciate Dorian Fuller's steady supply of reprints of his extremely reliable, well-researched articles, even if they are peninsular-o-centric instead of (properly!) Indus-Valley-o-centric.

NOTES

1. Scarborough is hampered in drawing further conclusions by the need to explain the "rapid rise" of the Indus state postulated by Possehl (1990) in earlier writings. In fact, excavations at Harappa between 1996 and 2000 have well-illustrated the gradual evolution of the society at that site from village to city (Kenoyer and Meadow 2000, and additional publications in press). These data are especially important because Harappa is the only urban Indus site for which the transition has been investigated by modern excavations of all the periods in question and for which extensive radiocarbon dating has been done. The radiocarbon dating program is key to this discussion, as the argument for an unusually rapid development of urbanized Harappan Phase society has been based in part on the great homogeneity of materials throughout different levels of

the early (undated) excavations at Mohenjo-daro during the 1920s and 1930s (Jansen 1989). With the addition of radiocarbon dates, this homogeneity is now seen as evidence for gradual indigenous evolution of the Indus Valley Tradition and perhaps social mechanisms encouraging continued use of "traditional" cultural symbols. However, the dates from Harappa do not necessarily negate Jansen's (1989) suggestions for a rapid building of Mohenjo-daro at the beginning of the urban Harappan Phase period rather than following the normal gradual growth from town to city. It is entirely possible that the two cities followed different life histories, with Harappa operating as a normal diversified city and Mohenjo-daro functioning more as a capital city or preeminent ceremonial site for the civilization. However, it may be extremely difficult to ever assess the life history of Mohenjo-daro given that the early levels there are at least 8 to 12 meters below the surface, under the current water table. In any case, these new data solve Scarborough's (2003:143) difficulties with having to postulate a rapid change in agricultural systems to explain the presumed "precipitous growth around 2500 BC" of the Harappan civilization.

2. In this chapter I focus on the agricultural side of the Harappan Phase food-production system for the Indus Valley floodplains. Studies of other aspects of the Indus Valley provisioning systems include research into domesticated animal management (Meadow 1996; Meadow and Patel 2003; Miller 2003), fishing (Belcher 1998, 2003), and procurement of other wild plants and animals for food and other purposes (Madella 2003; Miller 1991; Weber and Belcher 2003a). Research has also been done on the rather different agricultural and pastoral systems from adjacent regions associated with the Indus Valley Tradition, especially Baluchistan to the west (Meadow 1996; Lambrick 1964; Fairservis 1967; Costantini 1984; Costantini and Costantini Biasini 1985; Thomas 2003; Tengberg and Thiébault 2003), and Gujarat and beyond to the southeast (Bhan 1992; Fuller 2003; Fuller and Madella 2000; Reddy 1997, 2003; Weber 1996). Weber and Belcher (2003b), which updates and summarizes previous and recent research, is a particularly valuable source for Indus Valley Tradition plant use, animal use, and environmental data.

3. The exception to an increase in labor requirements would be if this is the period when the "chain-of-pots" is first invented, allowing both intensification and extensification on a previously unknown scale. A case might be made for the sudden and widespread appearance of a new pottery type at the end of the Harappan Phase, the "Pointed Base Goblet" (PBG), as the containers used in a "chain-of-pots" lifting device. They are currently hypothesized to represent "disposable" drinking cups created at a time when at least the city of Harappa was expanding rapidly. The scoring marks on the sides of these ubiquitous and hastily made vessels would provide attachment points for string tying the pots to the wooden wheel; however, the restricted mouths might not be ideal for the purpose. Experimental work coupled with use-wear analysis and distributional analyses of finds of these vessels would be interesting. At present there is no evidence for chain-of-pots lifting devices till thousands of years later.

REFERENCES

Belcher, W. R.

1998 Fish Exploitation of the Baluchistan and Indus Valley Traditions: An Ethnoarchaeological Approach to the Study of Fish Remains. PhD dissertation, Department of Anthropology, University of Wisconsin-Madison.

2003 Fish exploitation of the Indus Valley tradition. In *Indus Ethnobiology: New Perspectives from the Field*, edited by S. A. Weber and W. R. Belcher, 95–174. Lexington Books, Lanham, Maryland.

Bhan, K. K.

1992 Late Harappan Gujarat. *Eastern Anthropologist. Indus Civilization Special Number* 45(1–2):173–192.

Charles, M., C. Hoppé, G. E. M. Jones, A. Bogaard, and J. G. Hodgson

2003 Using weed functional attributes for the identification of irrigation regimes in Jordan. *Journal of Archaeological Science* 30(11):1429–1441.

Charles, M., G. E. M. Jones, and J. G. Hodgson

1997 FIBS in archaeobotany: Functional interpretation of weed floras in relation to husbandry practices. *Journal of Archaeological Science* 24(12):1151–1161.

Costantini, L.

1984 The beginning of agriculture in the Kachi Plain: The evidence of Mehrgarh. In *South Asian Archaeology 1981*, edited by B. Allchin, 29–33. University of Cambridge Oriental Publications No. 34. Cambridge University Press, Cambridge, UK.

Costantini, L., and L. Costantini Biasini

1985 Agriculture in Baluchistan between the 7th and the 3rd Millennium B.C. *Newsletter of Baluchistan Studies* 2:16–30 + tables.

Courty, M.-A.

1995 Late Quaternary environmental changes and natural constraints to ancient land use (Northwest India). In *Ancient Peoples and Landscapes*, edited by E. Johnson, 105–126. Museum of Texas Tech University, Lubbock.

Earle, T.

2000 Archaeology, property, and prehistory. *Annual Review of Anthropology* 29:39–60.

Fairservis, W. A.

1967 The origins, character and decline of an early civilization. *American Museum Novitates* 2302:1–48.

1986 Cattle and the Harappan chiefdoms of the Indus Valley. *Expedition* 28(2):43–50.

Flam, L.

1981 The Paleogeography and Prehistoric Settlement Patterns in Sind, Pakistan (ca.

4000–2000 B.C.). PhD dissertation, Department of Anthropology, University of Pennsylvania.

1986 Recent explorations in Sind: Paleogeography, regional ecology, and prehistoric settlement patterns (ca. 4000–2000 B.C.). In *Studies in the Archaeology of India and Pakistan*, edited by J. Jacobson, 65–89. Oxford and IBH Publishing, New Delhi.

Francfort, H.-P.

1992 Evidence for Harappan irrigation system in Haryana and Rajasthan. *Eastern Anthropologist* 45(1–2):87–103.

Fuller, D. Q.

2001 Harappan seeds and agriculture: Some considerations. *Antiquity* 75:410–413.

2003 Indus and non-Indus agricultural traditions: Local developments and crop adoptions on the Indian Peninsula. In *Indus Ethnobiology: New Perspectives from the Field*, edited by S. A. Weber and W. R. Belcher, 343–396. Lexington Books, Lanham, Maryland.

Fuller, D. Q., and M. Madella

2000 Issues in Harappan archaeobotany: Retrospect and prospect. In *Indian Archaeology in Retrospect, Vol. 2, Protohistory: Archaeology of the Harappan Civilization*, edited by S. Settar and R. Korisettar, 317–390. Manohar and Indian Council of Historical Research, New Delhi.

Habib, I.

1999 *The Agrarian System of Mughal India, 1556–1707*. 2nd rev. ed. Oxford University Press, New Delhi.

Hillman, G. C.

1984 Interpretation of archaeological plant remains: The application of ethnographic models from Turkey. In *Plants and Ancient Man—Studies in Paleoethnobotany*, edited by W. van Zeist and W. A. Casparie, 1–41. A. A. Balkema, Rotterdam.

Hoffman, M. A., and J. H. Cleland

1977 *Excavations at Allahdino II: The Lithic Industry*. American Museum of Natural History, New York.

Howard, A., and G. L. C. Howard

1909a The varietal characters of Indian wheats. *Memoirs of the Department of Agriculture in India, Botanical Series* 2(7):1–66.

1909b *Wheat in India: Its Production, Varieties, and Improvement*. Imperial Department of Agriculture in India, London and Calcutta.

Howard, A., G. L. C. Howard, and A. R. Khan

1910 The economic significance of natural cross-fertilization in India. *Memoirs of the Department of Agriculture in India, Botanical Series* 3(6):281–330.

Jansen, M.

1989 Some problems regarding the Forma Urbis Mohenjo-Daro. In *South Asian Archaeology 1985*, edited by K. Frifelt and P. Sørensen, 247–254. Curzon Press and the Riverdale Company, London, U. K., and Riverdale, Maryland.

1993 *Mohenjo-Daro. Stadt der Brunnen und Kanäle. Wasserluxus vor 4500 Jahren. City of Wells and Drains. Water Splendour 4500 Years Ago.* Bergisch Gladbach, Frontinus-Gesellschaft, Bonn.

Jones, G. E. M.

1984 Interpretation of archaeological plant remains: Ethnographic models from Greece. In *Plants and Ancient Man—Studies in Paleoethnobotany*, edited by W. van Zeist and W. A. Casparie, 42–61. A. A. Balkema, Rotterdam.

1987 A statistical approach to the archaeological identification of crop processing. *Journal of Archaeological Science* 14:311–323.

Jones, G. E. M., A. Bogaard, M. Charles, and J. G. Hodgson

1997 Distinguishing the effects of agricultural practices relating to fertility and disturbance: A functional ecological approach in archaeobotany. *Journal of Archaeological Science* 27(11):1073–1084.

Kenoyer, J. M.

1998 *Ancient Cities of the Indus Valley Civilization.* Oxford University Press and American Institute of Pakistan Studies, Karachi.

Kenoyer, J. M., and R. H. Meadow

2000 The Ravi Phase: A new cultural manifestation at Harappa (Pakistan). In *South Asian Archaeology 1997*, edited by M. Taddei and G. De Marco, 55–76. Istituto Italiano per l'Africa e l'Oriente (IsIAO) and Istituto Universitario Orientale, Naples, Rome.

Kenoyer, J. M., and H. M.-L. Miller

1999 Metal technologies of the Indus Valley tradition in Pakistan and western India. In *The Archaeometallurgy of the Asian Old World*, edited by V. C. Pigott, 107–151. University of Pennsylvania Museum Monograph 38. University Museum Publications, Philadelphia.

Lal, B. B.

1979 Kalibangan and the Indus Civilization. In *Essays in Indian Protohistory*, edited by D. P. Agrawal and D. K. Chakrabarti, 65–97. B. R. Publishing, Delhi.

Lambrick, H. T.

1964 *Sind: A General Introduction.* History of Sind, vol. 1. Sindhi Adabi Board, Hyderabad, Pakistan.

Leshnik, L. S.

1973 Land use and ecological factors in prehistoric north-west India. In *South Asian Archaeology*, edited by N. Hammond, 67–84. Duckworth, London.

Madella, M.

2003 Investigating agriculture and environment in South Asia: Present and future contributions of opal phytoliths. In *Indus Ethnobiology: New Perspectives from the Field*, edited by S. A. Weber and W. R. Belcher, 199–250. Lexington Books, Lanham, Maryland.

Meadow, R. H.

1996 The origins and spread of agriculture and pastoralism in northwestern South Asia. In *The Origins and Spread of Agriculture and Pastoralism in Eurasia*, edited by D. R. Harris, 390–412. Smithsonian Institution Press, Washington, DC.

Meadow, R. H., and A. K. Patel

2003 Prehistoric pastoralism in northwestern South Asia from the Neolithic through the Harappan period. In *Indus Ethnobiology: New Perspectives from the Field*, edited by S. A. Weber and W. R. Belcher, 65–94. Lexington Books, Lanham, Maryland.

Miller, H. M.-L.

1988 Preliminary Analysis of the Plant Remains from Tarakai Qila: Confessions of a Law Consumer. Unpublished M.Sc. thesis, University College London.

1991 Urban palaeoethnobotany at Harappa. In *Harappa Excavations 1986–1990: A Multidisciplinary Approach to Third Millennium Urbanism*, edited by R. H. Meadow, 121–126. Prehistory Press, Madison, Wisconsin.

Miller, L. J.

2003 Secondary products and urbanism in South Asia: The evidence for traction at Harappa. In *Indus Ethnobiology: New Perspectives from the Field*, edited by S. A. Weber and W. R. Belcher, 251–326. Lexington Books, Lanham, Maryland.

Morrison, K. D.

2000 *Fields of Victory. Vijayanagara and the Course of Intensification*. Munshiram Manoharlal Publishers, New Delhi.

Morrison, K. D., and C. M. Sinopoli

1992 Economic diversity and integration in a pre-colonial Indian empire. *World Archaeology* 23(3):335–352.

Oxford University Press

1955 Oxford Economic Atlas for Pakistan. Prepared by the Cartographic Department of the Clarendon Press, Oxford. Economic information compiled by the Intelligence Unit of *The Economist*. Geographical advisor: C. F. W. R. Gullick. Pakistan Branch, Oxford University Press, Karachi.

Percival, J.

1921 *The Wheat Plant; A Monograph by John Percival*. Duckworth, London.

Pollock, S.

1999 *Ancient Mesopotamia. The Eden that Never Was*. Cambridge University Press, Cambridge, UK.

Possehl, G. L.

1990 Revolution in the urban revolution: The emergence of Indus urbanization. *Annual Review of Anthropology* 19:261–282.

1998 Sociocultural complexity without the state: The Indus civilization. In *Archaic States*, edited by G. M. Feinman and J. Marcus, 261–291. School of American Research Press, Santa Fe, New Mexico.

Reddy, S. N.

1994 Plant Usage and Subsistence Modeling: An Ethnoarchaeological Approach to the Late Harappan in Northwest India. PhD dissertation, Department of Anthropology, University of Wisconsin-Madison.

1997 If the threshing floor could talk: Integration of agriculture and pastoralism during the Late Harappan in Gujarat, India. *Journal of Anthropological Archaeology* 16:162–187.

2003 Food and fodder: Plant usage and changing socio-cultural landscapes during the Harappan Phase in Gujarat, India. In *Indus Ethnobiology: New Perspectives from the Field*, edited by S. A. Weber and W. R. Belcher, 327–342. Lexington Books, Lanham, Maryland.

Scarborough, V. L.

2003 *The Flow of Power: Ancient Water Systems and Landscapes*. School of American Research Press, Santa Fe, New Mexico.

Shaffer, J. G.

1992 The Indus Valley, Baluchistan and Helmand traditions: Neolithic through Bronze Age. In *Chronologies in Old World Archaeology*, edited by R. Ehrich, 425–446. University of Chicago Press, Chicago.

Tengberg, M., and S. Thiébault

2003 Vegetation history and wood exploitation in Pakistani Baluchistan from the Neolithic to the Harappan period: The evidence from charcoal analysis. In *Indus Ethnobiology: New Perspectives from the Field*, edited by S. A. Weber and W. R. Belcher, 21–64. Lexington Books, Lanham, Maryland.

Thapar, B. K.

1973 New traits of the Indus civilization at Kalibangan: An appraisal. In *South Asian Archaeology (1971)*, edited by N. Hammond, 85–104. Noyes Press, Park Ridge, New Jersey.

Thomas, K. D.

2003 Minimizing risk? Approaches to pre-Harappan human ecology on the northwest margin of the Greater Indus system. In *Indus Ethnobiology: New Perspectives from the Field*, edited by S. A. Weber and W. R. Belcher, 397–430. Lexington Books, Lanham, Maryland.

Watson, A. M.

1983 *Agricultural Innovation in the Early Islamic World. The diffusion of crops and farming techniques, 700–1100*. Cambridge University Press, Cambridge, UK.

Weber, S. A.

1996 Distinguishing change in the subsistence and the material records: The interplay of environment and culture. *Asian Perspectives* 35(2):155–163.

1998 Out of Africa: The initial impact of millets in South Asia. *Current Anthropology* 39(2):267–274.

1999 Seeds of urbanism: Paleoethnobotany and the Indus civilization. *Antiquity* 73(282):813–826.

2001 Response. Harappan seeds and agriculture: Some considerations. *Antiquity* 75:413–414.

2003 Archaeobotany at Harappa: Indications for change. In *Indus Ethnobiology: New Perspectives from the Field*, edited by S. A. Weber and W. R. Belcher, 175–198. Lexington Books, Lanham, Maryland.

Weber, S. A. and W. R. Belcher

2003a Introduction. In *Indus Ethnobiology: New Perspectives from the Field*, edited by S. A. Weber and W. R. Belcher, xi–xx. Lexington Books, Lanham, Maryland.

Weber, S. A., and W. R. Belcher (editors)

2003b *Indus Ethnobiology: New Perspectives from the Field*. Lexington Books, Lanham, Maryland.

Wittfogel, K. A.

1976 [1957] *Oriental Despotism. A Comparative Study of Total Power*. Ninth printing. Yale University Press, New Haven, Connecticut.

Wright, R. P.

2000 Landscape, history and archaeology of the Harappa countryside. In *South Asian Archaeology 1997*, edited by M. Taddei. Istituto Italiano per l'Africa e l'Oriente (IsIAO) and Istituto Universitario Orientale, Naples, Rome.

Zohary, D., and M. Hopf

1988 *Domestication of Plants in the Old World*. Oxford University Press, New York.

Europe

The Barren and the Fertile

Central and Local Intensification Strategies across Variable Landscapes

TINA L. THURSTON

Archaeologists have long asked questions about the connections between political power, economics, and the role that state governments play in what we interpret as agricultural intensification attempts and all the ramifications—social, political, and economic—that may be inferred from them. Until the 1990s a fairly simple line was often drawn, connecting large-scale efforts to intensify production and the idea of top-down directives from authoritarian rulers. Recently, this idea has been questioned, largely via the consideration of ethnographic case studies, which sometimes illustrate the coercive or suggestive power of elites and sometimes indicate that labor mobilization and the building of substantial agricultural features can be traced to farming communities that are capable of imagining, creating, and maintaining large, complicated, and laborious intensification schemes without authoritarian commands. This has led some researchers to replace the top-down assumption with general skepticism about the extent of elite authority in organizing intensification.

Studies of current problems by sociocultural anthropologists, economists, political scientists, development experts, and others demonstrate that in between state coercion and total self-organization lies a large variety of processes: negotiation with the state, disgruntled or resigned semicompliance, rebellion, and other strategies. Since the ethnographic trajectories of such intensification sequences are so invariably complex compared to the simple either/or archaeological paradigm noted above, it is useful to study archaeological cases that have some ethnohistoric context, testing various scenarios against the physical and textual evidence.

The goal of this study is to argue that a superior approach to such problems is to combine two current paradigms, to take advantage of their individual strengths, and mitigate their weaknesses: what Stanish (this symposium) identifies as the *political economic core principle* and the *agency/historical ecology core principle*. The author's long-term field project, "Decentralized Societies and the Development of Secondary States," which began in 1992 in southern Scandinavia and continues today, has advocated for the general principle that by examining several scales of analysis simultaneously, both large-scale, statewide data and local (regional and subregional-scale) data, a more complete picture of past processes can be obtained. In addition, the study of site-specific data must be combined with overarching cultural and natural landscape analysis and, when possible, its interface with ethnohistoric records and relevant ethnographic data (Thurston 1997, 1999, 2001).

This contribution thus employs a cross-cultural perspective, using archaeological, ethnohistoric, and ethnographic data to explore both central elite strategies and the possibilities of local, rural agency in the process of cereal and animal intensification during a period of state building and unification. The study compares two areas under a single political authority yet with highly variable environments, both "barren" and "fertile," differing agroeconomic practices, disparate local sociopolitical conditions, and differing relations with the state. When we compare the archaeology with our knowledge of ethnohistorical accounts, what can be construed as local entrepreneurship and what as top-down directive? How are these two strategies manifest in recent ethnographic cases, and how can we relate this to archaeological problems?

In the recent literature pertaining to government-sponsored agricultural intensification and development projects, most authors acknowledge some general shared and somewhat predictable problems for economic and development planning in nations with diverse regions—sometimes with ethnic or language differences, sometimes with highly varied resources, sometimes with a variety of land-tenure systems. When planning or dictating activities in such areas the solution is often a specialized regional approach, applying different strategies for different areas (Bhalla and Khan 1979; Misra et al. 1985; Ndongko 1974:4). This has often been supported by archaeological evidence in many lesser-known and well-cited cases (Brumfiel 1980, 1983; D'Altroy and Earle 1985; Feinman et al. 1985) where the state has organized its various regions in quite different ways, analogous to Stanish's political economic model, in which rulers, leaders, or other political elites direct intensification attempts and strategize carefully to exploit production on a differential basis.

Yet ethnographers tell us that even as development schemes are hatched by central elites, differing peasant traditions and historical circumstances in various areas substantially impact even powerful centralized plans. For example, in Ethiopia several formerly autonomous polities and some tribal and chiefly areas were conquered by Menilek, who became emperor in the late nineteenth century (Ståhl 1973:4). These territories, and the farmlands they harbored, remained in the hands of an imperial family for many generations. By the late twentieth century, in some cases peasants had been forcibly turned into tenants; in others, groups retained corporate ownership of land.

Ruthless schemes were enacted in these conquered/incorporated regions to squeeze greater amounts of tribute or cash from smallholders. In some places they were dispossessed of their tenancies entirely so that land consolidation and intensification could take place according to elite wishes. In yet others, traditional land-tenure practices sheltered farmers from the worst of these state intensification attempts, especially where large kin groups held land corporately instead of individually (Ståhl 1973:5–8). Ethiopia is a good example but by no means isolated: the impact of *political history* on the outcomes of local-central interactions has been noted in many ethnographic cases (Bartlett 1980:552–553).

In the Ethiopian case we can thus see both the political economy model and the agency/historical ecology model in play simultaneously in different regions: agency + history creating differences in preexisting relationships that seem to have affected the amount of restraint that state authorities used while pushing their programs of intensification. These historical, contingent conditions include those within a region itself, such as the existence of long-established corporate land-tenure practices that increased group solidarity and also those that relate to conditions between local and state authorities. In this case the state seems to have postponed conflict with large, unified kin groups that might mobilize politically or foment rebellions against central authority. In areas where local farmers were not well organized and did not have "ancient" established rights, the state did not restrain itself at all. Both types of human-centered and human-created relationships had a strong effect on the outcome of political economic designs.

Few ethnographers would be surprised that two such models—elite coercion vs. local agency—would operate at the same time within the same state organization. In fact, a great deal has been written about what has been termed the difference between "policy-led intensification" and "autonomous intensification" (Birch-Thomsen 1999; Lele and Stone 1990). Archaeologists, as cited above, have sometimes been able to demonstrate, supportably, that different organization strategies for agriculture (and other institutions) are used by prehistoric states in differing regions. Yet it is difficult for many to fully support claims for the actual mechanisms that permit the state to sweep in, as did the Ethiopian imperial regime, or for those mechanisms that fend it off, notably kinship or other types of solidarity, strengthened by the ability of a group to claim it follows "ancient" or established ways of doing things. Yet it is clear in current and historic examples that this mind-set—that one's group has inalienable rights to some place or practice—can create formidable resistance.

In the following case study and discussion I will use my own research context, protohistoric in nature, to try and illuminate how such processes might be visible in the archaeological record. In late Iron Age Denmark, the context of my studies, there are enough fragmentary ethnohistoric documents to paint an overarching picture of "central state" intentionality regarding surplus, taxation, and agropastoralism. Yet in almost every other respect only archaeology can illuminate the process of centralized attempts at enacting these intentions. I argue that careful study of the archaeological material does indeed show that clearly differing regional strategies were employed. Furthermore, ethnohistoric records from two of the "conquered" or incorporated areas that made up parts of the state indicate the response of the regional elites and peasantry, giving us some idea about the "agency" aspects of the process often either lacking or only theoretically inferred in many archaeological sequences.

AGROPASTORALISM AND THE STATE IN PREHISTORIC AND PROTOHISTORIC SOUTH SCANDINAVIA

Paleoecological and archaeological studies of southern Scandinavia (Figure 6.1) have shown that human influence has impacted landscape formation for the last 5,000 years (Birks et al. 1988:229). Farming has been practiced in southern Scandinavia since the early Neolithic of that region (ca. 3200 BC), and dramatic episodes of clearing, expansion, and intensification can be seen in pollen cores, soil studies, and other ecological indicators. Beginning in the Neolithic and continuing until recent times, monumental earthworks and constructions dot the region, and prehistoric villages, fortifications, and towns have left their mark as well. Conversely, the natural landscape has also been a strong determinant in prehistoric agricultural and settlement strategies.

To put these factors in context, our project takes the perspective that environment is an influence on human behavior but not a deterministic one or necessarily a limiting one. We do agree, however, that some environments necessitate special strategies. While Denmark contains no deserts, rainforests, or other environments that some consider likely to strongly impact human subsistence strategies, there are some fairly extreme differences between the regions that we will examine. Ultimately, these differences partly lie behind later patterns of urbanization, market activity, and agriculture.

It is therefore important to consider both the physical and cultural landscape of the later Iron Age in southern Scandinavia, the period from about 500 BC to AD 1050, and into the early Middle Age, to about AD 1200. We can focus both on synchronic variation and diachronic change and attempt to distinguish the interrelation between changes in the agricultural landscapes

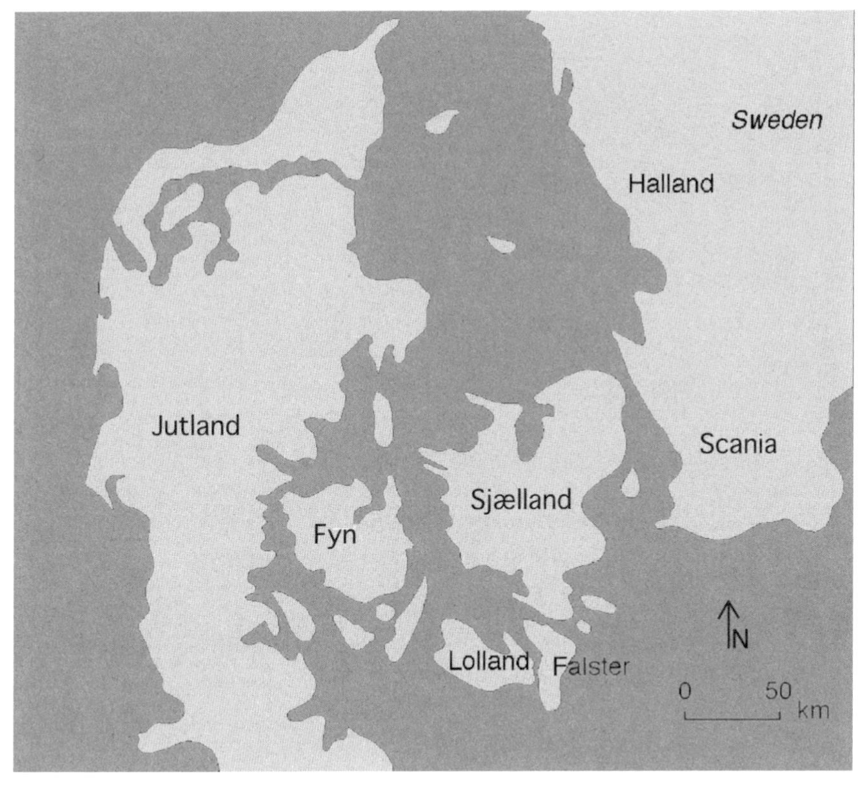

Figure 6.1. Denmark in the Iron Age: Jutland, the islands, Scania, and Halland.

and changes in sociopolitical organization during the unification of several regions into the state of Denmark.

Natural landscape conditions most affecting cultural practices include climate and temperature, geology and glacial effects, vegetation, soils, and topography. The cultural landscape includes the built landscape of settlements, both central places and hinterlands, monuments, roads, and fields but also the exploited nonanthropogenic features, such as bogs, shorelines, and forests, and the ideological use of springs, lakes, hills, and groves for ritual purposes. To address the cultural landscape, this chapter will examine archaeological features and historical data.

As noted, analysis on several scales is imperative to understanding both generalities and specificities of this case study: a general large-scale picture of southern Scandinavia and a closer discussion of two specific areas in which the author has conducted research (Figure 6.2). The first, Järrastads Herred, Scania, where the author carried out a five-year project from 1992 to 1996, is in an area with excellent soils and favorable conditions for cereal agriculture. The second area is Hassing Herred, Thy. This region lies in northern Jutland, with mostly poor soils formed in glacial outwash sands, on which heathland grew until late-nineteenth- and twentieth-century modifications rendered much of it into croplands. In historic and prehistoric times only a narrow strip of soil on the area's eastern shore was what might be considered "good farmland." The author has worked in this area since 1997 and is currently beginning a new three-year phase of research in the region.

These two regions were purposely chosen for comparative study because they are contrasting in their ecological nature. Such differing natural conditions suggest that different forms of agriculture were practiced, and different ways of life can be imagined. Additionally, and perhaps most interesting, each area had a distinct, prestate ethnopolitical identity, and the early ethnohistoric record indicates that both areas resisted incorporation into the state through bloody rebellions and uprisings during the time when the state was consolidating its power through the application of new taxes, laws, and land-tenure organization.

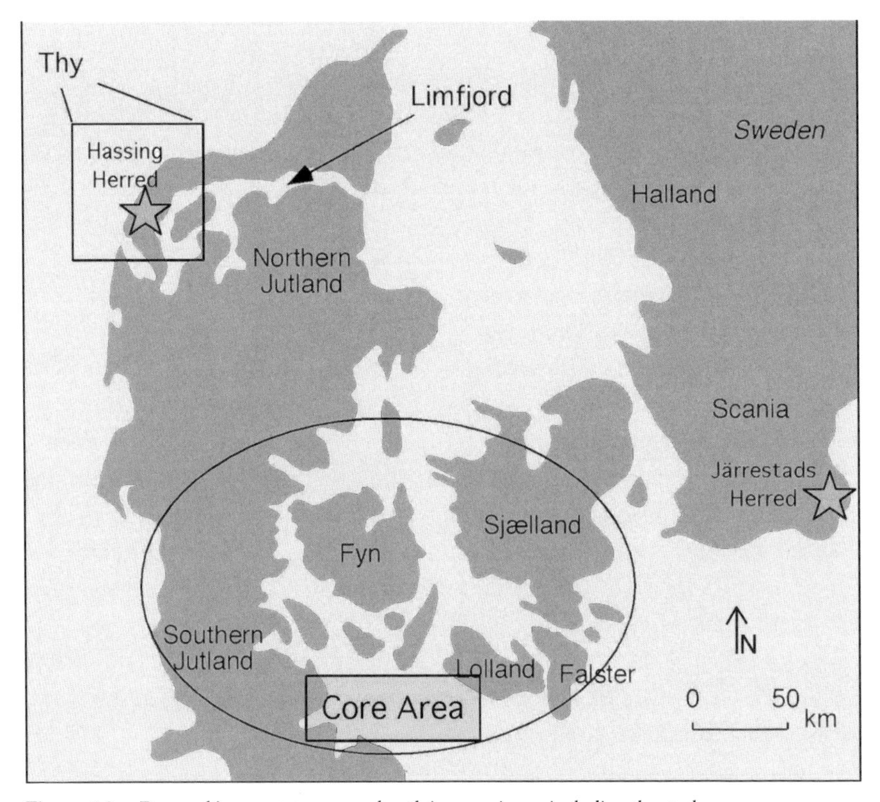

Figure 6.2. Denmark's core state area and outlying provinces, including the study areas.

One region was unsuccessful in its rebellion, which was quelled with several bloody episodes. The other achieved far more satisfactory results against state authority, although it was not able to completely forestall unification.

SPATIAL AND TEMPORAL BOUNDARIES OF THE STUDY AREA

The study area (Figure 6.1), which is sometimes referred to as "old Denmark," consists of the peninsula of Jutland, which lies above the Eider River, including Shleswig-Holstein, now part of northern Germany. It includes the four large islands Sjælland, Fyn, Lolland, and Falster, which lie to the east of Jutland, and 470-odd smaller islands that form an archipelago. Today's southern Sweden, or Scania, was the easternmost part of Denmark.

The Iron Age in southern Scandinavia is divided into several phases, and in most work pertaining to this region a culture-historic based chronology is used, which includes phases such as the pre-Roman, Roman, and Germanic Iron Ages, the Viking Age, and the early medieval. In this discussion I may refer to these terms, yet our project uses an entirely different parallel chronology that is based on political changes, which have corresponding social and economic shifts associated with them. Thus, we refer to the preintegration phase, the integration phase, and the consolidation phase, followed by the centralized state (Figure 6.3).

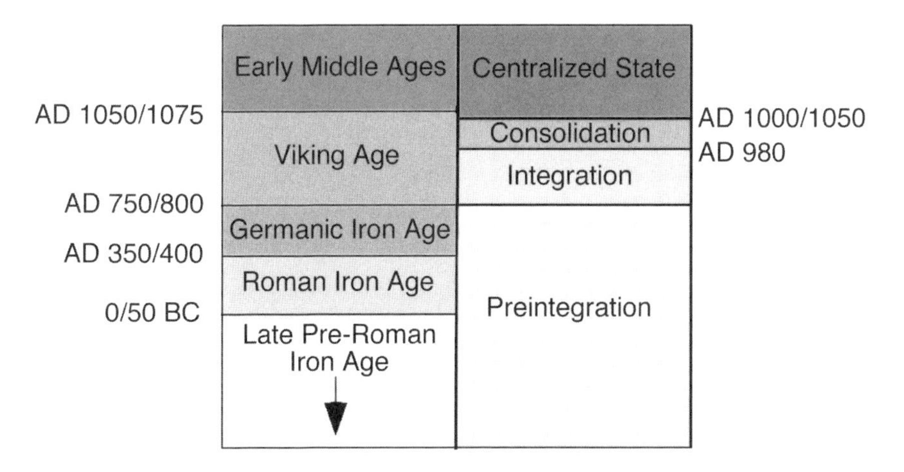

Figure 6.3. Chronology/phases for Danish state development.

THE NATURAL LANDSCAPE

Between 500 BC and AD 500, the Sub-Atlantic climatic minimum occurred, with its least favorable period occurring after 50 BC (Emanuelsson 1987). This period was considerably colder and wetter than modern times. After AD 500 the climate began to ameliorate, and by the late Iron Age in Denmark, AD 800–1050, temperatures were roughly similar to today's, ranging from 0 to 5 degrees centigrade in winter and 15 to 20 degrees centigrade in summer (Sømme 1968:118). Rainfall in western and central Jutland is about 750 mm per annum, with the islands and Scania reaching 1,000 mm (FAO 1981:Appendix 1). Deglaciated earlier than northern Scandinavia, the region that made up what would become Iron Age Denmark developed soil and vegetation earlier and is in the deciduous zone rather than the conifer belt, which lies directly to the north (Sømme 1968:119). Climate is classified as cool temperate Marine (FAO 1981:Appendix 9). The whole region lies in the Mesothermic temperate forest belt, with western Jutland consisting of boreo-atlantic oak and birch forest, and eastern Jutland, the islands, and Scania lying in the zone of Baltic beech forest (FAO 1981:Appendix 5).

Denmark's geomorphic structure falls into two types: old moraine, which escaped glaciation during the last Ice Age, and new moraine, which was covered with ice until the late Pleistocene, about 9,000 years ago (Sømme 1968:103). This dichotomy in turn affects the soils, vegetation, and landscape in the two regions (Figure 6.4).

The old moraine landscape, found on the western coast and central portion of Jutland, consists of glacial outwash plain bordered by salt marsh or Geest. Large deposits of outwash sand serve as parent material in much of Jutland. This outwash generated highly drained and low-nutrient soils (Sømme 1968:116). During parts of the Iron Age windblown sand eroding out of these areas also became a factor of soil formation, mixing with soils and sometimes covering entire settlements and their field systems, such as the village of Lindholm Høje in northern Jutland (Marseen 1959:66). This Viking Age village was completely covered with drift-sand during its occupation, forcing its abandonment. Situated in one of the few clay-soil areas of Jutland, its ridged field furrows and the tracks of a farm cart were perfectly preserved under the drifting dunes.

The new moraine is overlain by soils that are richer in clay and other nutrients and is the substrate for all the islands and the province of Scania. On Jutland new moraine is limited to a narrow strip on the east coast and small areas on the tip of the peninsula. In addition to these glacial contexts, soils were affected by the isostatic uplift occurring since the last Ice Age. The already poor west coast of Jutland is sinking, while the north and east is rising, resulting in

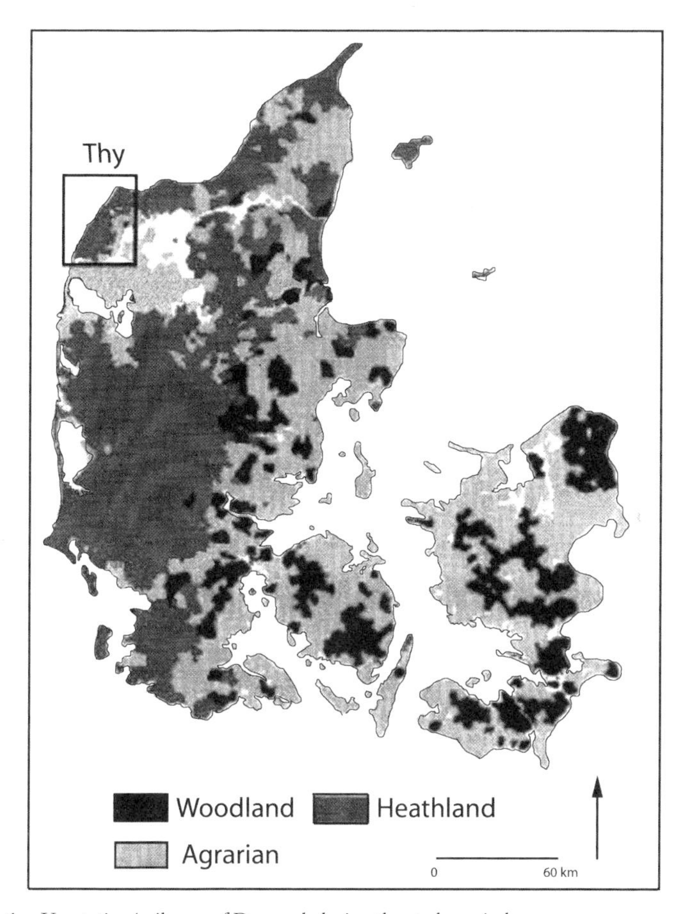

Figure 6.4. Vegetation/soil map of Denmark during the study period.

parent material of raised seabeds relatively rich in marine clays. More marine clays and sediments were deposited during several postglacial transgressions of the Baltic between 8000 and 4000 BC (Sømme 1968:114). These deposits added to the fertility of some soils.

As much as we will examine the agency paradigm in this discussion (which we will), these notable regional differences in geomorphology mean that soils are one of the primary factors relating to human subsistence and settlement strategies throughout prehistory, including the Iron Ages. Jutland's large, low-lying plain of over-drained outwash sand, plus the relatively high rainfall and moderate temperature, leads to the formation of several types of podzols, or sandy, nutrient-leached soils. Associated with these are histosols, soils with 75 percent or more organic matter, mainly peat (FAO 1974:24). The western coast of Jutland also contains large expanses of calcic fluvisols. These have been drained and reclaimed for farming since the nineteenth century (FAO 1981:67) but in the Iron Age were marshy hay meadows suitable for the cutting of animal fodder but not for cultivation. The very top of the Jutland peninsula and the eastern, Baltic, side has relatively large extents of eutric cambisols, soils far more favorable to agriculture (FAO 1974:37). The islands and the majority of Scania are also fertile, being largely made up of eutric cambisols and orthic fluvisols, both good agricultural media (FAO 1981:35).

Because of these soil conditions cereal agriculture on Jutland was limited and until early historic times relied on swiddening strategies, and the region was more suited to animal husbandry, while the islands and Scania were favorable for the production of wheat in climatically favorable times and for barley, rye, and oats during the Sub-Atlantic. These were the main grain crops of the Iron Age.

THE TRAJECTORY OF STATE POWER IN PREHISTORIC AND PROTOHISTORIC DENMARK

In Figure 6.3 one can note the traditional phase names on the left and the project's study phases on the right. *Preintegration* represents the era when several autonomous peer polities coexisted, *Integration* the long era during which slow unification and state formation occurred, and *Consolidation* a period of rapid shifts during which the state made decisive changes in the social, political, and economic orders. During the prestate era called the Roman Iron Age (RIA), Denmark consisted of several peer polities, best characterized as warlord chiefdoms, each with a chiefly center, surrounded by farming settlements (Hedeager 1992; Randsborg 1990). Concentrations of gold and prestige import goods on the island of Fyn (Hansen 1987; Randsborg 1990) presumably indicate the home of warlords powerful enough to monopolize importation of Roman items that were used to mark elite status and reward followers. By the end of the RIA and into the Germanic Iron Age (GIA), as smaller polities merged into larger, those elites who inhabited the island of Fyn appear, archaeologically, to be the wealthiest and most influential; early chronicles indicate that Fyn was the home of the first rulers of a coalescing, yet still decentralized, state.

From within Denmark, evidence is completely archaeological: apart from runestones, indigenous texts referring to contemporary and earlier events do not predate the twelfth century. *Germania* by Tacitus, a Roman historian whose surviving works document the era's imperial politics, is the primary source for this period. Recently his work has been scrutinized by archaeologists, who have found surprisingly accurate corroboration for the material culture he described (Hansen 1987; Hedeager 1992; Randsborg 1990). This is true of most Roman authors; in fact, although some of their perspectives are skewed, they are generally reliable and often sophisticated, almost ethnographic, in detail. Similar to the way in which Celtic archaeologists are able to avail themselves of Julius Caesar's descriptions of the Gauls and their social and political idiosyncrasies, most archaeologists and historians of Scandinavia and Northern Germany cautiously rely on descriptions of cultural practices described in Germania that do *not* leave archaeological traces.

Tacitus reported that chieftains had limited power and that leaders and followers had mutual bonds and obligations to each other. Rulers required assent from a powerful legislative assembly that met at predetermined times to debate the law and cast votes for or against the plans of warlords (Tacitus 1967 [AD 98]). Despite these "checks and balances," at the same time a vertical hierarchy operated within the war band, and social classes are evident. During warfare with Rome the chiefs would ally, and a single overlord was selected to lead collectively for short periods, but the support of peers was required (Lindow 1976:11). If an overlord seized too much power, the social code actively encouraged his assassination. Documented cases in Roman accounts specify that removal was aimed explicitly at those who were too "kinglike."

Beginning in the eighth century, contemporary *foreign* texts, such as the Frankish chronicles, allude to Danish elites in the Fyn/southern Jutland core, who successfully led allied groups against them. This may have bolstered core-area elites' growing power over former peers. Warfare simultaneously created a need for new sources of wealth, military manpower, and legitimation of leaders. The most visible solution was the outward trading/raiding expansion of the Vikings, eastward into Russia and southwestward to Europe. Yet this outward movement was far less important for state formation than the internal turmoil caused by changes at home. Here elites solved their problems of money and manpower by attempting the transformation of the autonomous, decentralized, and only loosely allied peer polities into a highly stratified, centralized state.

STRATEGIES OF ELITES FOR THE CONTROL OF DISPARATE REGIONS

Despite retaining decentralized characteristics, such states must still maintain a political economy and some kind of power over constituent parts. Economic, religious, and political decision making are channels through which centralized elites typically exert power. While elites in decentralized states may desire such power, it must be acquired slowly and indirectly or be disguised, for example, by using local officials as agents, maintaining illusions of local control.

In centralized states elites often intimidate and compel by dramatically displaying power, coercing labor, or encouraging belief in elite divinity, forcibly extracting support and obedience. Danish elites did not attempt this; it usually resulted in death. Similar to the strategy of Ethiopian elites in the mid-twentieth century, it is sometimes better to leave cohesive and powerful groups alone, while pursuing exploitation of others who are more helpless or less organized.

This does not imply that military or other violent solutions are not an option in such cases; they are simply costly and often not undertaken unless conditions change enough to force the government's hand. Such domination depends on direct coercion, but for controlling powerful entities that have, or believe themselves to have, inalienable rights, led by those who believe they are peers rather than subjects, hegemony is a better strategy: constructing political-economic conditions and sociocultural values that create a desire in subordinate groups to accept core primacy. Given the precarious nature of Danish rulers' power and the constant possibility of assassination either by elite rivals or one's own local constituency, it is not surprising that unification took a long time. Despite this, in a long, uneven, and difficult process the former peers were eventually subsumed into a unified, centralized state.

THE DUAL ROLES OF STATE POLITICAL ECONOMY AND REGIONAL AGENCY IN AGRICULTURAL INTENSIFICATION

To review the sequence: during the early part of the Viking Age a polity in the state's core area succeeded in unifying peers on southern Jutland and the large islands into a single kingdom (Figure 6.2). Markers of successful centralized elite directives appear during the 800s, in the form of massive bridges, complex systems of royal roads, royally administered markets, defensive earthworks, and large shipping canals that were cut through dry land at great labor costs. From runestones we know that different types of central officials oversaw activity in several towns in this core area.

In the areas of the map marked northern Jutland, Scania, and Halland, this was not the case. These areas remained remote and largely autonomous, even after the central state began to claim them as provinces in other runic inscriptions, something that can be observed by about the mid-tenth century. Partly, this can be traced to the long-standing political situation, in which allied peer polities sometimes operated in concert yet maintained autonomy. Another part of it can be assigned to ethnicity issues between the Danes and the neighbors they sought to control.

At first glance it may seem amusing and a little absurd that people inhabiting what are now southern Sweden and northern Denmark saw themselves as fundamentally different in many ways during the Iron Age. Yet these regions believed that they were ethnically different from each other and from the Danes of the central core area. Ethnicity is generally understood to be a self-defined concept based on beliefs about common descent and culture. It may be defined by kinship, eating habits, or historical experience as much as by appearance or language. The concept of ethnicity in archaeology is one that is sometimes difficult to nail down. It cannot always be tied to a "style," an artifact cluster, or a house type. Because of this we sometimes imagine that it is preferable to leave the concept out altogether rather than to venture into unknown territory and use it incorrectly. However, in our case study it is not impossible to address and may provide some models for other less-clear-cut cases.

That ethnicity is an important organizing structure in state politics is an established concept in cultural anthropology, as well as in a number of archaeological cases where it can be demonstrated (e.g., Brumfiel and Fox 1994). While current geopolitics are dominated by the concept of globalization, and "modern" political concepts were focused on the "bordered power container" of the nation-state (Giddens 1989:120), in other times and places ethnicity transcended many other sociopolitical categories within complex societies.

Ethnographers have no lack of evidence of its impact on the balance of power and the level of hostility or cooperation between region and center. Violent clashes can occur between minority groups and the ascendant group over trade, religion, language, culture, and domination vs. self-determination (Osaghae 1995:5). Battles over control of important economic resources are typical, especially when those who provide a nation's wealth feel marginalized (Osaghae 1995).

As if to further confound the archaeologist, ethnicity can have effects of an opposite nature in local affairs. When pressed by central authorities viewed as outsiders, or by any group considered to be outside their own, regionalized ethnic groups frequently use their common ethnicity for the "mobilization of local capital through self-help efforts" and an "unprecedented upsurge in the number and activities of ethnic unions of various complexions: 'development' unions, 'progressive' unions, 'hometown' associations, social clubs, community development associations, and cultural organizations" (Osaghae 1995:5). Reportedly, sometimes "new" forms of ethnic expression develop, but "the vast majority were old, sometimes moribund, associations that were invigorated to meet new challenges" (Osaghae 1995:6). This phenomenon—that ethnicity strengthens, revitalizes, or even initially forms most dramatically when challenges from "outside" appear—is a well-known aphorism in anthropology, a process sometimes referred to as tribalization (Whitehead 1995). Often tribalization occurs as a response by traditional peoples who are being subsumed within a nation-state (Cruz 1999). Ethnic unity movements may often take the shape of development, or intensification, of indigenous resources, be they industrial or agricultural, and often these movements indicate targeted strategies by locals to revive or bolster the health of local unity and identity (Stone 1998). As Southall (1988) expressed in the title of his exploration of the role of local centers in state intensification, "What else is development other than helping your own hometown?"

We can understand these identities if we examine what documentary evidence we have about northern Jutland and Scania during the preintegration phase of our study. North Jutland, and Thy (pronounced "tew") in particular, was the place of origin for a group that played a large role in the Roman accounts of the barbarian "migrations": the group the Romans recorded as the Teutones. In turn Scania was the homeland of the Langobards, who became known in later times as the Lombards. These "tribal" entities, as the Romans characterized them, were somewhat fluid in terms of the frequent creation and disbanding of alliances that were hard to keep track of—often, they traveled in large, shifting confederations as they roamed across Europe in Late Imperial times. This is not indicative of similarity in group identity but is instead an artifact of the long-standing alliance system that operated in their homelands for many centuries, at least.

During the period we refer to as the integration phase, when a slow, creeping unification was occurring (Thurston 1997) and external threats from the Franks united these groups, ethnic identity may have taken a backseat to creating a united front. By the late Viking Age the threat was mostly from would-be kings in Denmark, and an invigoration of old identities and solidarities may have well occurred.

THE POLITICAL AND ECONOMIC SEQUENCE IN SCANIA

Scania, the first autonomous/incorporated region to be examined, was a highly desirable territory. Around AD 1070 the chronicle of a German traveler to the Danish court described Scania as "fair" and rich, with a high population and "opulent" harvests and merchandise (Tschan 1959:191). Archaeological and paleobotanical evidence supports this: there were hundreds of large villages with populations of several hundred people each, and natural vegetation was almost entirely lacking. Since south-central Jutland, a large part of the core, was a sandy, agriculturally poor heath, Danish rulers had logical motives for incorporating Scania, but centuries of decentralization, combined with a strong local sense of ethnic identity, inhibited the process.

The earliest indigenous texts (Lund 1984; Scholz 1972) indicate that during the integration phase, Scanians loosely acknowledged the core-area Danes as *leaders* but probably not as *rulers*. As the core grew more centralized, settlement patterns did not change in Scania: there were no urban places, and the earlier chiefly centers continued to function, disconnected from central authority, unlike the core area, where signs of central authority are seen as early as AD 720 and increase through time. Neither the cultural landscape nor any institutions extant in the region exhibit centralization under broader authority. This does not mean that society did not function; rather, it reflects a system that people approved of and valued, one without central interference in established daily praxis or in higher-level political and economic events.

INCIPIENT AND MATURE CENTRALIZING TRENDS

In the mid-800s the state attempted to build infrastructure in Scania (Figures 6.5 and 6.6a). A group of seasonal export locations was established for collecting Scanian agricultural products and perhaps silver for taxes, and there is some evidence that marketplace activities accompanied the seasonal roundup of goods (Brattberg 1983; Ohlsson 1976; Rausing 1990). State control is evident in nearly identical names, administrative boundaries around each site, simultaneous founding, and near-perfect spacing for provincial control. Importantly, they are limited to Scania. They appeared when political unification existed, but systemic integration was incomplete. While this helped exploit Scania's wealth, it was a feeble attempt, since it added a method of wealth extraction but did not change preexisting political organization or regionalized institutions.

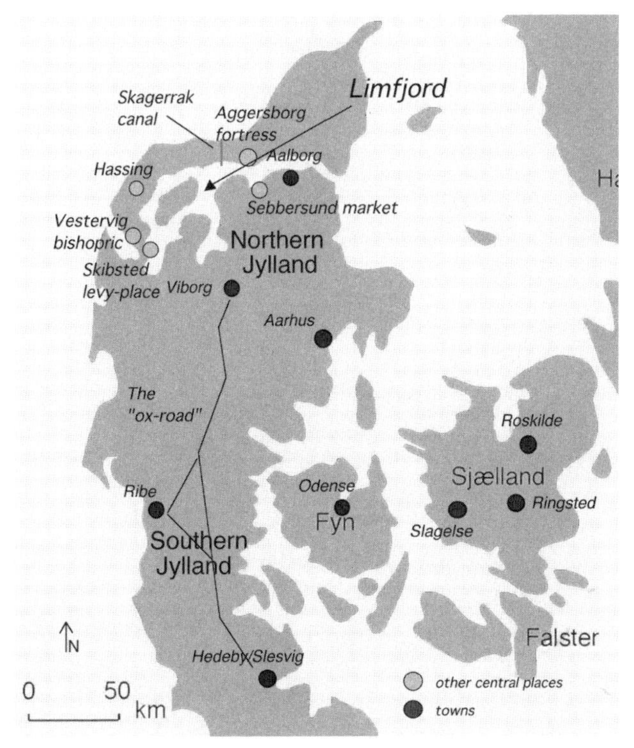

Figure 6.5. Towns in late Iron Age and early medieval Denmark, and some other places in northern Jutland mentioned in the text.

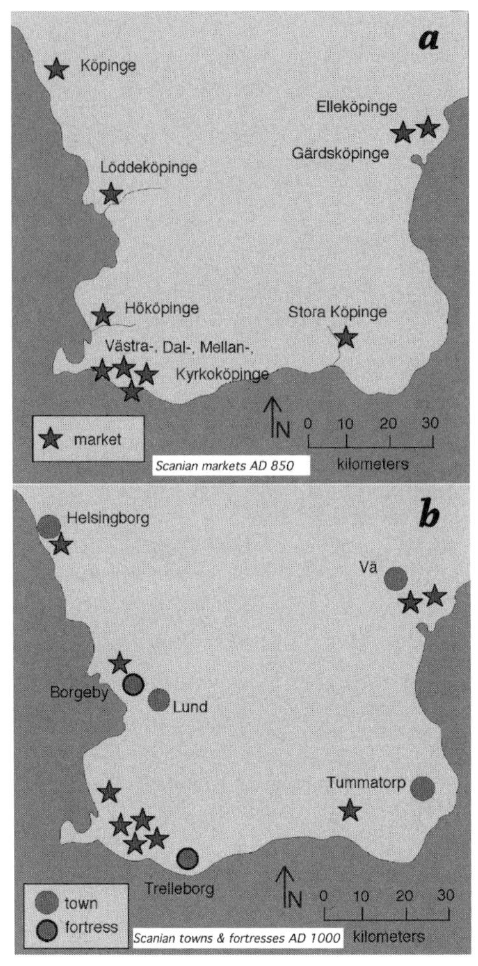

Figure 6.6. The development of Scanian markets, towns, and fortresses.

Then, between AD 980 and 1000, this pattern abruptly changed. Four evenly spaced administrative towns were founded in Scania by royal charter (Figure 6.6b). Law, taxation, coining, markets, industries, and important religious functions, previously practiced in rural locations, were all conflated at these new, urban central places, and redundant export markets were simultaneously abandoned, further indicating that they had not been as successful as desired. At the same time, the Scanian chiefly centers were abandoned. Also in this short period (dendrochronologically dated to the 980s) six large, nearly identical, centrally controlled fortresses, today called "Trelleborgs" (palisaded forts), were constructed around Denmark, two in Scania, one in North Jutland, and three in the general core region (Jacobsson 1995; Svanberg and Söderberg 1999).

AGRICULTURAL INTENSIFICATION AS A CENTRALIZING STRATEGY

Another strategy of the state may have been agricultural intensification and land reform. As will be seen, the course of intensification appears to be too large-scale, too simultaneous, and too ubiquitous to be completely assigned a local origin, although good arguments have been made that such conditions can be self-organized by farmers (Lansing and Kremer 1993). However, as will be discussed below, aspects of these changes strongly suggest otherwise, and these strategies, once in place, gave a great deal of economic and political advantage to the state, while reducing that of local people.

EVIDENCE FOR A STATE-SPONSORED MOVE TOWARD DISPERSED SETTLEMENT

The first of these possible state political economy strategies for agriculture is seen across Denmark but especially in Scania's rural hinterland. During the consolidation phase, the same 40 to 60 years that brought urbanization, military bases, and the disappearance of chiefly centers, thousands of *torps*, tiny, dispersed agricultural settlements, were founded, some of which appear to be laid out on identical, predetermined plans. Since it reduces labor and travel time, dispersed settlement often reflects intensification efforts (Drennan 1988:274).

These torp settlements were primarily carved out of already-used farmland, with some in new settlement areas. They were populated by people drawn from preexisting villages: as will be seen, older villages shrank by about the same amount of population that was needed to populate these new villages. While in other similar agricultural parts of Denmark, the density of these settlements is about .061 per square kilometer; in Scania it is .136 per square kilometer (Thurston n.d.), nearly double. This colonization effectively raised efficiency and yield on already-farmed land and opened up more arable land, so woodlands almost disappeared (Berglund et al. 1991:112). In addition to the labor saved in agriculture, the analysis of labor saved by conflation of various institutional services in one urban location—one of the four newly founded towns—also shows time savings, leaving farmers more time to spend on agricultural projects to raise yields (Thurston 1999).

EVIDENCE FOR FORCED SETTLEMENT REORGANIZATION

At the same time, there was an unusual restructuring of the older villages: they were internally reorganized and contracted as much as 60 percent, with excess population probably moving to the new torps. This involved the razing of each village, each of which had occupied the same locale for about 500 years, and the movement of previously loosely and somewhat haphazardly nucleated farms down to neat, side-by-side rows of houses along a village street, usually not farther than 250 to 500 meters from where the old farm had stood.

ADOPTION OF NEW INTENSIFICATION TECHNOLOGIES AND METHODS

Simultaneously, two-field crop/fallow rotation was replaced by the more intensive three-field system, and the low-productivity ard plow was replaced by the more productive moldboard plow.

DISCUSSION: POLITICAL ECONOMY OR AGENCY?

We do know, definitively, from early ethnohistoric texts, that at this time new taxes and regulations were put in place across Denmark, sparking much protest. It is quite conceivable that some of these changes observed in the agricultural landscape were locally organized strategies, as per the agency model discussed above. Farmers, pressed for higher taxes, may have adopted new crop rotations and technology; these were already in use in Western Europe and were probably not unknown. Farmers may also have pioneered new territory under duress to increase yields, although the much higher density of Scanian torps, compared to those in the islands, in what was already both the best and highest yielding land, is puzzling and suggests that some other factors, at least, were involved.

The deconstruction and rebuilding of extant villages, however, is a very strange phenomenon. Ethnographers working in all parts of the world have noted that such resettlement, when government sponsored, even over short distances, is highly disruptive and traumatic to farming people (de Wet 1993; Scudder and Colson 1982). Since the effect is similar cross-culturally, it is usually termed a cognitive effect, which we can argue would have operated 1,000 years ago much as it does today. The disturbance typically persists for a minimum of 18 to 24 months but usually can be observed for much longer. Even after years informants curse authorities that either forced or encouraged them to abandon their familiar, ancestral houses, outbuildings, and even pathways and, if possible, will eagerly move back to the village or farm, no matter how poor. There is also remarkable similarity, ethnographically speaking, in the way that governments knowingly exploit the havoc such moves create, since one symptom of uprooting is that long-held beliefs and practices are more easily lost or changed. It has even been suggested that this effect, as much as any agricultural "improvement," is behind such programs today (de Wet 1993). Since the practice of uprooting and moving people for purposes of control is at least as old as (and perhaps much older than) the textual evidence from the era of the neo-Assyrians of Iron Age southwest Asia, c. 900–600 BC, it is not impossible to hypothesize that this effect was anticipated by Danish rulers in the late first millennium AD.

Why should we link this change to political economy and not agency? In the earliest part of the historic period (ca. 1200), when written records became more common, it is clear that a system of rural taxation was in place with the following features: the state required farmers to site their actual households on closely packed regulated plots of land. The size of the house plot symbolized the size of the farmer's hectarage, which was spread in strips all over the village in what is called an "open field system." Thus farmers could not deceive tax collectors with vague or false descriptions of their holdings. The location of the farmhouses along a row, with fields behind, rather than out in the fields themselves, made census and taxation that much more efficient. Transfers of land through sale and inheritance would have to be updated from time to time, but for long periods the system would work with little adjustment.

Given the fact that it is highly unlikely that the total demolition, removal, and rebuilding of older villages was voluntary, combined with the later system of agricultural taxation, it is probable that the change we see in Scania's agricultural landscape during the consolidation phase denotes the beginning of this agricultural recording and taxation system. These "symbolic" landholding records were useful in times when even many state officials were illiterate. By the 1500–1600s this system was replaced with paper maps and accompanying land books that recorded every strip of land, who owned it, and what taxes were due.

Fortunately, we can say more about agency than the likelihood that people would not have disassembled their homes and farms willingly. Indigenous chronicles (Pålsson and Edwards 1986) indicate that in the eleventh century, specifically the 1080s, the Scanians met with the Danish king in the traditional assembly, demanding that "ancient" obligations and entitlements be restored, while refusing new taxes and increased military service requirements. A high church tithe had also recently been imposed, exacerbating their economic problems. When the king intimidated them by executing the loudest protesters, the Scanians quieted for a while, but by the 1180s they were once more demanding that "foreign" leaders be removed (core-area Danes) and taxes be repealed. They soon proclaimed independence, initiated violent attacks against both church and state representatives of central government, and elected a king from a pool of prestate elite (Andersson 1947). After a series of slaughters called the Scanian Uprisings, Scania lost its war with the Danes.

With an admittedly "opulent" economy and a satisfactory way of life, local Scanian elites and peasants alike had little reason to willingly pay more taxes to maintain a state and church that they found economically oppressive. The traditional system included some taxation but also a great deal of autonomy from the state. They also shared the Germanic tradition of rejecting or killing overbearing rulers, a freedom they probably still perceived, given their bold protests to the king (Thurston 2001). The bloody protohistoric uprisings were a continuation of a long-term attempt to remain disarticulated from the state. Why did it take 100 years, or five generations, for dissatisfaction to erupt into violence? Those who are prosperous are often politically conservative and less willing to endanger their lives and their holdings. They will usually attempt to negotiate as long as possible, to maintain their way of life before resorting to bloodshed. In the same 100 years the state's mechanisms for control also grew stronger. The Scanians apparently waited too long.

THE POLITICAL AND ECONOMIC SEQUENCE IN NORTH JUTLAND

There is much evidence supporting widespread, state-sponsored landscape change in Scania. However, it is not clear that elsewhere in Denmark similar sequences occurred. The work carried out by our project in North Jutland, which is continuing today, has begun to reveal how elite and non-elite central and local strategies differed from region to region. We are now examining how integration and consolidation in North Jutland, in a district called Thy, differed dramatically from Scania during the same relative time frame.

As noted above, North Jutland was home to the Teutones, and also at least one other distinct ethnopolitical group in the earlier preintegration phase, the neighboring Cimbri of Himmerland. During the migration of Germanic people across the Romanized continent several centuries earlier, both the Teutones and Cimbri were represented, both groups thus having origins as autonomous prestate polities. The same prestate social and political ideology as described in Tacitus, showing clear limits on elite power and the broad authority of farmers, would have been shared by these groups as they were in Scania. We fully expected to find archaeological evidence that Thy was at first highly resistant to the centralizing state.

Support for this hypothesis is again suggested by the protohistoric record: at approximately the same time that the Scanians were making initial demands for the restoration of old taxation levels and rights, yet were being intimidated into backing down by the terror of random executions, the North Jutlanders also rejected the new taxes and laws of the Danish king. The key difference was that the Jutlanders invaded Denmark's core area with a populist army and killed the king and his entourage of relatives, supporters, and advisers (Pålsson 1986). The chronicles, although admittedly on the side of the king, make it clear that this was their perceived right because of dissatisfaction with new taxes and restrictive laws. This was 100 years *earlier* than the Scanian Uprising, the much-delayed but eventual outcome of continuing high tax burden on rich Scanian farmers.

The North Jutlanders later put their support behind the slain king's brother, who was soon elected king of Denmark. This successor does not seem to have initiated any brutal retaliation on the rebels, as was done in Scania a hundred years later, and soon after, the North Jutlanders seem to be unified with Denmark. In the previously discussed Ethiopian example rulers were content to exploit the weak and leave those who are more organized alone. This factor may have been in play in the North Jutland unification sequence, especially after the show of force. While kinship was not the unifying factor, it is probable that other factors permitted the North Jutlanders to display such cohesive resistance. First, they had a strong ethnic bond that was many centuries old, if not older. Second, they were not "opulent" in either crops or merchandise. While some were certainly rich, and most may have been comfortable enough, they had far less to lose in the way of property than they did in the way of autonomy. A combination of differing internal relationships and differing relations between the state and region may have contributed to their continuing self-determination. This in turn may have discouraged the still decentralized and weak kingship to leave well enough alone and go after easier prey.

But why did the North Jutlanders soon after join with the state? One can imagine the costs of this relationship to both sides, but what were the benefits? At this point it is necessary to return to the environment.

We know from environmental reconstructions that the region was not agriculturally rich. The same chronicler who in 1070 called Scania "opulent" stated that Jutland was a sterile wasteland, "a salt land and a vast wilderness" (Tschan 1959:187). This German churchman also noted, however, that there were many cities and towns there, wherever there was a bay or inlet. This raises two issues: first, although farmland may have been lacking, trade and markets were not. This brings us to the main geomorphological feature of the region: the Limfjord (Figure 6.5). This huge inland fjord system is a natural trade route. It cuts through from the Baltic all the way to the Atlantic, with not only a quicker route but one that is safe and shallow. The alternate way, across the top of Denmark through the Skagerrak, is treacherous sailing in dangerous cross-currents, impassable because of storms at certain times of year, and much longer. Thus the movement of goods across the Limfjord was ideal and led to the growth of trading places. The royally founded Viking Age city of Alborg, which was established around AD 950, is a clear example of centralized presence, but several other nameless markets from this era have been surveyed and excavated that date to the 800s. Thus, the region saw the development of trading places and urbanization nearly 300 years before Scania's artificially introduced towns. Some of these had central elite presences in charge; it was not an unfamiliar presence even in a highly autonomous area. The benefits of state authority, in addition to the drawbacks, would have been well known.

Despite the fact that there was some wealth in the region and production for subsistence was not difficult, northern Jutland in general was less rich than Scania or the islands. Since they were poorer, did locals see advantage in a state urban/market system that the already-rich Scanians did not see? Jutlanders may have later assimilated into the state more easily because they were already used to state presence in trading places. Because some local elites were probably getting rich off the mercantile transport that passed through the Limfjord, local elites in North Jutland may have begun cooperating with and abdicating some of their power to the state much earlier.

It was also an ideal military seaway. Guarding the narrows at the eastern end of the waterway is the Viking Age royal fortress of Aggersborg, next to a regional royal estate from the same era. This part of the Limfjord is set in the small area of good soils found on the peninsula. Just to the west archaeologists have recently noted what is probably the remains of a state-built canal leading to the Skagerrak strait between Denmark and Norway, which was both a military target and threat. Similar canals have been found elsewhere in Denmark, built to enable the free movement of warships. Finally, the Limfjord is the site of at least one important national levying place where wartime fleets would gather. Thus, it is clear that whatever this region lacked in cereal, it made up for in essential strategic importance, both economic and military. We can surmise that North Jutland was the focus of elite attention during the state-building era, just a different kind of attention than Scania received.

Yet there was that eventual, successful uprising against the state by the North Jutlanders. It was sparked by a local event but due to the same general taxation and military obligation issues that caused protest elsewhere. The uprising was not an urban one but rural; what was being taxed if not grain production? This brings us to the second important point about Jutland's economy. The primary subsistence strategy throughout Jutland's prehistory was stock raising. There is evidence that in order to meet tax obligations, animals, and not cereals, were intensified. There is also evidence that the North Jutlanders were able to postpone this tax-related intensification by at least a century and maybe more, long past the era when the rest of Denmark was reorganizing to meet state demands.

AGROPASTORAL ECONOMIES IN JUTLAND'S "SALT WASTELAND"

Power and wealth in preintegration phase Jutland, during the Roman and Germanic Iron Ages, was always based in cattle and other livestock. Many archaeological excavations show that cattle were the primary product of the peninsula for many centuries. The west coast's wet, calcareous marsh meadows were ideal for hay cutting and stock grazing. Several Iron Age sites present good examples of this. The early Iron Age village of Hødde, excavated in southwestern Jutland, was completely dominated by cattle production. Its size ranged over time from 11 to 27 farms. It is possible to count stall spaces, which are well preserved in the architecture of the longhouses, and the number of simultaneously used stalls in combined households ranges from 188 to 460 over time. A small amount of cereal farming in infields was practiced for subsistence.

Another excavated Iron Age village, Grøntoft, was located on a sandy hill among vast expanses of heath (Vad Odgaard 1985:121–127). Cattle production was also the major activity here; pollen analysis work shows that some barley and spurrey were grown in a small "garden" field system around the village.

Elite settlements are found in a region of central Jutland belonging to the state's core and probably represent central and not local elite. These began to emerge in the middle Viking Age and reached their peak around AD 1000 to 1050, the consolidation phase. These illustrate how elite landowners built their wealth with cattle. At the sites of Vorbasse and Omgård, single elite estates, stalls show that there were simultaneous facilities for 100 cows and 80 cows, respectively. In addition, one large facility at Omgård with narrower stalls indicates extensive horse breeding with room for 30 horses. In northern Jutland, with identical conditions, local rather than central elite probably had similar wealth bases in stock.

Other evidence for the widespread importance of cattle is the prehistoric road that runs up two-thirds of the length of Jutland, referred to in early chronicles as the ox-road (also called the army road, showing its dual duties). This was a major route along which cattle were driven to markets, such as Ribe in southern Jutland. In Ribe, the earliest trading town in Denmark (ca. AD 720), many arts and crafts were plied, but large areas with meter-thick deposits of pure cow dung indicate holding pens for live cattle. At Ribe and other trading places cows were apparently a major commodity, although their passing does not leave much evidence, unless fortuitous features are preserved, such as these dung layers.

POLITICAL ECONOMY OR AGENCY? OR BOTH?

In Jutland, during and after the consolidation phase, from 1000 to 1200, torps, the "new" agricultural hamlets, *were* founded, not only on the richer east coast but on the heath of the west as well. Torp density in Jutland's better cereal farming areas was about .05 to .06 per square kilometer, similar to rich areas in the islands. In the poor heathland areas torp density was about .02 per square kilometer, much lower. In most of Denmark the linguistic element *torp* is combined either with a male genitive, such as Staffanstorp (Staffan's settlement) or with terrain names such as in Grustorp (gravelly torp). Interestingly, but perhaps not surprisingly, some are named for what was produced in them. The vast majority of these refer to animals, and almost all of these are in western and northern Jutland. Places called Kovstrup (cow torp) and similar abound and also many places referring to goats, sheep, pigs, and even a few to ducks or geese. Some refer to the production of linen (Hørup—flax torp). Thus, many of the "new" pioneering settlements in agriculturally poor parts of Jutland reflect animal intensification.

Probably, taxes were assessed in similar levels across Denmark. The existence of fewer torps in these cattle-production areas than in cereal-farming areas is quite easy to explain. Ethnographic examples indicate that animal intensification, up to a certain point, can be accomplished by putting more animals on the same land, with very little effort. Such changes in pastoralism often occur when new factors are introduced, such as access to markets or the need to generate new income (Bencherifa and Johnson 1990:394), both of which the Jutlanders were experiencing. Usually in such cases the idea behind the strategy is to get maximum yield on cattle with the least possible amount of investment of labor and capital (Bencherifa and Johnson 1990:397). One method is by territorial expansion to accommodate more cattle (Bencherifa and Johnson 1990:398), such as is seen with the pioneering torp settlements. Another common strategy is to load more cattle onto the same land, which eventually causes ecological stress but can be done quite easily as a short-term strategy. Farmers can wait to breed more animals, but this is often not even necessary. Instead, they will use cash picked up in nearby towns, either during seasonal labor or by family members who migrate there to make extra income. Both of these possibilities could occur in Jutland, where there were many marketplaces and numerous small and large towns in the late Viking Age and early medieval. This money was then invested in cattle (Bencherifa and Johnson 1990:399), and we know that cattle were bought and sold because such is clear from the traces found at Ribe. There are no special demands on labor during such an increase, as unskilled youths are often used to manage the larger herds.

For traditional agropastoralists in modern times, increasing incorporation into a national-scale economic system is a fact of life, both because of increased demands and the possibility for herders to generate economic rewards over and above what is due to the government. Indigenous farmers are completely aware of the changes that national policies subject them to, and these farmers are observed, first, to take advantage of opportunities for themselves that arise in such conditions and, second, to work toward increasing their own production efficiency and toward gaining knowledge about fairly distant market processes (Cruz 1999:380). Animals in most of these societies, as in Iron Age Jutland, have always been symbols of prestige, so this goes along with traditional systems rather than interrupting them (Bencherifa and Johnson 1990:400).

OTHER MEASURES OF THE AGENCY MODEL

So far we have looked at Jutland's differing sequence as a strategy of the state for exploiting differing resources. We have also looked at the probability that the methods used to produce income for new taxes was self-organized. But this combination of state demands and self-organized intensification is not the end of the story. In addition to being lower in density and reflecting activities of husbandry in their toponymy, all the torps we have investigated in North

Jutland during our many years of archaeological fieldwork appear to have been founded at least several generations later than those in Scania, at the very end and not the beginning of the consolidation phase.

In every case Scanian torps have late Viking ceramics (ca. AD 1000) scattered on the surface and in subsurface features, while those in Thy appear to be founded no earlier than 1100 and often as late as 1200. This means that the expansion and intensification of pastoral strategies was either not undertaken, or it was undertaken at a relatively low level during the time when other parts of Denmark, most notably Scania, were turned upside down by intensification efforts. Furthermore, in direct opposition to the sequence in Scania, in our many years of surveying, testing, and excavating in Thy, we have found no evidence of "contraction" or the demolition/reconstruction of older villages. In fact, in most cases Viking era houses lie immediately beneath early medieval ones, something that is almost never found in Scania. Thus, not only was North Jutland's intensification effort much later, but it was not accompanied by any large-scale uprooting and resettlement.

The fact that North Jutlanders gathered an indigenous army and disposed of the king and many of his key supporters in 1086 cannot be overemphasized. It is possible that in part, perhaps very *large* part, the lack of direct interference in agriculture and taxation-related settlement change seen archaeologically was not due to state disinterest, or to the lack of arable land, or to the ability to meet tax burdens merely by increasing numbers of cattle on the hoof, but to local resistance.

Lack of reprisal by the state after the North Jutlanders' act of regicide may indicate that early kings did not yet have authority to punish those with strong unity mechanisms, acting under the precepts of ancient social codes. Later kings, who responded quickly to the Scanian Uprising a few generations afterward, had consolidated their power and were able to impose previously unknown levels of coercion.

The North Jutlanders may have forestalled large increases in taxation through their show of force. The next ruler backed off from pressuring them, making unification with the state, with whom they were already profitably interacting, much easier. Quite soon after the rebellion, the region was rapidly incorporated into the state and fully integrated with the state system. By the time a few generations had passed, the state was the agent most capable of showing "force" through its summary crushing of the rebel Scanians. Later generations in North Jutland responded to tax demands with the above-described strategies, which may have been largely self-organized.

CONCLUSIONS: THE SURPASSING COMPLEXITY OF THE SOCIONATURAL LANDSCAPE

The substantivist-formalist debate surrounding economic anthropology of the 1960s stressed the differences between models that gave primacy to either the complete interdependence of social and natural factors and the predominance of culture in decision making *or* to "rational," maximizing choices carried out through individual or group agency. More recent ethnographic models have stressed a much more complex picture of the relationships between economic, social, and political behavior. Most societies do practice rational, agency-based decision making, but how it progresses is largely defined by culture. When differing cultural constructs, such as subcultures within a state, are subjected to similarly changing economic circumstances, the outcomes are likely to be shaped differently as well. One way to frame this process is through the concept of structuration (Giddens 1984), a more contemporary take on the importance of social structures but different from the old substantivist concept in that it stresses the idea of a recursive and dialectical relationship between structure and agency: people make decisions based on conditions and beliefs within their society, be it nation-state or subculture, but their actions also actively change society. In addition, there is more emphasis today on human behavior and the study of processes leading to such changes, the "historical" basis of change (even in prehistoric contexts), and the evolution of socioeconomic practices rather than on a concept of static cultural structures and institutions. This does *not* preclude the existence of overarching and comparative human responses to conditions and change, through which we can compare behavior across many cultures. These two perspectives, stressing historicity on the one hand and predictability on the other, must be combined and studied on different levels of interaction in order to more closely model past societies.

These complementary models more accurately reflect the almost infinite complexity involved in trajectories of change—a complexity that is daunting to many archaeologists—yet at the same time they permit us to construct sets of predictable and comparable strategies. I hope to have demonstrated that in many contexts a presupposition of complex relationships and behaviors should be interwoven with research design for archaeological fieldwork so that we may rationally maximize *our* ability to reconstruct the historically contingent past. As has been demonstrated, variation in human exploitation of the landscape, such as settlement and land-use choices, population density, and agriculture (Birks et al. 1988:209), can be traced in part to underlying ecological factors: climate, hydrology, topography, and soil fertility, as in Denmark. Yet, in the case of Denmark, it has been possible to also examine some specific relationships between the landscape and cultural systems, as well as between differing cultural systems themselves, and to reasonably examine intentionality and agency. Given that we know that central-state authorities in Denmark desired the incorporation and domination of both Scania and North Jutland, and that Scania and North Jutland actively resisted this trajectory, we have been able to contextualize regionalized evidence of intensification and hypothesize about whether some manifestations were agency-based local responses to changing demands and expectations, while others were top-down political economy directives from local elite, all operating at the same time and with different dynamics.

Only some archaeological sequences have the benefit of ethnohistoric data, yet at each juncture where historic data were relevant in the preceding case study, a number of other threads of evidence, purely archaeological, were there to be interpreted as well. Relevant ethnographic analogy can judiciously be used in the study of prehistoric or protohistoric sequences. If we ask questions of all data categories that presuppose complex interactions rather than deny them, our interpretations will more accurately reflect the past than the sometimes overly simplified models of earlier eras.

Acknowledgments

Grateful thanks go to J. H. Bech of the Museum for Thy and Vester Hanherred (National Museum of Denmark); to Lars Larsson of the Department of Prehistoric Archaeology, Lund University, Lund, Sweden; and to Doug Price, Gary Feinman, and Timothy Earle for our many conversations on this research. The fieldwork described in this chapter is directly supported by the National Science Foundation (SBR-98077729, SBR-0002371, SBR-0314407, SBR-0506074, and a previous dissertation improvement grant).

REFERENCES

Andersson, I.
1947 *Skånes Historia*. Norstedt and Sons, Stockholm.

Bartlett, P. F.
1980 Adaptive strategies in peasant agricultural production. *Annual Review of Anthropology* 9:545–573.

Bencherifa, A., and D. L. Johnson
1990 Adaptation and intensification in the pastoral systems of Morocco. In *The World of Pastoralism: Herding Societies in Comparative Perspective*, edited by J. G. Galaty and D. L. Johnson, 394–416. Guilford Press, New York.

Berglund, B., M. Hjelmroos, and E. Kolstrup
1991 The Köpinge area: Vegetation and landscape through time. In *The Cultural Landscape during 6000 Years in Southern Sweden—The Ystad Project*, edited by

B. Berglund, 109–112. Munksgaard International, Copenhagen.

Bhalla, G. S., and D. A. Khan (editors)

1979 *New Technology and Agricultural Transformation: A Comparative Study of Punjab, India, and Punjab, Pakistan*. United Nations Centre for Regional Development, Nagoya, Japan.

Birch-Thomsen, T.

1999 Animal traction and market conditions: A case study from south-western Tanzania and northern Zambia. In *Meeting the Challenges of Animal Traction: A Resource Book of the Animal Traction Network for Eastern and Southern Africa*, edited by P. Starkey and P. Kaumbutho, 33–39. Intermediate Technology Publications, London.

Birks, H. J. B., J. M. Line, and T. Persson

1988 Quantitative estimation of human impact on cultural landscape development. In *The Cultural Landscape: Past, Present, and Future*, edited by H. H. Birks, H. J. B. Birks, P. E. Kaland, and D. Moe, 229–240. Cambridge University Press. Cambridge, UK.

Brattberg, S.

1983 Studier Kring Skånes Köpinger—Vikingatida Marknadsplatser (Studies on Skånes Köpinges—Viking Age Marketplaces). *Limnhamnia 1983* [Yearbook]:83–102.

Brumfiel, E. M.

1980 Specialization, market exchange, and the Aztec state: A view from Huexotla. *Current Anthropology* 21(4):459–478.

1983 Aztec state making: Ecology, structure, and the origin of the state. *American Anthropologist* 85(2):261–284.

Brumfiel, E. M., and J. W. Fox (editors)

1994 *Factional Competition and Political Development in the New World*. Cambridge University Press, Cambridge, UK.

Cruz, M.

1999 Competing strategies for modernization in the Ecuadorian Andes. *Current Anthropology* 40(3):377–383.

D'Altroy, T. N., and T. Earle

1985 Staple finance, wealth finance, and storage in the Inka political economy. *Current Anthropology* 26(2):187–206.

de Wet, C.

1993 A spatial analysis of involuntary community relocation: A South African case study. In *Anthropological Approaches to Resettlement: Policy, Practice, and Theory*, edited by M. M. Cernea and S. E. Guggenheim, 321–350. Westview Press, Boulder, Colorado.

Drennan, R.

1988 Household location and compact versus dispersed settlement in prehispanic

Mesoamerica. In *Household and Community in the Mesoamerican Past*, edited by R. Wilk and W. Ashmore, 273–293. University of New Mexico Press, Albuquerque.

Emanuelsson, U.
1987 *Skånes vegetationshistoria* [Skåne's Vegetation History]. Svensk Geografisk Årbok.

FAO (Food and Agriculture Organization of the United Nations)
1974 *Soil Map of the World*, Vol. 1, *Legend*, UNESCO, Paris.
1981 FAO-UNESCO Soil Map of the World 1:5,000,000. In *Soil Map of the World*, Vol. 5, *Europe*, UNESCO, Paris.

Feinman, G. M., S. A. Kowalewski, L. Finsten, R. E. Blanton, and L. M. Nicholas
1985 Long-term demographic change: A perspective from the Valley of Oaxaca, Mexico. *Journal of Field Archaeology* 12:333–362.

Giddens, A.
1984 *The Constitution of Society: Outline of the Theory of Structuration*. Polity Press, Cambridge, UK.
1989 *The Nation-State and Violence: Volume Two of a Contemporary Critique of Historical Materialism*. Cambridge University Press, Cambridge, UK.

Hansen, U. L.
1987 *Romischer Import im Norden* [Roman Imports in the North]. Det Kongelige Nordiske Oldskriftselskab, Copenhagen.

Hedeager, L.
1992 *Iron Age Societies: From Tribe to State in Northern Europe, 500 BC to AD 700*. Translated by J. Hines. Blackwell, Oxford.

Jacobsson, B., E. Arén, and K. A. Blom
1995 *Trelleborgen*. Trelleborgs Kommun, Lund.

Lansing, J. S., and J. N. Kremer
1993 Emergent properties of Balinese water temple networks: Coadaptation on a rugged fitness landscape. *American Anthropologist* 95:97–114.

Lele, U., and S. W. Stone
1990 *Population Pressure, the Environment, and Agricultural Intensification: Variations on the Boserup Hypothesis*. World Bank, Washington, DC.

Lindow, J.
1976 *Comitatus, Individual, and Honor: Studies in Northern Germanic Institutional Vocabulary*. University of California Press, Berkeley.

Lund, N. (editor)
1984 *Two Voyagers to the Court of King Alfred: The Ventures of Ohthere and Wulfstan, Together with the Descriptions of Northern Europe from the Old English "Orosius."* William Sessions, York.

Marseen, O.
1959 Lindholm Høje. *Kuml 1959* [Yearbook]:53–68.

Misra, R. P., B. K. Becker, and T. D. Nguyen (editors)
1985 *Regional Development in Brazil: The Frontier and its People*. United Nations Centre for Regional Development, Nagoya, Japan.

Ndongko, W. A.
1974 *Regional Economic Planning in Cameroon*. Scandinavian Institute of African Studies, Uppsala.

Ohlsson, T.
1976 The Löddeköpinge investigation I: The settlement at Vikhogsvägen Lund. *Meddelanden från Lunds Universitets Historiska Museum* 1975–1976:59–161.

Osaghae, E. E.
1995 *Structural Adjustment and Ethnicity in Nigeria*. Scandinavian Institute of African Studies, Uppsala.

Pålsson, H., and P. Edwards (translators)
1986 *Knytlinga Saga: The History of the Kings of Denmark*. Odense University Press, Odense.

Randsborg, K.
1990 Beyond the Roman Empire: Archaeological discoveries in Gudme on Funen, Denmark. *Oxford Journal of Archaeology* 9(3):355–366.

Rausing, G.
1990 Löddeköpinge, Lund, and Lödde Kar. *Meddelanden från Lunds Universitets Historiska Museum* 1989–1990:143–148.

Scholz, B. W.
1972 *Carolingian Chronicles*. University of Michigan Press, Ann Arbor.

Scudder, T., and E. Colson
1982 From welfare to development: A conceptual framework for the analysis of dislocated people. In *Involuntary Migration and Resettlement*, edited by A. Hansen and A. Oliver-Smith, 267–287. Westview Press, Boulder, Colorado.

Sømme, A.
1968 *A Geography of Norden*. Heinemann, London.

Southall, A.
1988 Small urban centers in rural development: What else is development other than helping your own hometown? *African Studies Review* 31(3):1–15.

Ståhl, M.
1973 *Contradictions in Agricultural Development: A Study of Three Minimum Package Projects in Southern Ethiopia*. Scandinavian Institute of African Studies, Uppsala.

Stone, G. D.
1998 Keeping the home fires burning: The changed nature of householding in the Kofyar homeland. *Human Ecology* 26(2):239–265.

Svanberg, F., and B. Söderberg
1999 *Den vikingatida borgen i Borgeby. Arkeologiska studier kring Borgeby och Lödde-*

köpinge [The Viking Age Fortress in Borgeby. Archaeological Studies on Borgeby and Löddeköpinge]. Riksantikvarieämbetet, Malmö.

Tacitus, P. C.

1967 [AD 98] *Agricola, Germany, Dialogue on Orators*. Translated by H. W. Benario. Bobbs-Merrill, New York.

Thurston, T.

1997 Historians, prehistorians, and the tyranny of the historic record: Danish state formation through documents and archaeological data. In *New Approaches to Combining the Archaeological and Historical Records*, edited by S. Kepecs and M. Kolb, 239–263. Special guest-edited volume of the *Journal of Archaeological Method and Theory* (Vol. 4).

1999 The knowable, the doable and the undiscussed: Tradition, submission, and the "becoming" of rural landscapes in Denmark's Iron Age. *Antiquity* 73(281):661–671.

2001 *Landscapes of Power, Landscapes of Conflict: State Formation in the South Scandinavian Iron Age*. Kluwer Academic/Plenum Publishing, New York.

n.d. Infields, outfields, and broken lands: Agricultural intensification and the ordering of space in Iron Age Denmark. In *New Perspectives on Subsistence Intensification*, edited by T. L. Thurston and C. T. Fisher, forthcoming. Kluwer/Plenum Publishing, New York.

Tschan, F.

1959 *The History of the Archbishops of Hamburg-Bremen*. Columbia University Press, New York.

Vad Odgaard, B.

1985 A pollen analytical investigation of a Bronze Age and pre-Roman Iron Age soil profile from Grøntoft, western Jutland. *Journal of Danish Archaeology* 4:121–128.

Whitehead, N. L.

1995 The historical anthropology of text: The interpretation of Ralegh's *Discoverie of Guiana*. *Current Anthropology* 36(1):53–74.

Irrigation in Medieval Spain

A Personal Narrative Across a Generation

THOMAS F. GLICK

Irrigation is a privileged topic in anthropology because its practice both requires cooperation and generates conflict. It requires cooperation because of the way water flows, and it creates conflict because water is frequently a scarce resource. But the quality of water flow that produces the fine texture of allocation procedures is *stochasticity*. *Stochasticity* refers to the random quality both of temporal availability and of the volume of water that may be available at any given time (as opposed to *variability*, which refers to the differential flow of water over some period of time). Irrigators do not know from year to year, or even month to month, when water will arrive or how much there will be. Institutional responses to stochasticity have formed the core of my research on irrigation.

Irrigation did not become what Robert Merton used to call a "strategic research site" until after the publication, in 1957, of Karl Wittfogel's classic, but compromised, *Oriental Despotism*, which both made water control into a marker of political organization and provided a universalizing hypothesis that could be tested in the field (practically anywhere in the world because of the near universality of irrigation). Wittfogel's thesis linking water control (both irrigation and flood control) with centralizing, despotic states also proved stimulating to archaeologists, particularly those whose study areas were among the examples used by Wittfogel in his comparative study (e.g., Mesopotamia).

That Wittfogel was working out an idea that had been adumbrated by Karl Marx is well appreciated.[1] What isn't so well known is that Marx, in turn, was working out an idea that had been adumbrated by G. W. F. Hegel. In an appendix on the geographical foundations of history in his *Lectures on the Philosophy of World History*, Hegel notes that the early civilizations arose in fertile river valleys where agriculture was established and with it "the rights of communal existence" (Hegel 1975:158–159 [1837]). He goes on to say that soil fertility "automatically" brings about the transition to agriculture, "and this in turn gives rise to understanding and foresight." While Marx thought that the scale of these broad river valleys was too large to generate voluntary associations, Hegel implies they stimulated the communal solidarities that water management requires, although he does not specify irrigation. He, too, lacked a theory of mediation to explain how, exactly, agriculture of this type stimulates "understanding and foresight." (By the time Wittfogel wrote, by contrast, historians of science had described the relationship between water control and mathematics; and flood control obviously involved Hegelian foresight, as the famous Nilometer shows.)

There was a (literal) lack of fit, however, between the large scale of Wittfogelian hydraulic societies and the small-scale irrigation systems so attractive to anthropologists.[2] All of the latter displayed the salience of local control in irrigation and the relative ease with which peasants were able to resist external authorities. There are good reasons for this, the primary one (in my view) being that the stochastic nature of water supply requires a particularly close control of information if it is to be distributed effectively among multiple users; this distribution is best managed through very local information networks that are able to sense and transmit information about any one point in the system throughout the entire system, particularly in societies where kinship and water distribution networks are isomorphic.[3]

In 1970 I laid out three areas of scholarly research that till that time had pretty much been untouched, at least insofar as medieval Spain was concerned.[4] The first area of research was the operational structure of medieval Valencian irrigation communities, how they were administered, and how conflict was managed. The second area was the diffusion through the Islamic world of several key hydraulic techniques. The third was the parallel diffusion of Arab institutions of water allocation. I will discuss some aspects of the study of these institutions first and then address diffusion in the context of medieval archaeology.

MEDIEVAL IRRIGATION COMMUNITIES

In Christian Spain irrigation was organized either by municipalities or by autonomous communities of irrigators. The institutional distinctions between the two modes of organization, however, are not substantial enough to conclude that there was a connection between political control and irrigation. Wherever town councils administered an *acequia mayor*, their actions were virtually identical to those of the syndics of autonomous canal systems, and one could in fact argue that their discretionary authority was less because as representational entities they were further removed from daily operations than were officers of irrigation communities. (The latter were required to be members of the irrigation communities they represented; town councilors represented a variety of interests.)

The way I looked at irrigation systems was greatly influenced by the coordinated research being carried out at the same time as my own by my mentor, Arthur Maass (1917–2004). Maass devised an elegant methodology for characterizing and comparing irrigation systems: from the written ordinances of irrigation communities (followed up by field interviews) he established the operating procedures of irrigation locally.[5] From the operating procedures, he defined irrigator objectives (equity, equality, economic efficiency, justice, and local control). The trade-offs among objectives provide a basis for comparison (for example, equity may be sacrificed in the interest of efficiency, and vice versa). Finally, he assumed that irrigator objectives encoded the values of the broader culture. What he found, somewhat to his surprise, was that no matter what political system informed an irrigation system, or what kind of water law (whether appropriative, riparian, or administered) enshrined notions of optimal use, irrigators everywhere have been surprisingly successful in maintaining local control. Maass's "trade-offs" really relate to how stochasticity is handled. In Valencia variable debits do not matter because the equality rules (the irrigator takes the necessary amount of water and then cannot irrigate again until all other irrigators have had their turn) ensure that you will get your turn and that no one will take it away from you. In places where water is auctioned, variable flow is calibrated in the price of water (auctions are very sensitive to relative value).

Here we see very clearly the distinction between the Valencian system and that where irrigators are assigned a specific day and hour in permanently structured turns. The Valencian turn builds in a response to stochasticity and variability by assuring irrigators enough water whenever their turn may come up, but irrigators cannot be sure when they may be able to irrigate again. In a fixed-turn system the irrigator will get water—for example, every Tuesday at 10:00—but can never be sure how much will actually arrive at his turn-out. Uncertainty can be normalized to some extent, but it is a constant in irrigation.

Wittfogel lacked an adequate theory of mediation. In Wittfogel's analysis a certain constellation of physical features (aridity in the presence of large rivers liable to flooding) generate authoritarian political forms that, in turn, generate operating procedures at the local level (although he did not analyze them). In Maass's approach the social and economic factors that inform operating procedures are mediated by the objectives of the irrigators. Maass's view is quite similar to that of Jean Brunhes, who asserted that aridity generated specific psychological and social-psychological responses that, in turn, determined the specific nature of water rights. Brunhes (1902) inferred a psychological response to the abundance or shortage of water supply but offered no explanation of how that happened. He also lacked a theory of mediation.[6]

In sum, operating procedures reflect both the technological responses that specific physical features controlling water supply call into being and an integration of irrigator objectives with technological requirements. Social, economic, and political factors are external to the system. That is, they do not act directly to determine operating procedures but are mediated by, and expressed in, the objectives of irrigators.

The advantage of Maass's approach is that it views irrigation from the perspective of the irrigator. Medievalists are well aware that peasants have typically been mute. Irrigation may constitute something of an exception. The ordinances reflect arrangements worked out by irrigators in practice and embodied in legal terms by lawyers hired for the purpose. In litigation, however, it is almost always the peasant who is called on to define the custom.

Maass's "objectives" can also be invoked to characterize customary water law, both in substance and in the way it is applied. Customary water law draws on an unwritten archive of practice, long established and known to all. There was a civil law of water (of Roman inspiration) in medieval Spain too, but it was hardly ever pled in litigation. Oliver Wendell Holmes explained why. "It is the merit of the common law," he wrote in 1870, "that it decides the case first and determines the principle afterwards" (Holmes 1995:212 [1870]). In fact, in medieval litigation, principles underlying water rights were not specified at all unless the right was embodied in a written conveyance, like a royal charter. Otherwise, the rationale was only that tradition had it that way. There was a preference for time priority, particularly when countervailing evidence was lacking, but autonomous irrigation communities were "administered" systems, where irrigator rights and obligations were specified in ordinances ratified by the members of the communities in question rather than in abstract principles (although two monarchs enamored of Roman law—James I of Aragon and Alfonso X of Castile—both attempted to assert the public nature of water, against attempts to seigniorialize it). There was the further twist in the kingdom of Valencia that title to water was conferred, by royal fiat of James I, on landholders of whatever social class within the service areas of canals already in operation (that is, Muslim built) in the civil law code (*Furs/Fueros*) of the Kingdom of Valencia. Under these circumstances the universal strategy of litigants was to argue that whatever they had done was in conformity with custom, just as it was "in the time of the Sarracens."

It is this insight of Maass's—that ambient societal features are filtered through irrigator objectives, which in turn inform operating procedures—that virtually all historians have missed.[7] It is obvious that wherever it is practiced, irrigation constitutes an economic feature of the first order. But the fact that the kind of documents that irrigation generates are almost wholly focused on institutions and technical arrangements regarding the distribution of water, rarely alluding to economic features, has left medieval historians without an explanatory focus. Maass's method makes it possible to read irrigation systems for their social content but only through a close analysis of operating procedures, which in turn can be read for irrigation objectives, which constitute the rationale of allocation.

One of the salient features of the organization of these medieval irrigation communities is that the institutional responses developed to meet the requirements of variability and stochasticity in the flow of water had the unanticipated consequence of endowing them with the flexibility to respond to social as well as physical challenges. Thus interruption of service as a result of war (a social breakdown) or the destruction of a diversion in a storm (a technological breakdown) can be handled according to the same procedures and using the same administrative or executive mechanisms as established to control the normal consequences of variability of flow and stochasticity of the debit.

Once the operating procedures have been tested and found to work, they remain in place, cumulatively reinforced by the weight of precedent and custom. The institutional stability of irrigation systems is well appreciated.[8] If irrigation systems are ultrastable, how do they reflect short-term social, economic, or political change? They may absorb the changes through already established internal switching devices. Patterns of cultivation (including crop rotation, crop selection, and so forth) change over time. If under drought regimes there is an established priority of water allocation by crop, that principle of prioritization can obviously accommodate different crops and different economic valuation of crops. The most significant changes historically, nevertheless, have been in land tenure. Such changes may not directly affect operating procedures but may well affect the rights structure, in particular the progressive tendency to patrimonialize water (convert a public or semipublic water right into a private one).[9] Or, if the social distribution of water rights should be altered, concentrating rights in a particular class, its values will ultimately be reflected in irrigator objectives and operating procedures altered accordingly. Although operating procedures were surprisingly resistant to feudalization, feudal dues or work services (like corvées) may have been applied in specific irrigation systems (Pérez Picayo and Lemeunier 1990:156). In some places (Murcia, for example) municipal control, which was the predominant form of institutional organization through the fifteenth century, was weakened in the modern period when powerful interests became involved with the extension of irrigation systems; this did not always benefit local control by irrigators, particularly when those powerful interests were outsiders. On the other hand, the trend toward privatization or patrimonialization of water frequently strengthened the hand of irrigation communities (Pérez Picayo and Lemeunier 1985:36–37). In such changes local control is always the principal issue.

LITIGATION

From a social science perspective certainly, the records of litigation provide the most relevant source for the study of conflict in irrigation as a way of studying how conflict in irrigation systems was handled. It is surprisingly unstudied. Courtroom performance is important in itself. Witnesses attach specific emotional charges to some feature of institutional control or operating procedure. Maass's objectives are acted out.

Ditch Narratives

Considering how common a human activity irrigation is, there are surprisingly few narrative accounts of it. It appears rarely in novels: Blasco Ibáñez's *La Barraca* is seemingly exceptional. Novelist Stanley Crawford's *Mayordomo*, an account of his year's service as acequia commissioner in New Mexico, is unique. In Crawford's keen description everything that happens in a communal acequia is negotiated or rather expresses the salience of negotiation as a conditioning factor of the practice of irrigation. But there is a rich narrative vein in courtroom legal storytelling. Here are three stories. The first relates to interest-group politics on a canal, the others to the discretional authority of community irrigation officials.

1. Medieval Valencia. Downstreamers on the Moncada canal in the drought year 1416 have sued the acequiero for unlawfully giving preference to upstream irrigators. Both sides (upstream and downstream irrigators) parade witnesses. An upstream witness testifies that "he has seen and sees today that men of Puzol [a downstream village] complain they cannot have water and he sees many others, who are before Puzol in the turn, complain that the water of the canal does not suffice them. The witness believes that were it not for the good order and administration which the canal has it would not be able to supply half of its service area" (Glick 1970:81).[10] Here is a public performance in support of the established ordinances of the irrigation community. That a peasant says this publicly, in court, is a strong reaffirmation of existing legal arrangements, suggesting that the drought has affected the ability of downstreamers to appreciate the fairness of the system.

2. Medieval Valencia. An irrigation official (*sequier*; Castilian, *acequiero*) is patrolling his canal on an evening in the drought year of 1413 to prevent acts

contravening the regulations, we are told. When he comes to the Codinachs mill he finds the gates of the mill's return ditch illegally closed, backing up the water in the main canal and flooding agricultural fields. The following day he returns with a hook, to raise the gates, and he brings an assistant, a humpback named Aparici Romeu. When the official raises the gate, the miller's wife tells him to touch nothing, to which the official responds that she ill wasted the water, which "was of profit to other places as well as her own." One of the miller's sons asks to be shown the misdemeanor in question and while they inspect it, three other brothers appear, one of whom calls out to the assistant, Aparici, and says, "Halt, sir Humpback, that I might undo your hump!" A brawl ensues which finds the unfortunate humpback in the canal fending off the dagger of one of the sons. He would have been killed save for the hauberk he wears. (It is interesting that even more officials wore armor.) The sequier allowed that the brothers were about to kill his assistant without reason and that such "was not gallantry." The brother in the canal climbs out, draws his sword, and thrice repeats, "By God, were I not a gentleman I would slash your face!" The canal ordinances having failed the officer, he sues the miller's family in royal court (Glick 1970:68–70). The moral of this legal story, told in court by the aggrieved official, is to underscore a number of themes of medieval irrigation: that his discretionary authority had been thwarted and the ordinances of the irrigation community subverted, was clearly the main point. The lurid detail was meant to contrast the violence (especially of millers, whose behavior had fulfilled a social expectation regarding their poor character) with the pacific solution provided in the ordinances of the community.

3. New Mexico. Here is a New Mexican *mayordomo*, Cleofas Vigil, describing how the right-of-way functions operationally, as he recounts an argument with a newcomer over the mayordomo's right to free passage both in the ditch and on the rights-of-way (*bordos*) on either side:

> *Le dije yo* when I come to this place or go through this place *yo no vengo a ver te a ti, le dije. Yo voy a pasar esta acequia.* I patrol the ditch *y tu no me puedes prohibir de que voy a pasar por esta propiedad. No voy a pasar por tu propiedad, voy a pasar por la propiedad de la comunidad. Dijo*, you heard me, you call first. *Le dije* . . . call you, and let me tell you something . . . *ese atarque que tenes en esa acequia, le dije, quiero que lo quites de la acequia*. . . . I'll give you a week, *le dije. Si yo suvo en una semana, le dije* . . . we will see the judge. . . . *Al juez y meterle la regla de la asociacion de la acequia.*[11]

Cleofas explains that as mayordomo he has very wide latitude (that is, he exercises the discretional authority invested in him by the community) in the execution of the community's statutes, which include moving water around to maximize its use. "*El mayordomo puede hacer eso*," he emphasizes, "*si usa su juicio cabal*." His testimonial performance serves to mark important boundaries whose maintenance is vital for the integrity of the community of irrigators.

Litigation is integrative; it enhances cooperation through a public ritual (Smail 2003:87). In complex irrigation societies where higher levels of political organization (in Valencia, the city, some aristocrats, and the king) had an interest in the smooth functioning of agricultural production, courtroom performance could only enhance local control, particularly since the legal strategy was for each side to parade as many cultivator/irrigators as was feasible in order to demonstrate that the principles involved in their case were common knowledge. I have noted elsewhere (Glick 2003) that courtroom performance by peasants could also be a form of protest—for laying out the antinomies of conflict in such a way that the community as a whole could get its hands around it. In the medieval *huerta* of Valencia, where several of the irrigation canals were quite long and included a dozen or so villages along their course, although the community ordinances supposedly covered all member irrigators equally, in fact distribution problems relating to delivery acted to create informal moieties of upstreamers (*sobirans*) and downstreamers (*jusans*), and, on very long canals, an intermediate group (*medians*) also appeared (see Glick 1970:74–84). These were informal groups of irrigators who came together as pressure groups to represent their best hydraulic interests at times when it was perceived that other such groups were privileged by reason of their location. Because all rights holders were equals, as members of an irrigation community, the ordinances alone were unable to even out the inherent favorable position enjoyed by those located near the head of the canals over those located near the tail. That is to say, the physical layout itself was conducive to the formation of interest groups based on distance from the diversion dam: upstreamers sought to maximize their favorable location, downstreamers to challenge them to release more water.

Structural politicization, if I may call it that, the remedy for which frequently lay beyond the internal organization of the community, was handled at a higher level, especially in the court of the royal justice, which, in Valencia, was that of the governor.[12]

Customary practice also regulated upstream-downstream competition, not by appeal to community ordinances but by traditional practice (to borrow Daniel Smail's characterization), knowledge of which "was diffused among innumerable archives composed . . . not only of any available written acts but also . . . of sets of witnesses and their memories—or, to be more precise, the ways in which these witnesses chose to interpret their memories in courtroom settings" (Smail 2003:226). The social solidarities like the commons of an irrigation community, or—even more so—the informal communities of *jusans* and *sobirans*, were displayed through performance in court. When they testified in court they were giving a public display confirming the meaning of such a solidarity. The way the medieval justice system worked was that any collectivity, whether formally organized or not, could hire a lawyer and sue in court. There were cases where downstreamers sued upstreamers or one or the other group sued the commons of the irrigation canal or its officers, or where peasant irrigators sued nobles—all of whom had equal votes in the commons of the community.

How Courts Reflect Tradition

In water and irrigation law in the southwest United States, and more specifically in those areas where "Spanish rights" are valid, the relationship between customary operating procedures of presumably Spanish origin have perplexed lawyers and irrigation officials who have tried to negotiate customary practice in the corridors of civil law. The drift of water law in Spain over the past generation has been to abolish to the greatest extent possible any remaining special jurisdictions (such as the Tribunal of Waters of Valencia) and extend a uniform law code to irrigation generally. This trend is matched by a current rage for a particular kind of modernization that involves phasing out all traditional surface irrigation systems and replacing them with underground delivery conduits, a strategy in the interest of efficient use but one that has the by-product of destroying the social organization of the irrigation communities involved.[13]

In New Mexico, where there is a higher value for preserving the social organization of the community acequias (especially in areas where the community acequia is the basic unit of local governance, in the absence of any municipality), the trend in recent irrigation law has been the opposite: to press for civil recognition of traditional operating procedures and the rationales underlying them. The main case in point is that of the "tradition and customs hearings" in *New Mexico vs. Abeyta*, one of a series of recent suits in which the state sues all of the rights holders on a specific river or within a specific watershed in order to adjudicate their water rights. In this case a number of community acequias argued that their rights are in fact encoded in the customary practice of the irrigation community. Such arguments, which run counter to the standardizing rationale of civil law codes, have not gained much credence in the past. But here, a Special Master was appointed to assess the role of local custom and make recommendations accordingly. The testimony, by irrigators and mayordomos, has the same fineness of detail that medieval litigation reflects. Palemon Martínez, an irrigator and water commissioner, argued that if time priority should become the basis of irrigation turns among neighboring acequias, "it will disrupt the harmony they have traditionally enjoyed under a *repartimiento*, sharing to meet all needs."[14] The Special Master recommended that "declaration of customary allocations of water between and among acequias" be formally stipulated (that is, accepted as a basis for water rights claims), and the trial judge approved the recommendation. The rules of prior appropriation ("first in time, first in right") was set aside, and acequias were permitted to share water under customary rules, without regard to priority dates (see Rivera 1998:164–171).

Maass himself was not particularly sympathetic to the protection of customary practice for its own sake, perhaps because he overly identified with the mystique of twentieth-century engineering.[15] But the acequia movement in New Mexico is a prime example of irrigators marshaling their social values in order to preserve a key objective of such communities, namely local control. Their success in court corroborates Maass's findings.

THE ARCHAEOLOGICAL TURN

From the richly textured ritual space of the courtroom we come to archaeology, where relict canals are mute. For a decade or so I have been a sometime collaborator of a group of medievalists in Barcelona who practice what they call "hydraulic archaeology." I had read some of the classics of irrigation archaeology (Adams on Mesopotamia, Kosok on Peru, for example) at the time when I was working on medieval Valencia but was never able to make a connection between archaeology and the institutional approach that engaged Maass and myself—indeed, on field trips I used to drag him to mill sites where he would become overcome with boredom. But with the Catalans I found the connection, which was a particular way of looking at tribally organized space.[16]

In the 1980s and 1990s irrigation became a central focus of medieval archaeology, particularly as related to settlement patterns in Islamic Spain.[17] Extensive archaeology has become a preeminent form of medieval archaeology in Spain since its introduction in the 1980s. It was introduced from France under the stimulus of a broad-gauged discussion of *incastellamento*, the process by which the landscapes of Western Europe were rearranged in the wake of the feudal "revolution" of the eleventh century. Peasants were settled in villages in the district of a castle, whose function in the feudal system was to control peasants. The early practitioners of extensive archaeology in Spain, however, worked on Islamic Spain, where there was no feudalism. The strategy was first to identify castles and their "dependent" villages and then to determine what the function of the castle might have been, given the lack of a feudal rationale.[18]

The paradigm that emerged had a number of associated features: a castle (*hisn*; pl. *husun*), generally located on high ground controlling a mountain valley; groups of six to ten tribally organized villages (*qura*; s. *qarya*) in close proximity; and an irrigation system serving the villages.[19] The castles that

Guichard privileged as the central organizing element of peasant village complexes were not huge defensive structures (although they too existed in al-Andalus) but smaller, defensive structures that could not have been permanently garrisoned. These castle-refuges, as Guichard came to call them, could only accommodate small, temporary garrisons, and their huge walled perimeters (called *baqar* in Arabic; *albacar* in the Romance vernaculars) were large enough to accommodate all of the townsmen and their cattle in times of danger. That they were in fact refuges cannot be doubted inasmuch as Christian chroniclers record the Muslim population so enclosed when facing attack.[20]

There were problems with the model, chiefly of a functional nature. While the function of the castle-refuge seems clear enough, no one has been able to identify with any clarity the enemy against whom the villagers were defending (see Barceló 1995:10–41). Guichard has shown that fear of external Christian attack could not have a motive because, in Castellón Province, for example, the *husun* face the interior, not the putatively hostile borders.

The *hisn/qarya* complex quickly became paradigmatic in the sense that workers coming along after the French surveys now began research by identifying castles and their associated villages in whatever zone was studied (Bazzana et al. 1988). But no sooner was the paradigm in place than it began to change. Not only were there examples of castral units where the castle was an afterthought, but more significant, there were large areas typified by village complexes, irrigated to be sure, but without castles (Martínez Enamorado 2003; Kirchner 1998). It soon became clear that water was the defining element of settlement design. In a reflective article, Barceló asked, how would the course of extensive archaeology differ if water mills rather than castles were its organizing feature. Given the differential way in which gristmills were integrated into feudal and nonfeudal societies, it can be argued they encode a broader spectrum of social information than do castles.

"HYDRAULIC ARCHAEOLOGY"

The expression was coined by Miquel Barceló and his group at the Autonomous University of Barcelona to characterize the survey approach they developed. The method has been used to study small valley irrigation systems, some deserted, many still occupied, that served tribal irrigators, in limits set by the water supply and the cultivable area, which were found to match an optimal population size. The standardization is so great that the design of the irrigation system becomes the archaeological signature of a particular kind of peasant social organization.

Fieldwork in Guajar Faragüit (Granada) and three sites in Mallorca in 1985 and 1987 revealed patterns that suggested that water systems had been designed there prior to their construction and that the design established the articulation between the water source, the course and slope of the canal network, localization of storage or regulating tanks, emplacement of mills, and the layout of agricultural parcels, taking into account possibilities of future extension of the system (Kirchner and Navarro 1993; this study is the major methodological statement of the Barcelona group). A base map is established using enlarged aerial photographs on which the outlines of the irrigated space and the residential zone are drawn with markers. The network of canals is then walked and measured, with all results transferred to the enlarged photograph. The resulting maps are morphologically very detailed but have to be corrected for the distorting effects of aerial photography. So the next step was the drawing of a planimetric map. At the same time the residential zones of settlement associated with the irrigated spaces are reconnoitered. This phase includes the collection of surface ceramic sherds, some of the residential zones having either disappeared without a trace or having been absorbed into present-day urban neighborhoods.

Once the maps are completed, available medieval documentation is searched for any relevant material. Note that in spite of the fact that the objective is to study Arab and Berber tribal settlements, the only written documentation usually available is that left by the Christian conquerors of those places. Land division books (*Libros de repartimiento*), which record the allocation among Christians of parcels formerly held by Muslims, reveal all kinds of topographic and material features together with indices of ethnicity in the form of place-names based on tribal affiliation. The two registers, archaeological and historical, are both incomplete and therefore are complementary (Barceló 1992:458–459). Mills listed in documents with no specific topographical clues can be located physically within canal systems, for example (Kirchner and Navarro 1993:132–134).

Moreover, later transformations of the original design, after the Christian conquest, are also reflected in documents of the most varied types: municipal records, land surveys, ordinances of irrigation communities, and so forth. Using such records as guides, it is possible to identify later additions to the design (e.g., by abrupt connections with the original perimeter or the distinctive morphology of newer terraces with respect to older ones; Kirchner and Navarro 1993:135). A further problem that makes such analysis even more complex is that continued use of the irrigation systems over hundreds of years, many until the present, has involved the steady repair, reconstruction, or replacement of original elements with new ones—whatever was required to keep the water flowing and the crops growing. What does not change, however, is the original articulation of the various operative elements of the system: qanats, main canals, and distribution tanks—all basic to the original hydraulic design—cannot be altered without jeopardizing the functioning of the system (Kirchner and Navarro 1993:136).

The first objective of hydraulic archaeology is to establish the original design of the system, which is done by walking the entire canal network, measuring its components, and mapping them. Gravity flow by its very nature supposes that the irrigated space will be delimited by "lines of rigidity," which result from the establishment of channels for the controlled flow of water.[21] Thus any irrigated space has an upper line of rigidity, described by the main canal or canals for water distribution, above which the water cannot flow because of slope characteristics. The lower line of rigidity is that established by the lowest slope favorable for water flow—a gully, a river, the floor of a valley—or at least the lowest point at which the debit of water available will permit irrigation. A concrete amount of water will permit the irrigation of some maximum area, dependent also on the type of crops planted and the frequency of irrigation.[22] The lower line of rigidity is less stable and can be modified in response to changes in the water supply, changing patterns of land use, and population growth. Modifications to the original design tend to be easily detectable in the structures of the irrigation system itself. The modifications that turn up in these studies tend to be those that least compromise the original design, typically the introduction of hydraulic appurtenances such as mills or tanks, which require no alteration in the gradient or trajectory of the main canal.

Sometimes long after the original system was built, canals were lengthened or new water sources were added. The population explosion of the eighteenth century, for example, triggered the terracing of new fields all over eastern Spain and the consequent expansion of very old irrigation systems. If in a given valley, new water sources are located at a higher altitude than the original catchment, a new line of rigidity would be established above the former one. In reality, however, tapping a new water source implies a new design and the creation of a new system, with its water delivery grid in some instances linked into the older one.

The salience of design in the physical layout of a hydraulic system lends force to Elinor Ostrom's (1990:90) insistence on design in the crafting of institutions for hyperstable systems. These features include clearly defined boundaries (who is in the unit and what rights they have); appropriation rules restricting time, place, technology, and quantity of resource units must be congruous with local conditions; individuals affected by operational rules can participate in modifying them; an accountable system of monitoring; graduated sanctions for offenses; conflict resolution mechanisms; minimal recognition (by outside powers) of rights to organize (= local control). Normal tribal social organization encompasses most of these requirements, or their equivalents, as in the case of stock herding, and can be easily appropriated for the organization of irrigation.

TAIL-END MILLS

Mill emplacement was another key element of the scholarly debate over irrigation. It called attention, first, to overall socioeconomic organization, because for Barceló and his group it is axiomatic that mills were privileged in feudal society (the milling of grain was a widespread feudal monopoly) but were either communal or artisanal in nonfeudal societies, such as that of Arab and Berber tribes. This structural distinction is reflected in the differential siting of mills. Tail-end mills become significant markers of tribal social organization.[23] The tail-end siting keeps conflict with irrigators to a minimum, allows for self-monitoring, and (according to Barceló 1981) represents a lower social value with respect to agriculture. Usually these are small systems with one main channel and the mill located at the tail end.[24]

Helena Kirchner (1997) has described an elegant variation of the tail-end mill principle in an irrigation system in Coanegra, a valley in Mallorca, where there is one long canal, evidently divided into sections, each of which is ended by a gristmill. Place-name and physical evidence suggest that each segment served a tribal segment or clan.[25] This is a significant finding because it also suggests a related institutional feature, namely a preference of tribal groups for self-monitoring. There are also instances where the mill is located completely outside of the irrigation network, right on the banks of a stream, for example. There has been no comparable survey of feudal mills in eastern Spain, but there is enough evidence to suggest that at least in some seigniorial domains, the mill was located at the head of the hydraulic system and irrigation was fed via the mill's tailrace in plots referred to as *subtus rego* (Glick 1979:98). In the large peri-urban huertas of eastern Spain (Valencia, Murcia), which were quite well developed when, in the thirteenth century, the Christians captured them, mills were located preferentially either at the heads of canals or on secondary canals dedicated only to milling. The rationale, however, was similar. Concentrating the mills made monitoring easier and more efficient and reduced the possibilities for conflicts with agricultural irrigators.

Mills are included in the original site design. Kirchner gives an interesting example of a clan settlement at Felanitx, Mallorca, with an irrigated space of 8.1 hectares and no mill. The layout suggests that prior to settlement a choice had to be made between using available water for agriculture or to sacrifice one-third of the irrigated space in order to support a mill (Glick and Kirchner 2000:313n,110).

For this reason the modifications that turn up in these studies tend to be those that least compromise the original design, typically the introduction of hydraulic appurtenances such as mills or tanks that require no alteration in the gradient or trajectory of the main canal, or else, alterations in the area irrigated owing to the lengthening of the main canal or the addition of new water sources if they are located higher up than the original catchment, permitting the establishment of a new line of rigidity above the former one.

Such an approach is very attractive to historians. Most of the rich detail of the *Libros de repartimiento* is useless without actually walking the places described and filling in the missing spatial articulation.[26] Hydraulic archaeology represents a challenge to historians to make better use of the documentary data by articulating them geographically and morphologically, allowing us to read the function of such elements from their structure. And it embodies a running critique of classical archaeology, which has ignored the space of work in deference to residential space, as Barceló (1992) has put it.[27]

DIFFUSION OF TECHNIQUES

When the Islamic Empire (Umayyad, then Abbasid) constituted a more or less contiguous area stretching from Spain to northern India, at least nominally under the same rule, it created a vast medium for the diffusion of ideas and techniques, preeminently from east to west. There is not much analytical difference between the way ideas and techniques diffuse. It is logical to view the diffusion of the style of cultivation that the Arabs called "Indian Agriculture" (*filaha hindiyya*) and that of the Indian way of doing arithmetic (*hisab al-hind*) as analogous processes. Indian agriculture referred specifically to a set of cultivars grown under monsoon conditions in India that the Arabs acquired and pushed westward, where, particularly in the Mediterranean basin with its summer droughts, they could not be grown without irrigation.[28] Therefore once the new agricultural package left India and began working its way westward, through Persia, it picked up technical elements (such as the *qanat* and the *noria*, sometimes called "Persian wheel") required to cultivate the new crops, forming a distinctive medieval Arab-Muslim tool kit that included a variety of crops.

Tribal settlement, with its built-in predilection for small valleys, ensured, moreover, that a vast range of microclimates or microhabitats would be settled. The result was a very flexible agriculture that was always prepared to maximize the genetic plasticity of, for example, the common roster of cereals, which quite easily adapted, through both natural and human selection, to a broad range of microclimates. As Peregrine Horden and Nicholas Purcell (2000:262) note, this kind of flexibility creates a situation where the microhabitats themselves function as gardens for the acclimatization of new varieties vastly outweighing new crop introduction by members of the elite.[29]

A key piece of hydraulic hardware was the qanat, or filtration gallery, where a tunnel skims the surface of a water table and conducts it laterally to a canal or aqueduct. It is a Persian invention that diffused westward in classical times and then intensified throughout the Islamic Empire. Spanish qanats had been known before the advent of extensive archaeology, particularly the monumental system supplying water to Madrid, but the small-scale use of the technique was unrecognized. Fieldwork by Barceló in Mallorca (Barceló et al. 1986) turned up a plethora of very small qanats, some as small as a meter or two in length. Through this research it became clear that these small-scale, backyard qanats were part of a standard peasant tool kit, unlike the monumental qanats, which required formal engineering.

Another component of Arab and Berber water systems was the noria. There is no argument over the ubiquity of the noria, but there has been one over the date of its introduction. Sonia Gutiérrez (1996) sees them used in very early neo-Muslim settlement, so early that there does not seem to be enough time allowed for the diffusion to take place. The elements of a tool kit, almost by definition, do not travel alone. Norias used as a primary irrigation element—as in La Mancha—make family farms highly productive, with marketable surpluses, always in association with a storage tank (*alberca*). But the noria also appears as a complementary element integrated into systems consisting mainly of surface canals.[30]

INSTITUTIONAL DIFFUSION

Irrigation layouts cannot be read for ethnicity. Markers of ethnicity in studies of medieval irrigation systems are either tribal names associated (in historical records and in toponyms) with a given village; or, if the institutions of water distribution are documented or if the system in question is still functioning, then often Arabisms in technical terminology of irrigation may permit the identification of its paternity. At first I had surmised that such small-scale systems were generic. But when Arabized terminology (for irrigation turns, water measures, and so forth) is documented, these systems can be traced to specific areas of the Islamic world. If they are associated with a tribal place-name, then one knows where to look. If not, it's a matter of luck. Two examples:

1. In the huerta of Murcia agricultural parcels were classified in thirteenth-century documents by area (*tahulla*) and by a unit of valuation (*alfaba*). Both terms are Arabisms, but the meaning of *alfaba* was not understood. A colleague recommended a book about irrigation in a Saharan oasis, and there, to my amazement, the origin of the *alfaba* was explained (the author was unaware of the Spanish Arabism or its use in Murcia). It is a measure of the capacity of a particular parcel to produce wheat, both the usage and the concept being peculiar to Zanata Berbers. The presence in the huerta of Murcia of a place-name derived from a Zanata settlement there (Zeneta) further solidifies the identification.[31]

2. In the valley of the Huete River in the Ebro Valley irrigation turns were described in medieval documents by two Arabisms seemingly confined to that area. One of them, *alfetma*, which turns up as a place-name in Murcia (Alfeytami), means "thread of water" (meaning a proportion of the debit), whose Spanish calque, *hilo* or *hila* (Catalan, *fil*; Valencian, *fila*) is widely diffused. If the term has a particular hearth in the Arab world, it has not yet been identified. But the other term, *azaira*, seems to be of Tripolitanian origin, although the argument is somewhat inferential (it involves the standardization of one of the canonical prayer times, the *zuhr* or afternoon prayer, into a general unit with a time value of four hours) (Glick and Teixeira 2002–2003).

Barceló's (2004:37, 160) comparative study of Yemeni settlement in Yemen and al-Andalus legitimizes this kind of philological approach because it provides a framework for the migration of tribal groups who practiced irrigation in both their hearth and their new site(s) in al-Andalus, and makes obvious the significance of associated place-names. The study areas were ancient hearths of the Banu Ru'ayn in Yemen and in successive, segmented settlements in al-Andalus, including Fahs Ru'ayn (Málaga), al- Ru'ayni (Almería), Beniroaym (Vall de Gallinera, Alicante), Caria Benirroym (Ibiza), Binirroi, Binirohaym (Mallorca), Biniarroi (Menorca).

ASCERTAINING THE DESIGN PRINCIPLES IN YEMEN

This analytic approach magnifies the match between the design (as environmental perception) and the settlement morphology that eventually emerges:

> Water distribution, which is constrained by slope needs, forces the planning of each parcel slope, and the exact path of all channeling, from the outset. We are not therefore referring to spaces of imprecise limits, nor are they in any way different from the spaces irrigated from permanent water collecting [i.e., in al-Andalus]. It might additionally be hoped that there is an initial coherence between the capacity of each asdad [diversion dam], the volume of water flow, the surface to be irrigated and the irrigation time after the stocking of water in the *sayl* and the water needs—quantity and frequency of irrigation—of the cultivated plants. [The coherence between these different variables is typical of any hydraulic system (Anderson and Maass 1971).] The existence of this coherence is not improvised and is surely a decisive factor, providing an explanation for regularity of size, if this regularity is confirmed. . . . The inclusion of [terraces] with an irrigation perimeter implies morphologically compact groups, in which the slopes must be planned from the very beginning. (Barceló et al. 2000:38)

Some Features of Tribal Settlement

Place-name illustrates the geographical patterns of segmentation; the fact that tribes, both Arab and Berber, were still segmenting in al-Andalus (certainly through the tenth century and perhaps even later) makes it possible to link this process to site selection for new settlements and original water-system design because all were undergirded by a perception of the optimal size of a clan. Barceló (2004) calibrates this size both demographically and in standard areas of irrigated space. In Banu Ru'ayn alquerias in al-Andalus this is a maximum of around 25 persons.

> Even this population size makes little sense if not placed in the context of the other alquerias that form part of the network and of the higher regional level. So that to set with the great precision the range of sustainability—that is, the quota of fixed production of plants in a settlement—of the whole network is a solid indication not only of the detail of survival but how it had been conceived [conceptualized] and the way in which risks had been confronted, reducing them with the introduction of a calendar not subject to the rainfall regime. (Barceló 2004:40)

The uniformity of irrigated spaces was quite consistent between the Ru'ayn homeland in Yemen and their new settlements in al-Andalus: Banu Ru'ayn settlements in Yemen around 10 ha each matched nicely with the average size of such settlements in al-Andalus: 10.2 ha, 7.7, 8.1, 8 (all Mallorca); 10.4 (Ibiza); 18, 28.8, 27.6 (Liétor) (Barceló 2004:92).[32]

The migration of the Ru'ayn from Yemen to al-Andalus, passing through Berber, North Africa, was both generational and presumed the formation, along the way, of alliances with Berber groups also in the throes of migration to Spain. In this view the immigration of clans is only possible "with the context of a migration which, arriving in Tangiers, had already been going on for almost 40 years, to try alliances with *imazigen* [Berber] tribes and clans" (Barceló 2004:51). This in turn suggested that the disposition of the Ru'ayn among other clans "seems to indicate a network of settlements that should be studied as such: distances, range of resources, technical solutions, sizes, distribution of water sources, local roads and their connection with axes of . . . a regional road network" (Barceló 2004:43).

The modalities of irrigation in Yemen and Spain were quite different. Banu Ru'ayn agriculture in Yemen was based on runoff systems, with rainwater (*sayl*) channeled through wadis into diversion dams (Barceló 2004:87). These were relict sites, where the diversion dam and canal networks could be reconstructed but where no indication of the institutions of allocation could be known. The study suggests that peasant tool kits were sufficiently broad and flexible to permit new settlement under a considerable range of hydraulic possibilities, all of them subject to the same adjustment to "optimal" clan size.

Practicing hydraulic archaeology there with the same method developed in Spain, the team establishes the carrying capacities of 10-ha units formed by tribal segments, from which the size of migrating groups can be estimated (Barceló 2004:115).

POLITICAL DIMENSIONS OF TRIBAL MIGRATION AND SETTLEMENT

Barceló's surveys of these small, tribally settled spaces turned up some patterns that were politically suggestive. The mixture of Berber and Arab clans (identified by place-names), living in closely grouped villages sharing the same water source, also suggests the operation of a system of alliances. So in the Vall de Gallinera (a small valley in the province of Alicante) he describes a number of alquerias (one of which was a Banu Ru'ayn settlement) arrayed along a gully, "the complex result of some hydraulic designs and some specific organizations of peasant work, and that the whole thing required, in order to have [been] possible, alliances among groups in consonance with the congruence among the sizes of the initial populations and the spaces parceled out and managed" (Barceló 2004:30–31). Moreover, there are cases of specific Berber and Arab clans, seemingly paired, that appeared in neighboring villages in areas widely separated from one another.

The following list summarizes some results of these tribal irrigation studies:

1. The irrigation potential of a specific water source appears to have been the crucial component of site selection for immigrating Berber and Arab tribal segments in the eighth and ninth centuries.

2. There is a correlation between the carrying capacity of such sites and the optimal size of a tribal segment. Optimization does not imply maximization. Optimal use may have embodied an allowance for a surplus of water or building or not building a water-powered gristmill, depending on how much arable land was needed, and other such factors.

3. Segmentation occurs when the carrying capacity is inadequate for the increased population size; the segments that shear off found new settlements with the same hydrological characteristics and the same irrigable area as the old settlements.

4. Tribal norms of governance, under these conditions, are sufficient to encompass the management of water allocation and to mediate minor, everyday conflicts.

5. There is a preference for self-monitoring.

6. The main factors affecting the need for specialized officials are canal length and the need to measure water allocations by time.

7. There is a general fit between "tribal values" and Maass's notions of equity, justice, and local control. "Efficiency," as such, is not an issue except when the optimal carrying capacity is reached or the clan must decide whether there is enough water to install a mill and, if so, where to install it (i.e., within the irrigation system or completely detached from it).

8. The standard areas of peasant work as established by the methods of hydraulic archaeology can be used to infer the size of immigrating tribal segments.

9. The standard array or composition of networks of clan-based villages suggests a political process of alliance formation. Place-name evidence suggests that such alliances were patterned, with two groups (Berber/Berber, Arab/Arab, and Berber/Arab) coexisting in close proximity in more than one site.

A final reflection on the two registers—historical and archaeological—of irrigation that, over the length of a career, I have sought to align. In the final analysis technology is a mental process; it pertains to whatever device may be used to mediate between human society and its environment. In this context the operating procedures of irrigation systems (turns, rotations, and the like) are no less technologies than are diversion dams, canals, holding tanks, qanats, or norias that make up irrigation layouts. Integrated peasant tool kits contain both conceptual and material sets that are necessarily complementary ways of working through the problems related to tribal settlement linked to a particular approach to water use.

NOTES

1. See, in particular, Peet (1985:3–20).

2. Wittfogel's theory is valid in those cases where irrigation was from large rivers where flood control was also an issue, under conditions of aridity. In the case studies written in the 1960s *contra* Wittfogel, the study areas did not meet these conditions; most were of small-scale systems in semiarid climates where flood control was not an issue.

3. On kinship and irrigation the locus classicus is Berque (1955:153–159). See also Jolly (1997:60–90).

4. See Glick (1970, 1988, 2004a).

5. Maass and Anderson (1978:1–10, 366–369, 375–376 [local control], and 395–397 [justice]). Maass wrote the text, Anderson the accompanying computer simulation.

6. See Maass's comment on Brunhes in Maass and Anderson (1978:399–400).

7. See Instituto de Estudios Almerienses (1989). This meeting was marked by the excitement of a new field and by the interdisciplinary foci of the participants. While many of the papers presented useful data, however, there was a general inability to deal analytically with institutions of water allocation, nor has any corrective emerged since.

8. See Ostrom (1990:69–82) on Spanish irrigation systems as hyperstable, self-organizing "common pool resources," using my and Maass's data. On viewing hyperstability in irrigation in historical perspective see Glick (1995:66–67): institutions are presented (particularly when litigated) as having a more coherent structure over time than they actually have. To establish precedent, lawyers had to argue that there was no institutional change over time; see example in Glick (1970:238), where a particularly perceptive medieval water lawyer explains the illogic of such pleading.

9. On the patrimonialization of water in the Canary Islands and San Antonio, Texas, see Glick (1972).

10. On scripted court performances see Smail (2003:93).

11. This bit of legal storytelling is from an interview, which is another kind of ritual performance. But the event recalled had, by this time, become common knowledge in the interviewee's community.

12. The governor of the kingdom was a mixed judicial and executive officer that later evolved in the position of viceroy. Most cases were heard by the lieutenant governor.

13. For an interesting example of how local communities are negatively affected by the loss of traditional irrigation systems there is the case of the San Juan Ditch in San Antonio, Texas; see Torres (1997:109) and Rivera (2003).

14. *New Mexico vs. Abeyta*, "Report of the Special Master," US District Court for the District of New Mexico (typescript), 1993, p. 68.

15. Even though he wrote a famous critique of it (Maass 1951).

16. See Glick (2002, 2004b) and Glick and Kirchner (2000), three recent syntheses that integrate the institutional and archaeological perspectives.

17. Early statements were Bazzana and Guichard (1981) and Barceló (1981).

18. The term *incastellamento* as a term describing the general process is misleading. I prefer "directed settlement nucleation," following Horden and Purcell (2000:80).

19. Many of these villages, especially in the Valencian region, have place-names in Beni-, "sons of," reflecting clan organization; see Glick (1995:31–35).

20. On *hisn/qarya* complexes see Glick (1995:18–29).

21. There is a connection between the rigidity of hydraulic networks and institutional hyperstability. As Maass himself recognized, water rights were structured in such a way as to control, limit, or prevent the physical expansion of irrigation systems (Maass and Anderson 1978:369). For the perspectives of Maass and Barceló compared see Glick (1995:73).

22. The maintenance of an equilibrium between irrigated surface and volume of water is not solely guaranteed by the principles of proportional distribution and equity, virtually universal in norms of social distribution of water. The efficacy of local management of a hydraulic system is especially dependent on the ability of an irrigation community to restrict the expansion of the area under irrigation.

23. Once you have the original design, you can assess the carrying capacity of the unit, and with that information together with the identity of the settlers provided by

place names, you can deduce irrigator objectives (to use Maass's term). The trade-off is between equality and justice (tail-end communal mills) and economic efficiency (front-end feudal mills).

24. For a tail-end mill in a medieval *hisn/qarya* complex in Wadi Zgane, Morocco, see Powers (2002:119).

25. In the case of Coanegra the four *alquerías* all have names relatable to Berber segments. The pattern of a long canal divided into clan-based sectors is also clear in the case of the Alquibla Canal in the huerta of Murcia. The canal changes names four times along its course, the last two sectors being Benicotó and Benicomay, toponyms indicative of tribal segments (Torres Fontes 1971:21).

26. Without a field component drawn from extensive archaeology the *Libros de repartimiento* yield generalizations that are interesting in the aggregate but shed scant light on how specific irrigation systems worked in practice or on how residential zones were articulated with the space of work. For such an exercise, involving reading the *Repartimientos* in chronological order to detect large-scale pattern changes, see Glick (1995:chap. 6; 1999).

27. An appropriate proof of this contention is the ongoing discovery of classical and dark-age mill sites and mill stones. Their supposed absence, which Marc Bloch famously attributed to the abundance of slave labor for milling flour by hand in Imperial times that acted to dissuade the construction of water-driven gristmills, was in reality owing more to the disinterest of classical archaeologists (see Wikander 2000).

28. In the agricultural belts surrounding most cities in the medieval Middle East, such as Baghdad, the Indian style was likewise practiced.

29. Martínez Enamorado (2003:114–115) supposes that in the mountains of Málaga, where Berbers and neo-Muslims vastly outnumbered Arabs, even when acclimatization of varieties by Arab elites is documented, as in the case of the *safari* pomegranate, the innovation quickly spread to Berber cultivators and became absorbed into the peasant repertory.

30. See examples in Glick and Kirchner (2000:Figures 7-4, 7-6, 7-11), all from the Balearic Islands.

31. The book referenced is Marouf (1980). For an analysis of the semantic issues that permit the identification with Murcia see Glick and Kirchner (2000:319–320).

32. Cf. Paul Trawick's (2001:369) notion that uniformity of parcel size is in turn related to proportionality, and thus the standardization or irrigation time and water consumption are "inherent features of this technology" (terracing).

REFERENCES

Anderson, R. L., and A. Maass
[1971] *A Simulation of Irrigation Systems: The Effect of Water Supply and Operating Rules*

on Production and Income on Irrigated Farms. U.S. Department of Agriculture, Economic Research Service, Washington, D.C.

Barceló, M.

1981 *Arqueología medieval: En las afueras del 'medievalismo.'* Crítica, Barcelona.

1992 Historia y arqueología. *al-Qantara* 13:457–462.

1995 Los *husun*, los *castra* y los fantasmas que aún los habitan. In *El agua. Mitos, ritos, y realidades*, edited by J. A. González Alcantud and A. Malpica Coello, 10–41. Anthropos, Barcelona.

2004 *Los Banu Ru'ayn en al-Andalus: Una memoria singular y persistente*. Al-Baraka, Granada.

Barceló, M., M. A. Carbonero, R. Martín and G. Roselló-Bordoy

1986 *Les aigües cercades (Els qanat[s] de l'illa de Mallorca)*. Institut d'Estudis Baleàrics, Palma de Mallorca.

Barceló, M., H. Kirchner, and J. Torró

2000 Going around Zafar (Yemen), the Banu Ru'ayn field survey: Hydraulic archeology and peasant work. *Proceedings of the Seminar for Arabic Studies* 30:27–38.

Bazzana, A., P. Cressier, and P. Guichard

1988 *Les châteaux ruraux d'al-Andalu*. Casa de Velázquez, Madrid.

Bazzana, A., and P. Guichard

1981 Irrigation et société sans l'Espagne orientale au Moyen Âge. In *L'homme et l'eau en Méditerranée et Proche Orient*, edited by J. Metral and P. Sanlaville, 115–140. Maison de l'Orient, Lyons.

Berque, J.

1955 *Structures Sociales du Haut-Atla*. Presses Universitaires de France, Paris.

Brunhes, J.

1902 *L'Irrigation dans le Péninsule Ibérique et dans l'Afrique du Nord*. C. Naud, Paris.

Crawford, S.

1988 *Mayordomo: Chronicle of an Acequia in Northern New Mexico*. University of New Mexico Press, Albuquerque.

Glick, T. F.

1970 *Irrigation and Society in Medieval Valencia*. Harvard University Press, Cambridge, Massachusetts.

1972 *The Old World Background of the Irrigation System of San Antonio, Texas*. Texas Western Press, El Paso.

1979 *Islamic and Christian Spain in the Early Middle Ages*. Princeton University Press, Princeton, New Jersey.

1988 *Regadío y sociedad en la Valencia medieval*. 1st ed. Del Cenia al Segura, Valencia.

1995 *From Muslim Fortress to Christian Castle: Social and Cultural Change in Medieval Spain*. Manchester University Press, Manchester, UK.

1999 Reading the *Repartimientos*: Modeling settlement in the wake of conquest. In *Christians, Muslims, and Jews in Medieval and Early Modern Spain*, edited by M. D. Meyerson and E. D. English, 20–39. University of Notre Dame Press, South Bend, Indiana.

2002 Tribal landscapes of Islamic Spain: History and archeology. In *Inventing Medieval Landscapes: Senses of Place in Western Europe*, edited by J. Howe and M. Wolfe, 113–135. University Press of Florida, Gainesville.

2003 Thin hegemony and consensual communities in the medieval crown of Aragon. In *El feudalisme comptat i debatut: Formació i expansió del feudalisme català*, edited by M. Barceló, G. Feliu, A. Furió, M. Miquel, and J. Sobrequés, 523–538. Universitat de València, Valencia.

2004a Sistemes agrícoles islàmics de Xarq al-Andalus. In *Història agrària dels Països Catalans*, edited by J. M. Salrach, 45–89. Fundació Catalana per a la Recerca, Barcelona.

2004b *Regadío y sociedad en la Valencia medieval*. 2nd ed. Biblioteca Valenciana, Valencia.

Glick, T. F., and H. Kirchner

2000 Hydraulic systems and technologies of Islamic Spain: History and archeology. In *Working with Water in Medieval Europe: Technology and Resource-Use*, edited by P. Squatriti, 267–329. Brill, Leiden and Boston.

Glick, T. F., and S. Teixeira

2002–2003 Azaira, Alhetma: Two medieval Arabisms reflecting the allocation of irrigation water. *Suhayl: Journal for the History of the Exact and Natural Sciences in Islamic Civilization* 3:213–219.

Gutiérrez, S.

1996 El aprovechamiento agrícola de las zonas húmedas: La introducción del arcaduz en el sureste de al-Andalus (siglos VIII y IX). In *Agricultura y regadío en al-Andalus*, edited by L. Cara Barrionuevo and A. Malpica, 7–19. Instituto de Estudios Almerienses, Almería.

Hegel, G. W. F.

1975 [1837] *Lectures on the Philosophy of World History*. Cambridge University Press, Cambridge, UK.

Holmes, O. W.

1995 [1870] Codes, and the arrangement of the law. In *Collected Works of Oliver Wendell Holmes*, edited by Sheldon M. Novick, 2:212–221. University of Chicago Press, Chicago.

Horden, P., and N. Purcell

2000 *The Corrupting Sea*. Blackwell, Oxford.

Instituto de Estudios Almerienses

1989 *El agua en zonas áridas: Arqueología e historia*. 2 vols. Instituto de Estudios Almerienses, Almería.

Jolly, G.

1997 La maîtrise lignagère dans la vallée de l'Azzaden (Haut Atlas, Maroc): Vision historique et spatiale. In *Jacques Berque, la Méditerranée, la Haut Atlas*, edited by C. Bromberger, 60–90. Université de Provence, Aix-en-Provence.

Kirchner, H.

1997 *La construcció de l'espai pagès a Mayurqa: Les valls de Bunyola, Orient, Coanegra i Alaró*. Universitat de les Illes, Palma de Mallorca.

1998 Redes de alquerías sin *husun*. Una reconsideración a partir de los asentamientos campesinos andalusíes de las islas orientales. In *Castillos y territorio en al-Andalus*, edited by A. Malpica, 450–469. University of Granada, Granada.

Kirchner, H., and C. Navarro

1993 Objetivos, métodos y práctica de la arqueología hidráulica. *Archeologia Medievale* 20:121–151.

Maass, A.

1951 *Muddy Waters: The Army Engineers and the Nation's Rivers*. Harvard University Press, Cambridge, Massachusetts.

Maass, A., and R. Anderson

1978 *And the Desert Shall Rejoice: Conflict, Growth, and Justice in Arid Environments*. MIT Press, Cambridge, Massachusetts.

Marouf, N.

1980 *Lecture de l'espace oasien*. Sindbad, Paris.

Martínez Enamorado, V.

2003 *Al-Andalus desde la periferia. La formación de una sociedad musulmana en tierras Malagueñas (siglos VIII–X)*. CEDMA, Málaga.

Ostrom, E.

1990 *Governing the Commons: The Evolution of Institutions for Collective Action*. Cambridge University Press, Cambridge, UK.

Peet, R.

1985 Introduction to the life and thought of Karl Wittfogel. *Antipode: A Radical Journal of Geography* 17:3–20.

Pérez Picayo, M. T., and G. Lemeunier

1985 *Agua y coyuntura económica: Las transformaciones de los regadíos murcianos (1450–1926)*. Geo-Crítica No. 58 University of Barcelona, Barcelona.

1990 Los regadíos murcianos del feudalismo al capitalismo. In *Agua y modo de producción*, edited by M. T. Pérez Picayo and G. Lemeunier, 150–187. Crítica, Barcelona.

Powers, D.

2002 *Law, Society, and Culture in the Maghrib, 1300–1500*. Cambridge University Press, Cambridge, UK.

Rivera, J. A.

2003 Restoring the oldest water right in Texas: The mission San Juan Acequia of San Antonio. *Southwestern Historical Quarterly* 106:367–395.

1998 *Acequia Culture: Water, Land, and Community in the Southwest.* University of New Mexico Press, Albuquerque.

Smail, D. L.

2003 *The Consumption of Justice: Emotions, Publicity, and Legal Culture in Marseille, 1264–1423.* Cornell University Press, Ithaca, New York.

Torres, L.

1997 *Voices from the San Antonio Missions.* Texas Tech University Press, Lubbock.

Torres Fontes, J.

1971 *Repartimiento de la huerta y campo de Murcia en el siglo XIII.* Consejo Superior de Investigaciones Científicas, Patronato "José María Quadrado," Academia "Alfonso X el Sabio", Murcia.

Trawick, P.

2001 The moral economy of water: Equity and antiquity in the Andean commons. *American Anthropologist* 103:361–379.

Wikander, O. (editor)

2000 *Handbook of Ancient Water Technology.* Brill, Leiden.

Oceania

Agricultural Intensification

A Polynesian Perspective

Patrick V. Kirch

The far-flung Polynesian islands were not by nature predisposed to become foci of agricultural innovation and intensification; human intervention and intentionality made them so. The islands are far removed from the original centers of plant domestication in the Old and New Worlds; their isolated and highly endemic floras mostly lack even wild taxa suitable to domestication. Thus the agricultural systems[1] that flourished on these islands at the time of first contact with Europeans had been in a real sense transported by the Austronesian-speaking voyagers who over many generations expanded eastward from their ultimate homeland in island Southeast Asia. As Peter Bellwood (2004) avers, the dispersal of the Austronesians ranks as one of several great linguistic diasporas that were fundamentally driven by agriculturally fueled demic expansions. In settling the Polynesian archipelagoes (between about 900 BC and AD 1200), one branch of the Austronesians brought with them a diverse array of domesticated plants (tuber and root crops, tree crops, narcotic and medicinal plants), along with the cultural concepts (ethnobotanical and agronomic knowledge, tool kits) necessary to transfer cultivation systems from one island to another. A key dryland crop, sweet potato (*Ipomoea batatas*), was later added to the Austronesian cultigen suite through contact between Eastern Polynesia and South America (Hather and Kirch 1991; Yen 1974a)

Following initial discovery and colonization, adaptation to a diversity of island environments (high islands and atolls; tropical to subtropical, and even temperate climates) led to further reassortment of crops and technology (Yen 1971, 1973). Nonetheless, with the notable exception of temperate South Island (New Zealand) and the Chathams, agricultural systems were successfully transferred to every island colonized by Polynesians. Moreover, these systems—varying in their particular mix of crops and agronomic technologies—were over the course of prehistory further changed through expansion, innovation, and intensification, eventually to display a remarkable range of variation, as witnessed in the ethnohistoric and ethnographic record (Barrau 1965a, 1965b). What is worth remarking here is that the hierarchically elaborated and politically complex social formations—the famous *chiefdoms*—for which Polynesia is anthropologically famed (Goldman 1970; Sahlins 1958; Service 1967) all owed their existence to agricultural economies more often than not highly intensive in the use of both landscape and labor.

Polynesia offers fertile ground for studying, from an archaeological perspective, the processes of agricultural intensification, as these are sedimented in the physical record of landscape and technological changes and in relation to parallel changes in other aspects of economy, population, and sociopolitical organization. Effectively, the prehistoric sequences of the many Polynesian islands and archipelagoes offer a series of "comparative experiments" in which the outcomes of agricultural change may be compared and contrasted with respect to similarity and difference in a range of potentially significant variables (isolation, island size and age, edaphic and hydrologic parameters, and so on). Of course, the "experiments" are completed and cannot be re-run; nor can the contingencies of history be wholly discounted. Still, the opportunity posed by the possibilities of historical comparison of multiple evolutionary pathways all ultimately stemming from common cultural origins has much to commend it (Kirch and Green 2001).

Archaeologists of Polynesia began to tackle the archaeological record of agricultural systems and agronomic change toward the close of the 1960s, consonant with the theoretical and methodological influence of the "New Archaeology" (i.e., processualism). At first included within the purview of "settlement pattern" surveys (e.g., Green 1961), terraces, field boundaries, and other archaeological vestiges of agricultural systems soon began to attract attention in their own right (e.g., Kirch 1977; Leach 1979; Yen et al. 1972). After nearly four decades of field research, writing, and discussion we have learned a great deal about ancient Polynesian agriculture (Kirch 1991), despite continuing disagreement over even fundamental matters (Leach 1999). Without trying to mask such healthy debate, in this review I attempt to canvas some key aspects of agricultural intensification in Polynesia, particularly with an eye to broader theoretical implications.

MODES OF INTENSIFICATION

The problem of intensification in agriculture is tied up intimately with the Malthus-Boserup debate over the linkage between population, resources, and technology. Indeed, the anthropological engagement with intensification was heavily stimulated by Ester Boserup's first monograph (Boserup 1965; Spooner 1972) and by her model of a linear sequence of intensification beginning with forest fallow and ending with multicropping (see Johnston 2003:Table 1 for a succinct summary of the Boserup model). But just what is meant by "intensification" has not always been clear or consistent. Both Kathleen Morrison (1994:115) and Patrick Kirch (1994:17) adopted Harold Brookfield's definition as the most appropriate for archaeological considerations: "the addition of inputs up to the economic margin . . . measured by inputs only of capital, labour and skills *against constant land*" (Brookfield 1972:31, emphasis added). Thus, for example, the mere *expansion* of an agricultural system with constant technology and labor inputs—even though this may appear quite impressive from the landscape perspective—does not count as intensification per se. Even with Brookfield's clear definition, however, there is the considerable problem of whether intensification can readily be detected in the archaeological record. As will be seen, this depends in part on the *mode* of intensification, for despite Boserup's overly simplistic model of a linear progression, it is clear that intensification can proceed by at least two quite distinctive pathways.

It is also useful, again following an insight of Brookfield's (1984), to distinguish between *intensification* and *innovation* in agriculture. To be sure, the two concepts are often linked, in that sequences of intensification frequently (but not always or necessarily) follow on innovations, as in the development of new irrigation techniques or more efficient farming tools. Furthermore, there is a critical difference between innovation and intensification, as pointed out by Brookfield, that whereas "intensification is always burdensome, and is adopted from necessity," innovation "offers the hope of advantage" (1984:35). This is because intensification typically requires increased labor inputs, whereas innovation may (or may not!) allow for a reduction in labor.

To return to the issue of modes of intensification, Boserup's model (1965) defines a linear progression of successive stages in the reduction of fallow length, beginning with (1) long fallow (15–25 year fallow period) in climax forest, and continuing with (2) bush fallow with 6–10 year rotation, (3) short fallow with 1–2 year rotation, (4) annual cropping with fallow limited to a few months, and finally (5) multicropping (two or more crops per year) with fallow completely absent. This sequence is accompanied by progressive innovations in technology (from dibble stick to plow, and possibly irrigation), by increased frequency of weeding (from none to highly intensive), and in the use of green manure, mulch, or other fertilizers. In this sequence there is also a steady progression in the labor inputs. This kind of classic Boserupian mode of intensification I have called *cropping cycle intensification* because it is necessary to distinguish it from a quite different form of intensification, which following Blaikie and Brookfield (1987) is termed *landesque capital intensification* (Kirch 1994:19).

In landesque capital intensification, labor inputs occur heavily up front, in the construction of terraces, canals, and similar agricultural infrastructure, with the goal making *permanent* modifications to the landscape (hence the notion of "capital") in a manner that allows for increased yields per unit area. Once constructed, however, such permanent agricultural infrastructure often has the capacity not only to increase yields per area but to do so with relatively low continuing labor inputs in maintenance, weeding, and so forth. In my study of the "wet and dry" agricultural systems of Futuna Island in Western Polynesia (Kirch 1994) I showed how the pondfield irrigation systems require relatively little maintenance after planting, in part because the water inhibits weedy growth. When this fact is combined with the much higher yields produced by taro (*Colocasia esculenta*) when grown in pondfield conditions (Spriggs 1982), the great advantage of landesque capital intensive irrigation over short-fallow or annual cropping (in which high labor inputs for weeding and mulching are necessary) becomes clear.

Less remarked in the literature are some fundamental differences in the ways in which cropping cycle and landesque capital forms of intensification interact with the biogeochemical regimes that are critical to nutrient status and hence to sustainability of agricultural production. Cropping cycle intensification is likely to be progressively *extractive* of soil nutrients over time, requiring increased inputs not only of labor but of green manure or other forms of exogenous nutrient input (or fallowing with nitrogen-fixing plants) if productivity is not to be significantly diminished. Recent investigations of the Kohala field system in Hawai'i document such progressive depletion of nutrients under cropping cycle intensification (Vitousek et al. 2004).[2] The construction of irrigated pondfields, in contrast, creates artificial microecosystems in which nutrients are continually supplied hydraulically, and in at least certain cases nitrogen may be fixed through algae that grow in the fields. Thus irrigation systems, once constructed, may be much better at sustaining high yields over the long term.

For the archaeologist these distinctions between cropping cycle and landesque capital intensification also have real implications with respect to the potential visibility of a temporal sequence of agricultural change in a particular regional archaeological record. Landesque capital intensification, by definition requiring substantial and permanent modifications to the landscape, will always be likely to leave highly visible traces, even centuries or millennia after abandonment. Indeed, in some places ancient terrace systems have been cleared of forest regrowth and put back into cultivation, as on Aneityum Island (Spriggs 1982). Cropping cycle intensification, on the other hand, may be more difficult to detect in the archaeological record. In some sequences the late stages of cropping cycle intensification do involve permanent modifications to the landscape, such as field walls or boundaries, rendering them archaeologically detectable. A classic case in Polynesia is the Kohala field system on Hawai'i Island (Kirch 1984:181–192; Ladefoged et al. 2003; Rosendahl 1994). Lacking such structural modifications, however, a sequence progressing from forest fallow to bush fallow to short fallow may be evidenced only by more subtle and difficult-to-interpret indications, such as changes in regional pollen spectra (decrease in arboreal taxa and increases in grasses and/or weeds), in rates of soil erosion and sedimentation, and in the frequency and size of microscopic charcoal particles in sediment catchments. The problems attendant on such indirect indicators of agricultural activity are at the core of recent debates over the interpretation of agricultural change in the Pacific (Leach 1999).

To sum up the above discussion, in thinking and writing about various kinds of agricultural change it is desirable to make several terminological distinctions, as diagrammed in Figure 8.1. First of all, we must distinguish agricultural innovation from intensification. Moreover, innovations themselves can be of several types. While it may be most common to think of *technological* innovation, a very important role has been played by *genetic* innovations. Ever since humans began domesticating plants in the early Holocene, they have continued to modify the genetic basis of their crops, primarily through simple "Burbankian selection." A fine example from Oceania is the selection for gigantism in the fruits and nuts yielded by certain tree crops, as shown ethnobotanically by Yen (1974b). Lepofsky et al. (1998) demonstrated that this process of selection for increased nut size is detectable in archaeobotanical remains of *Canarium* almond from the early Lapita site of Talepakemalai, in western Melanesia. The remarkable diversity of cultivars of such key crops as taro, breadfruit, or sweet potato, as documented ethnobotanically in Polynesia (Handy 1940; Wilder 1928; Yen 1974a), indicates that Polynesian horticulturalists (and here the term *horticultural* is indeed apt) were constantly attuned to the somatic variability in their planting stocks.

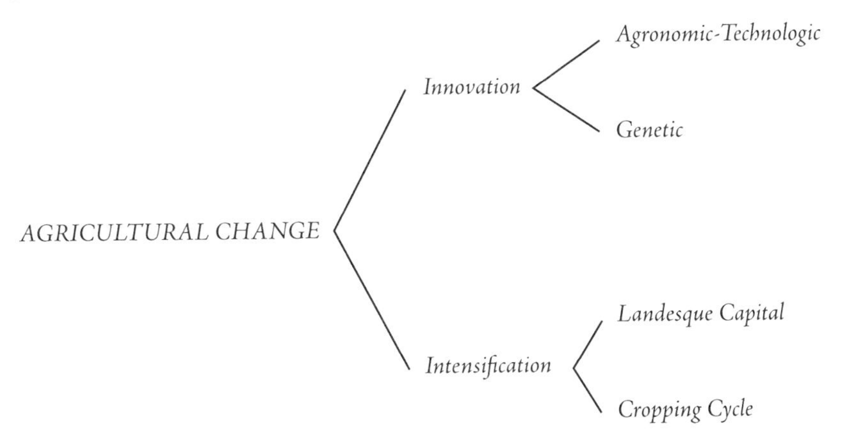

Figure 8.1. A taxonomy of agricultural change (from Kirch 1994:Figure 5).

Referring again to Figure 8.1, we must secondly distinguish such agricultural innovations from intensification proper, recalling that the latter is marked by increased labor inputs per unit area. Here again we will wish to differentiate between cropping cycle intensification—the classic sequence defined by Boserup—and landesque capital intensification, which results in permanent investments in physical infrastructure. Certainly in the Pacific, if not more generally throughout the preindustrial world, these two pathways of intensification have rather different implications for the economies in which they were embedded.

INTENSIFICATION IN THE POLYNESIAN ARCHAEOLOGICAL RECORD

I turn now to selected examples of agricultural intensification in Polynesia, as these have been recorded and studied archaeologically. This review is by no means exhaustive but merely illustrates some of the range of variation in modes and sequences of intensification. Documenting sequences of agricultural change in the archaeological record is challenging, requiring interdisciplinary collaboration and a diversity of evidential sources (e.g., classic field archaeology, geomorphology, pedology, palynology, archaeobotany). I also include in this section a discussion of arboriculture and storage as examples of agricultural innovation, in contrast with intensification.

Landesque Capital Intensification

In Polynesia, as indeed throughout much of island Melanesia, systems of landesque capital intensification are almost exclusively associated with taro, *Colocasia esculenta*, an aroid of Indo-Pacific origins with a remarkably broad adaptability to a range of edaphic and hydrologic conditions. Thus, while various taro cultivars were important components of swidden and dryland gardens, even tolerating relatively arid conditions (as on Easter Island or in leeward regions of the Hawaiian Islands), they flourish under hydromorphic soil conditions. Indeed, taro responds so well to inundated planting media that its highest yields are in fully irrigated, pondfield contexts. Whereas taro grown in traditional (nonfertilized) dryland gardens has yields estimated at between 7 and 15 tons per ha, yields under irrigation are regularly estimated at between 20 and 40 tons per ha, with extremes up to 58 tons per ha (Kirch 1994:Table 10; Spriggs 1982). Moreover, fallow length in such pondfield irrigation systems can be quite low, in the Futunan case of one to three years between each cycle of three to seven taro crops (Kirch 1994:155).

Given such high and sustainable productivity, it is not surprising that various agronomic systems for wet taro cultivation were developed throughout

Polynesia. These included not only true pondfield irrigation (defined as bunded terraces or pondfields holding standing water, and generally fed by one or more irrigation canals) but also reticulate systems of ditched "garden islands" in naturally swampy terrain and even pits excavated to tap the thin freshwater lens of coral atolls (Kirch 1984:Table 19). In parts of New Zealand, for example, extensive prehistoric systems of reticulate drainage ditches have been identified in alluvial floodplains, probably for taro cultivation (Barber 1989, 2001; Jones 1986, 1991). It is the terraced, irrigated pondfield systems, however, that have attracted the greatest attention from archaeologists. Although where they occur on alluvial flats such pondfields were often constructed of earthen bunds and embankments, in steeper valleys and on colluvial slopes Polynesian pondfield terraces were usually faced with stone retaining walls, and irrigation ditches were also stone-lined. In some cases stratified sequences of terrace wall construction and reconstruction have been revealed through excavation (Allen 1991, 1992; Riley 1975; Yen et al. 1972).

Discriminating between terraces that were irrigated from those that were used for slope retention or erosion control, but not irrigated, can pose problems for the archaeologist. Through excavation and examination of soil micromorphology, however, such discrimination is possible as a result of the characteristic reduction-oxidation regimes that develop under conditions of inundated taro cultivation. Such "redox" profiles were first noted by Yen et al. (1972) in prehistoric pondfield terraces of the upper Makaha Valley, O'ahu, and were systematically described by Kirch (1977).[3] A recently studied example, from the Waikolu Valley on the windward coast of Moloka'i Island, is illustrated in Figure 8.2. In this stratigraphic section two distinct periods of pondfield cultivation are evidenced (Layers III and VI), separated by a high-energy flood deposit (Layer V). Charcoal of endemic *Pritchardia* palm incorporated into the lower pondfield soil was radiocarbon dated to cal AD 1240–1280 (Kirch 2002:45–46). The pondfield soil itself is a very dark colored clay loam rich in organic matter, while the underlying oxidized horizon is streaked throughout with reddish-brown limonite tubes that formed around the rootlets of taro plants. Such limonite tubes form through a process of precipitation of ferrous and manganous ions, which were mobilized in the upper reductive pondfield horizon.

Kirch and Lepofsky (1993) addressed the origins and development of pondfield irrigation in Polynesia using a combination of direct archaeological and historical linguistic evidence. Two hypotheses had been raised to account for irrigation in the Pacific: (1) it had been developed in Southeast Asia and diffused to the Pacific, or (2) irrigation technology was independently innovated in the islands. The historical linguistic evidence strongly supports the latter hypothesis, for while there is a suite of widespread Oceanic words for crop plants and methods of swidden cultivation, implying a considerable

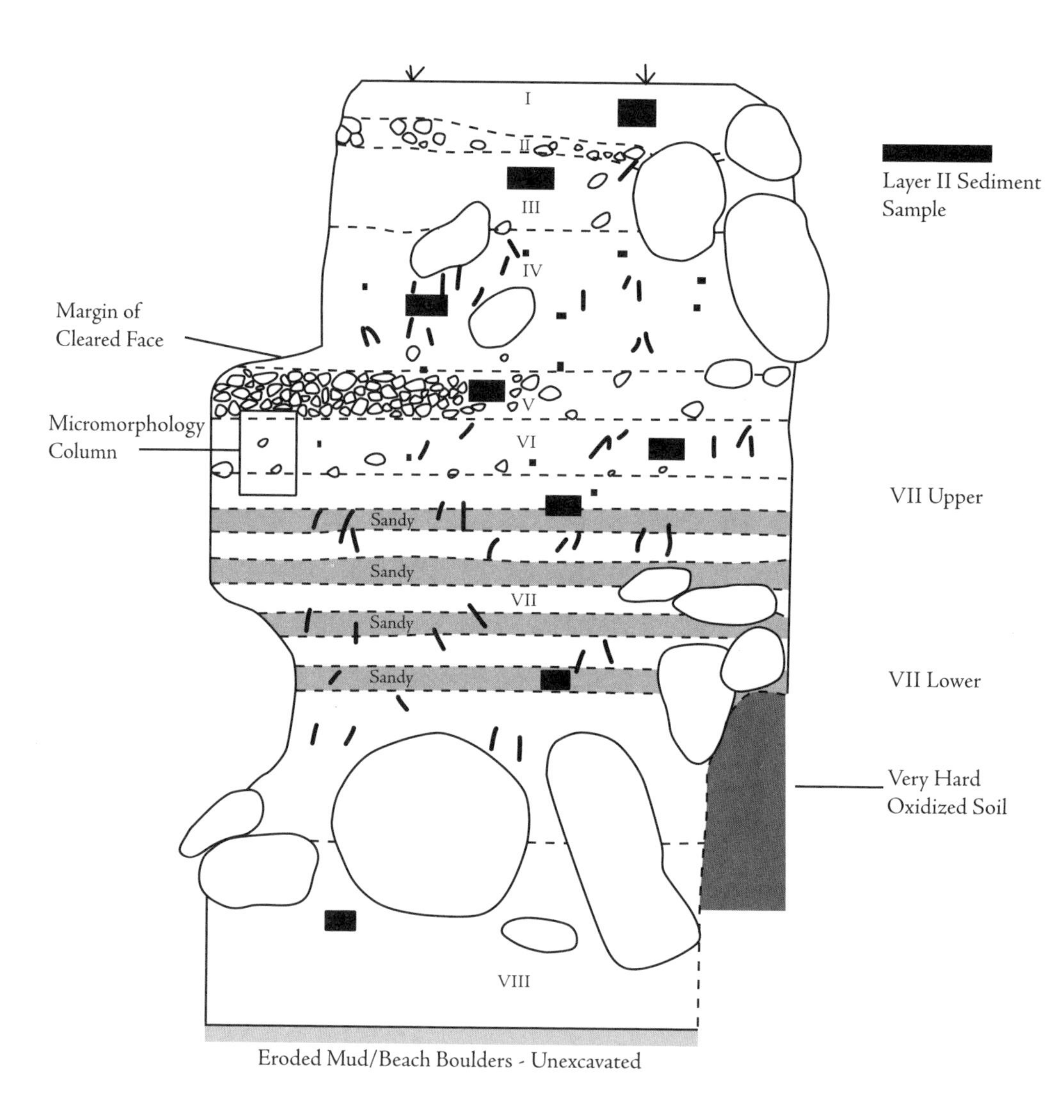

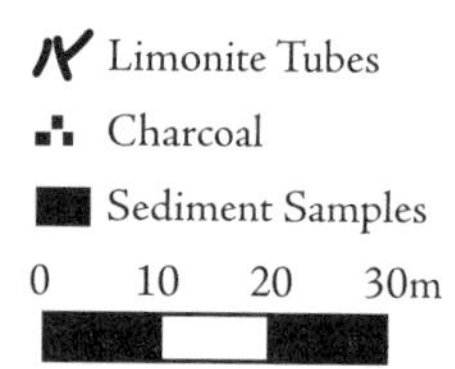

Figure 8.2. Stratigraphic section through a pondfield irrigation complex in Waikolu Valley, Moloka'i Island (from Kirch 2002:Figure 11).

antiquity for horticultural practices, there is no corresponding set of widely shared terms specifically referring to irrigation. Rather, words for *terrace*, *irrigation canal*, *pondfield complex*, and the like appear to be local innovations within particular island groups. To the extent that they have been excavated and dated, irrigation complexes in such islands as Futuna, Hawaii, and the Society Islands likewise indicate that such systems were constructed in the more recent periods of prehistory. This is not, however, to suggest that the earliest Polynesian horticulturalists did not understand the ability of taro to thrive in hydromorphic conditions. Rather, we suggest that until island populations had expanded to sizable numbers and begun to put pressure on food systems, wet taro cultivation was confined to nonintensive plantings in naturally swampy coastal plains and alluvial bottomlands. Thus only later in island sequences did large-scale irrigation arise, as independent innovations from a common knowledge of taro's agronomic propensities.

By far the most extensive archaeological work on pondfield irrigation has focused on the Hawaiian Islands, with important studies of sites on Kaua'i (Earle 1978, 1980), O'ahu (Allen 1987, 1991; Dega and Kirch 2002; Kirch and Sahlins 1992; Yen et al. 1972), Moloka'i (Kirch 2002; Riley 1975), and Hawai'i islands (Tuggle and Tomonari-Tuggle 1980). Jane Allen's synthesis of 46 radiocarbon dates from windward agricultural contexts (1992:Figure 4) indicates that major investments in irrigation infrastructure began around AD 1100 and continued for several centuries, thus corresponding to the Expansion period in Hawaiian prehistory (Kirch 1985). Most, if not all, of the large-scale irrigation works that covered the floors and lower colluvial slopes of windward, and some better-watered leeward, valleys throughout the island appear to have been built during the Expansion and subsequent Protohistoric periods. Some systems continued to be constructed in the first decades after European contact, as in the upper Anahulu Valley of O'ahu Island (Kirch and Sahlins 1992).

Ethnohistoric accounts stress the close linkage between irrigation and chiefship, which is not surprising given the ability of such systems to yield high levels of surplus; in the Anahulu Valley such surplus production has been estimated at between 50 and 70 percent (Kirch and Sahlins 1992:2:172). While many Hawaiian irrigation systems were small in scale, and may have been farmed by only one or a few household groups, the largest systems covered as much as 53 ha and were fed by canals from 1.0 to 3.7 km in length.

The long-term history of pondfield irrigation elsewhere in Polynesia is less well studied. Kirch (1994) and Anne Di Piazza (1990) carried out detailed ethnoarchaeological research on traditional Futunan irrigation but only limited excavations; these suggest a later prehistoric development of irrigation technology, as in Hawaii, but this hypothesis remains to be fully tested. In the Society Islands Lepofsky (1994, 1995) studied the small-scale irrigation systems of the ʻOpunohu and other valleys on Moʻorea, putting their development into a general temporal framework for agriculture and land use. Much work remains to be done, however, in these and other island groups where irrigation was important, such as the Marquesas (Addison 2001) and Austral Islands.

Cropping Cycle Intensification

As noted above, historical linguistic evidence from Proto-Polynesian and even older interstages of Austronesian languages such as Proto-Oceanic leaves no doubt that the first settlers of the tropical Pacific islands possessed an extensive suite of root, tuber, and tree crops and practiced both classic shifting cultivation and arboriculture (Kirch and Green 2001:128). The Proto-Polynesian language, for example, contains robustly reconstructed words for swidden garden (PPN **maqala*), fallow (**talu*), to weed (**qua-talu*), and many other terms associated with gardening (Kirch and Green 2001:Table 5.2). In its pioneering phases on most islands, however, the practice of shifting cultivation within a long-fallow context would not be expected to leave substantial archaeological evidence. However, as shifting cultivation expanded over time and as fallow length shortened, or as climax forest was progressively replaced with secondary growth, such changes might predictably be evidenced in the paleoenvironmental record of sedimentation and pollen spectra. Likewise, as there were few natural sources of ignition in the humid forests of tropical Pacific islands, dramatic increases in the deposition of microscopic charcoal particles can be expected to correlate with the expansion and/or intensification (i.e., reduction of fallow length) of dryland cultivation systems.

There is now substantial paleoenvironmental evidence from a number of islands for (1) forest clearance and the replacement of indigenous climax forest with secondary growth, and in some cases by terminal fern or grasslands; (2) increased rates of erosion and alluvial sedimentation, sometimes dramatic; and (3) significant increases in charcoal influxes to depositional contexts, all following on initial Polynesian settlement. Some of the best documented cases include Tikopia (Kirch and Yen 1982), the Lau Islands (Bayliss-Smith et al. 1988), Futuna (Kirch 1994), Atiu (Parkes 1997), Mangaia (Kirch 1996; Kirch et al. 1992), Moʻorea (Lepofsky et al. 1996; Parkes 1997), Rapa Nui (Flenley et al. 1991), and Oʻahu in the Hawaiian Islands (Athens 1997; Athens and Ward 1993). To be sure, shifting cultivation was probably not the exclusive cultural practice responsible for this suite of environmental changes, but it was likely to have been the most significant. Nonetheless, it is unfortunately true that such evidence provides at best only an indirect proxy for agricultural practice, and lacking other kinds of archaeological indicators it may be impossible to discriminate intensification sequences from long to short fallow (Leach 1999).

On some Polynesian landscapes, however, the intensification of cropping cycles did result in permanent traces and modifications that allow us to detect actual change over time. Far and away the best documented of these landscapes is the Kohala Field System situated on the northern part of Hawai'i Island, which has seen intermittent archaeological study over more than three decades (Kirch 1984:181–192; Ladefoged et al. 1996, 2003; Newman n.d. [1972]; Rosendahl 1994). This vast system of reticulate fields, marked by stone-and-earth walls or berms (running with the contours) and cross-cut by stone-lined trail systems (running up-slope), may have originally covered as much as 80 km^2 on the leeward flank of the Kohala mountains; some 55 km^2 remain intact (Figure 8.3). Radiocarbon dates indicate that the system developed throughout the Expansion and Protohistoric periods of the Hawaiian cultural sequence (ca. AD 1200–1800). More important, Ladefoged et al. (2003) have documented a relative chronology of field boundaries in several study areas within the system, convincingly demonstrating the progressive subdivision of field units into smaller and smaller parcels. Their evidence supports the interpretation of a shift from "reliance on a relatively low-density expansive agricultural productive system to use of an increasingly intensified one" (Ladefoged et al. 2003:937). Here, then, is a compelling case of progressive cropping cycle intensification that at the ethnohistoric endpoint consisted of short fallow intercropping of sweet potato, dryland taro, and other crops (with sugarcane grown as windbreaks on the field boundaries). Recent and ongoing research as part of the Hawaiian Ecosystem Biocomplexity project has also shown that the Kohala system both coincides with a particular zone of favorable edaphic conditions within the Kohala leeward biogeochemical nutrient gradient and that successive cropping within the zone has measurably lowered nutrient status within the fields (Vitousek et al. 2004).

Field systems with similar archaeological remains of walls and other agronomic features are documented from other parts of Hawai'i, Maui, and Moloka'i islands, where edaphic and climatic conditions were favorable to intensive dryland cultivation of sweet potato and taro (Allen 2001; Kirch 2002; Schilt 1984). In Kahikinui, Maui, an extensive upland zone between 400 and 900 m elevation was densely occupied and cultivated from the late Expansion through Protohistoric periods, although no regular grid work of fields was developed (Coil and Kirch, in press). Rather, cultivation took advantage of natural swales or basins in the undulating aa lava flow surfaces and of patches of slightly older lavas, which were blanketed in tephra ash deposits. As in Kohala, this intensive agrarian landscape coincides with a particular edaphic zone with elevated cation base saturation and higher available phosphorus (Kirch et al. 2004).

Outside of the Hawaiian Islands, examples of cropping cycle intensification that led to permanent landscape modifications are certainly known, although less well studied. In New Zealand, dryland cropping systems for sweet potato

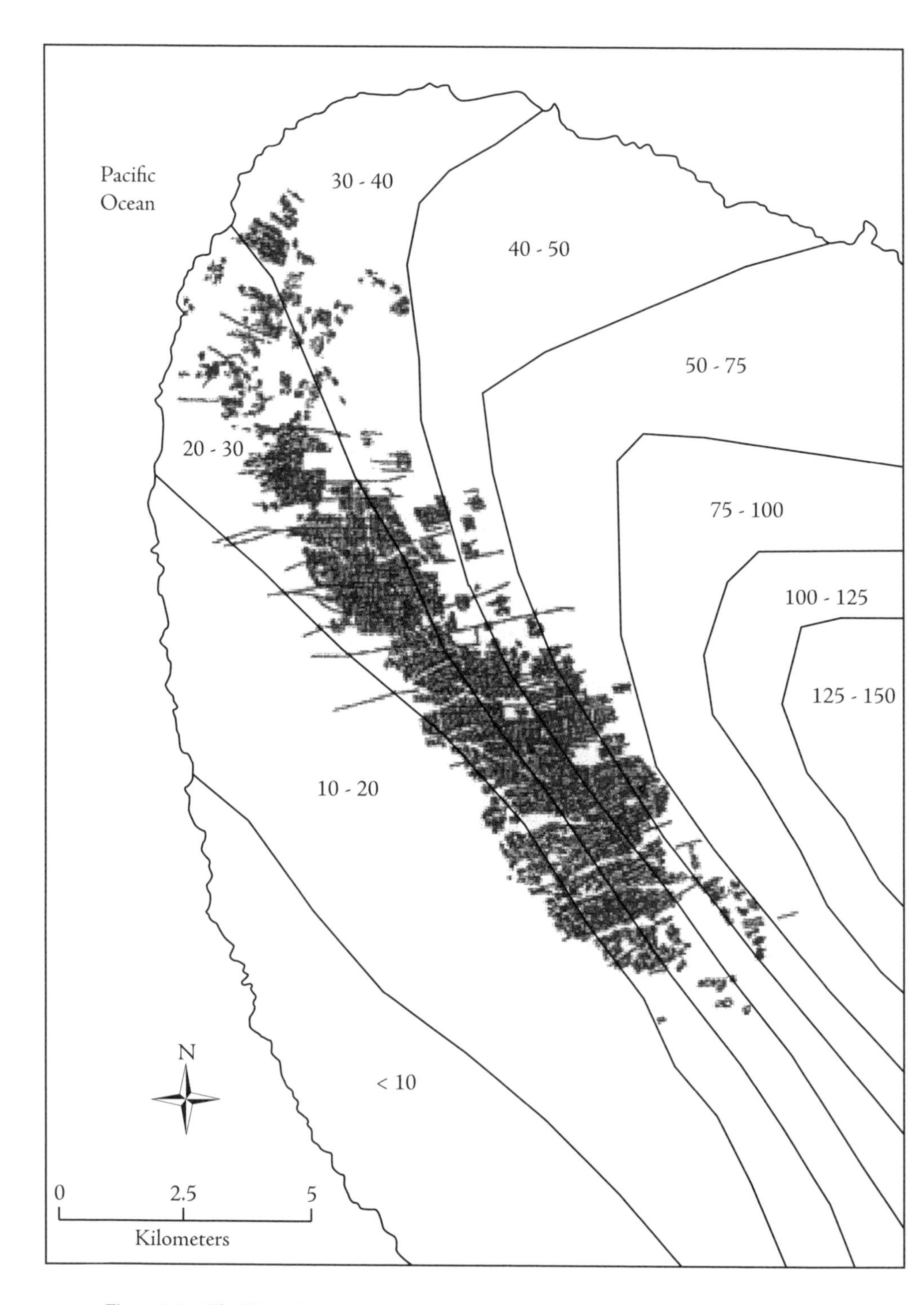

Figure 8.3. The Kohala Field System of northern Hawai'i Island, showing major field walls and trails as defined by aerial photography, in relation to annual rainfall zones (in inches) (from Ladefoged et al. 2003:Figure 2).

that resulted in stone-walled field systems have been noted in various parts of North Island, such as Palliser Bay (e.g., Leach 1979) and surrounding the fortified volcanic cones at Pouerua (Sutton 1990:Figure 3). In this temperate climate however, Polynesian agronomic systems were forced to adapt to conditions that were at best marginal to the limited array of crop plants introduced from the tropical islands. The resulting systems were dominated by sweet potato, which was well adapted to the temperate conditions and whose tubers could be stored in semisubterranean pits (Leach 1984). Nothing better reflects the intensity of late prehistoric New Zealand horticulture than the density of volume of such *kumara* storage pits, which were incorporated into a fortified village settlement pattern (Irwin 1985).

In the tropical core of Western Polynesia, intensive short-fallow cropping based primarily on yams (*Dioscorea alata* and other species) was described by the early European explorers in Tonga (Beaglehole 1969:252). What may have been a similar system on the karst island of Alofi left an archaeological landscape of field walls and trails, now covered in climax forest regrowth, that has yet to be closely studied (Kirch 1994:237–241, Figure 98). In parts of Samoa intensive short-fallow cropping was closely integrated with dispersed household settlements, marked by a dense grid of stone-walled fields and plots, as in the Mt. Olo tract on 'Upolu Island (Jennings and Holmer 1980).

At the farthest southeastern extreme of Polynesia, isolated Rapa Nui (Easter Island) offers another case where, on ethnohistoric as well as paleoenvironmental evidence, the original forest cover of the island was progressively reduced as cropping cycles were intensified under a horticultural system based dominantly on sweet potatoes and taro. Although no field systems similar to that of Kohala have been recorded, the Rapa Nui landscape is covered in a myriad of small-scale architectural features related to intensive cultivation, such as mounds, "stone mulch gardens," and stone-lined pits (*manavai*) (Stevenson 1997; Wozniak 2001). Whether the Rapa Nui population pushed the intensification of its agricultural subsistence base beyond its "carrying capacity," thus precipitating ecological and social disaster, has been hotly debated (Bahn and Flenley 1992). It does seem likely that on this isolated and circumscribed island the relations between population, production, and sociopolitical organization were tightly bound and that the collapse of the enormously labor-intensive statue cult was linked to an over-taxing of agricultural possibility. But the details of this fascinating sequence remain to be worked out.

ARBORICULTURE AND STORAGE AS INNOVATION

Having stressed the distinction between *intensification*, involving increased inputs of labor or landesque capital (based on labor) to constant land, as against *innovation*, it is worth briefly describing some Polynesian agricultural systems whose trajectories of change depended heavily on the latter. These systems followed a pathway of agricultural change that emphasized and elaborated a subsystem present in early Oceanic cultivation, namely *arboriculture* or tree cropping. Anaerobically preserved remains of at least 24 taxa of nut, fruit, and seed-bearing tree crops from the early Lapita site of Talepakemalai, dated to c. 1600–1200 BC (Kirch 1989), demonstrate that the immediate ancestors of the Polynesians had an extensive arboricultural array at their disposal. Although tree crops remained important to some degree in a majority of Polynesian agricultural systems, only certain islands were to develop production systems in which arboriculture was foremost.

Among these were the Marquesas and Society archipelagoes, which developed production systems dominated by the tree cropping of breadfruit (*Artocarpus altilis*). To be sure, in both island groups there was considerable development of irrigated taro terracing in the inland valleys, and the coastal plains of the Society group also supported ditched "raised bed" cultivation of *Colocasia esculenta*. However, in both archipelagoes, and particularly the Marquesas, ethnohistoric sources clearly document the primacy of breadfruit (Handy 1923; Wilder 1928). Here we have an example of agricultural innovation in two aspects: (1) *genetic innovation* in the selection for a diversity of cultivars and (2) *technological innovation* in the development of subterranean semianaerobic fermentation storage facilities. Wilder (1928) recorded some 32 distinct cultivars of breadfruit on Tahiti and Mo'orea, including a large number of seedless varieties that were propagated vegetatively. By developing the capacity to store large quantities of fermented breadfruit paste, the islanders were able to buffer both seasonal variation in yields and the adverse effects of longer-term environmental perturbations such as cyclones and drought. In the Marquesas the subterranean storage of breadfruit reached its technological apogee, with pits often located in fortified positions or in close association with elite residences and having capacities as large as 200 cubic meters or more (Linton 1925; Kirch 1984:133).

A second example of agricultural innovation comes from the Polynesian Outlier of Tikopia, geographically situated in the far Eastern Solomon Islands (Kirch and Yen 1982). Here there is also a dominant emphasis on arboriculture, in this case developed as a complex form of multistory cropping involving an understory of perennial *Cyrtosperma* aroids and bananas, shaded by such tree crops as *Artocarpus*, *Metroxylon*, and *Cocos*. Archaeological and paleoenvironmental evidence hints that this system developed over time out of an earlier phase of classic shifting cultivation involving firing of the original climax forest. The Tikopia system seems to be closely mimicked by a similar "agro-forest" production system known from the island of Kosrae, in Micronesia (Athens et al. 1996).

INTENSIFICATION AS PROCESS

To conclude this review, I now turn to issues of process, or explanation, which by their nature are theoretical and often contested. It is in this realm, however, that the Polynesian cases may exhibit their greatest import, insofar as they represent a series of "parallel experiments" in agricultural change, all stemming from the same historical base.

Intensification and Population

At a fundamental level there is no doubt that agricultural change and population are linked phenomena. Growing populations require expanded food supplies; expansion and intensification of an agricultural system require labor inputs, often stimulating population increase. The problem is teasing out cause and effect in what is, essentially, a tightly bound system of reciprocal interactions.

Population sizes and densities in Polynesia, prior to the ravages resulting from the introduction of Old World diseases to which there was effectively no resistance, are a matter of some dispute (Kirch 2000). All indications, however, are that the tropical and subtropical volcanic ("high") islands were heavily populated, with densities ranging from lows of perhaps 50 persons per km^2 of arable land in more marginal localities (such as Kahikinui on leeward Maui) to highs of as many as 242 persons per km^2 in the case of ethnographically documented Tikopia. In a general sense the Polynesian data confirm the strong correlation between high levels of population density and intensive agricultural production.

What we would really like to know, however, is how population and agricultural intensification track each other over longer time periods. Unfortunately, good quantitative data on either side of the equation are extremely difficult to obtain. From all of the Polynesian cases we have considered, only that of Hawaii comes close to providing the necessary data. Although there is much debate concerning the absolute size of the Hawaiian population at the point of initial contact with Europeans in 1778–1779, archaeologists are reasonably confident concerning the overall shape of the archipelago's population growth curve from Polynesian discovery c. AD 700–800 until the time of contact (Dye and Komori 1992; Kirch 1985). The curve is best described as two discrete segments: (1) an exponential growth segment from initial human settlement until a population peak was achieved c. AD 1400–1500; followed by (2) a late period marked by dramatic reduction in growth rates and possible fluctuation of the total population around a relatively stable mean. Most important, the phase of exponential population growth, especially during a three- to four-century-long period between about AD 1100 to 1400/1500 (the Expansion period of the Hawaiian cultural sequence) is characterized by both expansion and intensification of the islands' agricultural systems. Thus in Hawaii, at least, dramatic population increase and agricultural intensification do track each other closely. To move beyond this level of abstraction, and assert that either population or intensification is an "independent" variable driving the other, would be to push the data too far.

Intensification and Social Production

The economics of any society—but particularly of complex and hierarchically differentiated societies—cannot be reduced to a minimal equation linking population numbers to resources, the fundamental importance of demographic pressure notwithstanding. Sahlins (1972) wrote of this problem when he described the contradiction between the "domestic mode of production," with its inherent antisurplus ideology, and the continual demands for extradomestic surplus that are generated by the larger political economy. Thus, in Sahlins's words, the ideology of an antisurplus domestic economy "can only be 'a disarray lurking in the background,' always present and never happening" (1972:313). Brookfield, always a clear thinker on such definitional matters, stressed the necessity of disaggregating production for use (autoconsumption by the household units of production) from what he termed *social production* (1972:37–38). Whereas production for use (closely linked to demographic pressures) may be reasonably well modeled according to the principle of least effort, social production is another matter. As Brookfield writes: "Inputs may be wildly uneconomic when measured by calorific returns, yet wholly reasonable when measured against social returns" (1972:38). Brookfield went so far as to claim that in the Pacific social production could account for cases of intensification where none would be expected on the basis of population numbers alone (1972:44).

At the time of their common origins in Ancestral Polynesia (ca. 500 BC to AD 300), Polynesian societies were organized as heterarchical household units with asymmetric exchange relations between them, a social structure that inherently encourages competition for prestige and control over land and other resources (Kirch and Green 2001). Also present were suprahousehold kin aggregations with positions of hereditary leadership, especially the position of "priest-chief" or Proto-Polynesian **qariki*. Such sociopolitical structures were arguably a sufficient force to continually exert pressure on local production systems so as to generate surpluses above the minimal level of household needs, especially for competitive feasts. After the dispersal out of the Ancestral Polynesian homeland in the first millennium AD, and the subsequent settlement of the Eastern Polynesian islands and archipelagoes, specific Polynesian societies followed distinctive pathways of sociopolitical evolution. In the larger and more resource-rich islands, however, these typically moved in the direction of increased hierarchy and social stratification. At the time of European intrusion in the late eighteenth century, the most elaborated societies, in terms of rank and status distinctions, were those of Tonga, the Society Islands, and most particularly, Hawaii (Goldman 1970; Sahlins 1958). In the case of Hawaii one may argue that the course of sociopolitical evolution had crossed a fundamental threshold, marked by the emergence of kingship and an "archaic state." In these latter societies social production had become a major force in their political economies, marked not only by competitive feasting on a grand scale (Kirch 2001:179) but by production to underwrite specialization, as in the case of warriors, priests, canoe-makers, artisans of other sorts, and so on.

Futuna, in Western Polynesia, provides an ethnoarchaeologically documented case of social production in an island where underlying environmental patterns have divided the island into two sectors, one concentrated on landesque capital and the other on cropping cycle intensification (the two chiefdoms of Sigave and Alo, respectively; Di Piazza 1990; Kirch 1994). In the irrigation chiefdom of Sigave the relatively high calorific production of the taro pondfields is channeled into maintaining large pig herds, pigs being the most highly valued prestige food, essential for competitive feasts called *katoaga*. The dryland chiefdom of Alo, which has only limited areas in pondfields but more extensive dryland gardens, channels significant labor into the swidden cropping of paper mulberry (*Broussonetia papyrifera*), which in turn is converted into large sheets of decorated bark cloth, a necessary component of interhousehold exchanges (to celebrate birth, marriage, death, and other rites of passage). Pigs and bark cloth, along with various other prestige goods, flow against each other in the exchange cycle (Kirch 1994:196, Figure 84). Moreover, in the competitive intervillage feasting cycle taro and yams—the quintessential wet and dry crops—are symbolically ranked against each other, with an emphasis on the *quantity* of taro and the *quality* of yams (Di Piazza 1990:161). The immense efforts that go into the production of a successful *katoaga* feast underscore the force that social production can have on agricultural efforts. In one major feast that I witnessed in the Alo chiefdom in 1974, the food prestations included between 3,500 and 4,000 whole yam tubers, 300 small-to-medium-sized pigs, 30 large hogs, and large quantities of bark cloth. All of this was distributed to visiting guests from other villages, in turn setting up higher expectations for return feasting at some future date.

In the case of protohistoric Hawaii, in which kingship and a true class differentiation between chiefs and commoners had emerged, social production took on a distinctively "predatory" nature. Various Hawaiian proverbs speak to the boundless propensity of chiefs to extract surplus from the commoners: "The chief is a shark that travels on the land," consuming all in sight (Pukui 1983). The relationship between land, labor, and chiefship in late precontact Hawaii, however, may be best summed up with this proverb: "The land remains the land because of the chiefs, and prosperity comes to the land because of the common people" (Pukui 1983:125). That is to say, the chiefship holds the land while the commoners work it; in Marxian terms the Hawaiian elite had managed to transform the ancient Polynesian system of landholding by kinship groups into a territorial system of chiefly patronage (Kirch and Sahlins 1992). Lacking significant food storage (unlike the Marquesas or Society Islands) and a magazine economy, the Hawaiian elite moved peripatetically, extracting food and other productions from the many territorial units under their control. This exaction system became ritually encoded in the annual *Makahiki* rites dedicated to the deity of dryland agriculture, Lono, in which the Lono priests accompanied by the king's warriors visited each territory in succession, demanding specified levels of surplus goods (Valeri 1985). In short, in protohistoric Hawaii social production had evolved to an incipient system of taxation.

What is particularly intriguing about the Hawaiian case is that the evolution of such systems of taxation, legitimized by ritually encoded ideologies of propitiating agricultural deities, and corresponding to a hierarchical system of rank and land rights, were most strongly developed in that part of the archipelago dominated by cropping cycle intensification, not by irrigation. Ethnohistoric sources, as well as archaeology, tell us that these sociopolitical institutions emerged in the large-scale polities of Hawai'i and Maui islands during the last two to three centuries prior to European contact. After the conquest of the westerly islands and their vast holdings of irrigated pondfields, by Kamehameha I in 1795–1804, this system was extended archipelago-wide. Hence the origin of the proverb, "a taxing of small fields by the Hawai'i chiefs" (Pukui 1983), referring to the manner in which the Hawai'i Island chiefs demanded their due from even the smallest land parcels.

Intensification in Polynesian Sociopolitical Evolution

As various groups of early Polynesians dispersed out of their ancestral homeland in the Western Polynesian archipelagoes in the first millennium AD to discover and settle the far-flung islands and archipelagoes that make up the Polynesian Triangle, their societies began to differentiate in response to founder effects and adaptation to environmental differences, even as they continued to share many aspects of the common cultural history (Kirch and Green 2001). The subsequent evolution of their sociopolitical structures, both in Western and Eastern Polynesia, followed not a single trajectory but rather a number of alternative pathways, influenced by degree of isolation, environmental conditions and resources, population growth, and the playing out of structural contradictions inherent in their social organizations, not to mention all the messy contingencies of human agency and history. All of this cannot be reduced to a single or simple explanatory model, but certain consistent trends are discernible. One of these is the tendency, over time, toward intensification of the islands' production systems. Indeed, the only clear cases of *dis*intensification in the prehistoric record for Polynesia would appear to be those of southern New Zealand and the Chatham Islands, where true temperate conditions forced the abandonment of horticulture altogether and a dramatic shift to a hunting-and-gathering economic mode. Otherwise, within tropical and subtropical Polynesia all of the sequences we have teased out of the archaeological record testify to varying modes and degrees of intensification, sometimes accompanied by significant innovation.

But I have repeatedly stressed that intensification operates in contrastive modes, especially between landesque capital intensification and cropping cycle intensification, with very different implications for local political economies. Anthropologists have often made much of the linkage between irrigation and sociopolitical structure, as in the famous "hydraulic hypothesis" of Wittfogel (1957; but see also Adams 1966; Downing and Gibson 1974; Earle 1980), and in Polynesia there can be no doubt that chiefs preferred to base their economic power on wetland cultivation whenever possible. Yet a comparative analysis of prehistoric intensification in several Polynesian islands (Kirch 1994) strongly suggests that the most powerful links between intensification and chiefship did not occur in those regions most favorable to irrigation. Rather, as in the Hawaiian case just mentioned, but also in Futuna, the Society Islands, and Rotuma, decidedly hierarchical and—most notably—militarily aggressive political structures seem to have emerged in areas marked by cropping cycle intensification. There are presumably several reasons for this pattern. One is the "burdensome" nature of increased labor inputs, as one progresses along the pathway of cropping cycle intensification, in contrast with the ability of irrigation systems to respond through a process of involution (Geertz 1963). At a certain point the ability of intensive dryland systems to continue to yield surplus begins to decline, especially if these systems are extracting nutrients at a high rate. This point may come relatively late in the sequence of sociopolitical evolution, however, after the elite cadres have already instituted relatively oppressive systems of surplus and tribute exaction. In such an event there are at least two possible outcomes. One is a revolt of the overburdened producers and throwing off of oppressive chiefs, a response in fact recorded in Hawaiian oral traditions (Sahlins 1972). The other, also well documented in Polynesia, is predatory territorial expansion. The wars for hegemonic control that raged between Maui and Hawai'i in the Protohistoric period, and ultimately led to the conquest and incorporation of the rich irrigated islands of Moloka'i, O'ahu, and Kaua'i, can perhaps be understood in this light. Similarly, the political histories of Futuna, the Society Islands, and Rotuma document instances of territorial expansion moving from the "dry" to the "wet" (Kirch 1994:317–321). The case of Tonga may also be salutary, for on these makatea islands without possibility of irrigation and noted ethnohistorically for their highly intensive yam-based field systems, arose another of Polynesia's most hierarchical and territorially aggressive polities (Kirch 1984:217–242). Finally, we might invoke the much-debated case of Rapa Nui, where the materialization of rank and power reached an apogee in the unique statue cult, all founded on an agricultural base of short-fallow sweet potato cultivation (Yen 1988). Yet the history of Rapa Nui seems to show all too well the limits to cropping cycle intensification when pushed too far, as not only the economic system but the very fabric of Rapa Nui society was torn asunder in late prehistory.

CONCLUSION

The chiefdoms of Polynesia were embedded in economies based on intensive modes of agricultural production, themselves founded on a suite of crops, domestic animals, and agronomic techniques originally introduced to the islands from ancestral homelands to the west. Over the past 30 years archaeological study of the direct and indirect traces of Polynesian agricultural practice, combined with insights from ethnography and ethnoarchaeology, have greatly advanced our understanding of the economic histories of these island societies. While most island sequences show a trend toward increased intensification over time, this trend did not follow identical pathways. In particular, it is essential to distinguish between two main modes of intensification: landesque capital intensification, involving taro irrigation, and cropping cycle intensification, which centered primarily on sweet potato and yams. These modes had entirely different labor requirements, productive capacities, and implications for sustainability. The rise of hierarchical, territorially expansive sociopolitical structures appears to correlate closely with sequences of cropping cycle intensification. Other sequences of agricultural change in certain Polynesian islands did not emphasize either of these modes but depended more on genetic and technological innovations, especially in the development of arboricultural systems.

NOTES

1. There is something of a terminological debate in the literature regarding Oceanic cultivation systems and whether the term *agriculture* or *horticulture* is more appropriate (e.g., Leach 1999). Certainly much Polynesian cultivation can be properly characterized as horticulture from the perspective of individual plant tending, yet the extensive monocropping of taro or sweet potato seen in some systems was clearly "agricultural" in extent and intensity. Here I use the label "agricultural systems" as a

general cover term for any form of Polynesian cultivation, including horticulture and arboriculture, as well as more intensive monocropping.

2. Throughout this chapter I use *Hawaii* to refer to the archipelago as a whole and *Hawai'i* to refer to the island of that name.

3. Essentially identical redox soil profiles likewise develop in irrigated rice systems, as noted by Kawaguchi and Kyuma (1977) for Southeast Asia.

REFERENCES

Adams, R. McC.

1966 *The Evolution of Urban Society*. Aldine, Chicago.

Addison, D. J.

2001 Irrigation in traditional Marquesan agriculture: Surface survey evidence from Hatihe'u Valley, Nuku Hiva. In *Pacific 2000: Proceedings of the Fifth International Conference on Easter Island and the Pacific*, edited by C. M. Stevenson, G. Lee, and F. J. Morin, 267–272. Easter Island Foundation, Los Osos.

Allen, J.

1987 *Five Upland 'Ili: Archaeological and Historical Investigations in the Kane'ohe Interchange, Interstate Highway H-3, Island of O'ahu*. Department of Anthropology Report 87-1, Bernice P. Bishop Museum, Honolulu.

1991 The role of agriculture in the evolution of the pre-contact Hawaiian state. *Asian Perspectives* 30:117–132.

1992 Farming in Hawai'i from colonization to contact: Radiocarbon chronology and implications for cultural change. *New Zealand Journal of Archaeology* 14:45–66.

Allen, M. S. (editor)

2001 *Gardens of Lono: Archaeological Investigations at the Amy B. H. Greenwell Ethnobotanical Garden, Kealakekua, Hawai'i*. Bishop Museum Press, Honolulu.

Athens, J. S.

1997 Hawaiian native lowland vegetation in prehistory. In *Historical Ecology in the Pacific Islands: Prehistoric Environmental and Landscape Change*, edited by P. V. Kirch and T. L. Hunt, 248–270. Yale University Press, New Haven, Connecticut.

Athens, J. S., and J. V. Ward

1993 Environmental change and prehistoric Polynesian settlement in Hawai'i. *Asian Perspectives* 32:205–223.

Athens, J. S., J. V. Ward, and G. M. Murakami

1996 Development of an agroforest on a Micronesian high island: Prehistoric Kosraean agriculture. *Antiquity* 70:834–846.

Bahn, P., and J. Flenley

1992 *Easter Island: Earth Island*. Thames and Hudson, London.

Barber, I. G.

1989 Of boundaries, drains, and crops: A classification system for traditional Maori horticultural ditches. *New Zealand Journal of Archaeology* 11:23–50.

2001 Wet or dry? An evaluation of extensive archaeological ditch systems from far northern New Zealand. In *Pacific 2000: Proceedings of the Fifth International Conference on Easter Island and the Pacific*, edited by C. M. Stevenson, G. Lee, and F. J. Morin, 41–50. Easter Island Foundation, Los Osos.

Barrau, J.

1965a L'humide et le sec: An essay on ethnobiological adaptation to contrastive environments in the Indo-Pacific area. *Journal of the Polynesian Society* 74:329–346.

1965b Histoire et préhistoire horticoles de l'Océanie tropicale. *Journal de la Société des Océanistes* 21:55–78.

Bayliss-Smith, T., R. Bedford, H. C. Brookfield, and M. Latham

1988 *Islands, Islanders, and the World.* Cambridge University Press, Cambridge, UK.

Beaglehole, J. C. (editor)

1969 *The Journals of Captain James Cook: The Voyage of the* Resolution *and* Adventure, *1772–1775*. Cambridge University Press, Cambridge, UK.

Bellwood, P.

2004 *First Farmers: The Origins of Agricultural Societies.* Blackwell Publishing, Oxford, UK.

Blaikie, P., and H. C. Brookfield (editors)

1987 *Land Degradation and Society*. Methuen, London.

Boserup, E.

1965 *The Conditions of Agricultural Growth: The Economics of Agrarian Change under Population Pressure*. Aldine, Chicago.

Brookfield, H. C.

1972 Intensification and disintensification in Pacific agriculture: A theoretical approach. *Pacific Viewpoint* 13:30–48.

1984 Intensification revisited. *Pacific Viewpoint* 25:15–44.

Coil, J., and P. V. Kirch

in press Ipomoean landscapes: Archaeology and the sweet potato in Kahikinui, Maui, Hawaiian Islands. In C. Ballard, P. Brown, and M. Bourke, eds., *Sweet Potato in the Pacific*. Oceania Monographs, Sydney.

Dega, M., and P. V. Kirch

2002 A modified culture history of Anahulu Valley, O'ahu, Hawai'i, and its significance for Hawaiian prehistory. *Journal of the Polynesian Society* 111:107–126.

Di Piazza, A.

1990 Les jardins enfouis de Futuna: Une ethno-archéologie de l'horticulture. *Journal de la Société des Océanistes* 91:151–162.

Downing, T. E., and M. Gibson (editors)
1974 *Irrigation's Impact on Society*. Anthropological Papers of the University of Arizona No. 25. Tucson.

Dye, T. S., and E. Komori
1992 A pre-censal population history of Hawaii. *New Zealand Journal of Archaeology* 14:113–128.

Earle, T.
1978 *Economic and Social Organization of a Complex Chiefdom: The Halelea District, Kaua'i, Hawaii*. Museum of Anthropology, University of Michigan, Ann Arbor.
1980 Prehistoric irrigation in the Hawaiian Islands: An evaluation of evolutionary significance. *Archaeology and Physical Anthropology in Oceania* 15:1–28.

Flenley, J., S. King, J. Jackson, C. Chew, J. Teller, and M. Prentice
1991 The late Quaternary vegetational and climatic history of Easter Island. *Journal of Quaternary Science* 6:85–115.

Geertz, C.
1963 *Agricultural Involution: The Processes of Ecological Change in Indonesia*. University of California Press, Berkeley.

Goldman, I.
1970 *Ancient Polynesian Society*. University of Chicago Press, Chicago.

Green, R. C.
1961 Moorea archaeology: A preliminary report. *Man* 61:169–173.

Handy, E. S. C.
1923 *The Native Culture in the Marquesas*. The Museum, Honolulu.
1940 *The Hawaiian Planter, Vol. 1. His Plants, Methods, and Areas of Cultivation*. Bernice P. Bishop Museum Bulletin 161. Honolulu.

Hather, J. and P. V. Kirch
1991 Prehistoric sweet potato (*Ipomoea batatas*) from Mangaia Island, Central Polynesia. *Antiquity* 65:887–893.

Irwin, G.
1985 *Land, Pa, and Polity*. Monograph 15. New Zealand Archaeological Association, Auckland.

Jennings, J., and R. Holmer (editors)
1980 *Archaeological Excavations in Western Samoa*. Bernice P. Bishop Museum, Honolulu.

Johnston, K. J.
2003 The intensification of pre-industrial cereal agriculture in the tropics: Boserup, cultivation lengthening, and the Classic Maya. *Journal of Anthropological Archaeology* 22:126–161.

Jones, K. L.
1986 Polynesian gardening and settlement in two river catchments of the eastern North Island. *New Zealand Journal of Archaeology* 8:5–32.

1991 Maori settlement and horticulture on the Rangitaiki Plains, Bay of Plenty, New Zealand. *New Zealand Journal of Archaeology* 13:143–175.

Kawaguchi, K., and K. Kyuma

1977 *Paddy Soils in Tropical Asia: Their Material Nature and Fertility*. University of Hawaii Press, Honolulu.

Kirch, P. V.

1977 Valley agricultural systems in prehistoric Hawaii: An archaeological consideration. *Asian Perspectives* 20:246–280.

1984 *The Evolution of the Polynesian Chiefdoms*. Cambridge University Press, Cambridge, UK.

1985 *Feathered Gods and Fishhooks: An Introduction to Hawaiian Archaeology and Prehistory*. University of Hawaii Press, Honolulu.

1989 Second millennium B.C. arboriculture in Melanesia: Archaeological evidence from the Mussau Islands. *Economic Botany* 43:225–240.

1991 Polynesian agricultural systems. In *Islands, Plants, and Polynesians: An Introduction to Polynesian Ethnobotany*, edited by P. A. Cox and S. A. Banack, 113–134. Dioscorides Press, Portland, Oregon.

1994 *The Wet and the Dry: Irrigation and Agricultural Intensification in Polynesia*. University of Chicago Press, Chicago.

1996 Late Holocene human-induced modifications to a central Polynesian island ecosystem. *Proceedings of the National Academy of Sciences, USA* 93(11):5296–5300.

2000 *On the Road of the Winds: An Archaeological History of the Pacific Islands Before European Contact*. University of California Press, Berkeley.

2001 Polynesian feasting in ethnohistoric, ethnographic, and archaeological contexts: A comparison of three societies. In *Feasts: Archaeological and Ethnographic Perspectives on Food, Politics, and Power*, edited by M. Dietler and B. Hayden, 168–184. Smithsonian Institution Press, Washington, DC.

Kirch, P. V. (editor)

2002 *From the 'Cliffs of Keolewa' to the 'Sea of Papaloa': An Archaeological Reconnaissance of Portions of the Kalaupapa National Historical Park, Moloka'i, Hawaiian Islands.* Archaeological Research Facility, University of California, Berkeley.

Kirch, P. V., J. Flenley, D. Steadman, F. Lamont, and S. Dawson

1992 Ancient environmental degradation. *National Geographic Research and Exploration* 8:166–179.

Kirch, P. V., and R. C. Green

2001 *Hawaiki, Ancestral Polynesia: An Essay in Historical Anthropology*. Cambridge University Press, Cambridge, UK.

Kirch, P. V., A. Hartshorn, O. Chadwick, P. Vitousek, D. Sherrod, J. Coil, L. Holm, and W. Sharp

2004 Environment, agriculture, and settlement patterns in a marginal Polynesian landscape. *Proceedings of the National Academy of Sciences, U.S.A.* 101:9936–9941.

Kirch, P. V., and D. Lepofsky

1993 Polynesian irrigation: Archaeological and linguistic evidence for origins and development. *Asian Perspectives* 32:183–204.

Kirch, P. V., and M. Sahlins

1992 *Anahulu: The Anthropology of History in the Kingdom of Hawaii.* 2 vols. University of Chicago Press, Chicago.

Kirch, P. V., and D. E. Yen

1982 *Tikopia: The Prehistory and Ecology of a Polynesian Outlier.* Bernice P. Bishop Museum Bulletin 238, Honolulu.

Ladefoged, T. N., M. W. Graves, and R. P. Jennings

1996 Dryland agricultural expansion and intensification in Kohala, Hawai'i Island. *Antiquity* 70:861–880.

Ladefoged, T. N., M. W. Graves, and M. D. McCoy

2003 Archaeological evidence for agricultural development in Kohala, Island of Hawai'i. *Journal of Archaeological Science* 30:923–940.

Leach, H. M.

1979 Evidence of prehistoric gardens in eastern Palliser Bay. In *Prehistoric Man in Palliser Bay*, edited by B. F. Leach and H. M. Leach, 137–161. National Museum of New Zealand Bulletin 21, Wellington.

1984 *1,000 Years of Gardening in New Zealand.* Reed, Wellington.

1999 Intensification in the Pacific: A critique of the archaeological criteria and their application. *Current Anthropology* 40:311–339.

Lepofsky, D.

1994 Prehistoric Agricultural Intensification in the Society Islands, French Polynesia. PhD dissertation, Department of Anthropology, University of California, Berkeley.

1995 A radiocarbon chronology for prehistoric agriculture in the Society Islands, French Polynesia. *Radiocarbon* 37:917–930.

Lepofsky, D., P. V. Kirch, and K. Lertzman

1996 Stratigraphic and paleobotanical evidence for prehistoric human-induced environmental disturbance on Mo'orea, French Polynesia. *Pacific Science* 50:253–273.

1998 Metric analyses of prehistoric morphological change in cultivated fruits and nuts: An example from island Melanesia. *Journal of Archaeological Science* 25:1001–1014.

Linton, R.

1925 *Archaeology of the Marquesas Islands.* Bernice P. Bishop Museum Bulletin 23. Honolulu.

Morrison, K. D.

1994 The intensification of production: Archaeological approaches. *Journal of Archaeological Method and Theory* 1:111–160.

Newman, T. S.

n.d. [1972] *Hawaiian Fishing and Farming on the Island of Hawaii in A.D. 1778.* Division of State Parks, State of Hawaii, Honolulu.

Parkes, A.

1997 Environmental change and the impact of Polynesian colonization: Sedimentary records from central Polynesia. In *Historical Ecology in the Pacific Islands: Prehistoric Environmental and Landscape Change*, edited by P. V. Kirch and T. L. Hunt, 166–199. Yale University Press, New Haven, Connecticut.

Pukui, M. K.

1983 *'Olelo No'eau: Hawaiian Proverbs and Poetical Sayings*. Bernice P. Bishop Museum Special Publication 71, Honolulu.

Riley, T.

1975 Survey and excavation of the aboriginal agricultural system. In *Prehistory and Human Ecology in a Windward Hawaiian Valley: Halawa Valley, Molokai*, edited by P. V. Kirch and M. Kelly, 79–115. Pacific Anthropological Records 24. Bernice P. Bishop Museum, Honolulu.

Rosendahl, P. H.

1994 Aboriginal Hawaiian structural remains and settlement patterns in the upland agricultural zone at Lapakahi, Island of Hawai'i. *Hawaiian Archaeology* 3:14–70.

Sahlins, M.

1958 *Social Stratification in Polynesia*. University of Washington Press, Seattle.

1972 *Stone Age Economics*. Aldine-Atherton, Chicago.

Schilt, R.

1984 *Subsistence and Conflict in Kona, Hawai'i*. Department of Anthropology Report 84-1. Bernice P. Bishop Museum, Honolulu.

Service, E.

1967 *Primitive Social Organization: An Evolutionary Perspective*. Random House, New York.

Spooner, B. (editor)

1972 *Population Growth: Anthropological Implications*. MIT Press, Cambridge, MA.

Spriggs, M.

1982 Taro cropping systems in the Southeast Asian-Pacific region. *Archaeology in Oceania* 17:7–15.

Stevenson, C. M.

1997 *Archaeological Investigations on Easter Island. Maunga Tari: An Upland Agricultural Complex*. Bearsville Press, Los Osos.

Sutton, D. G. (editor)
1990 *The Archaeology of the Kainga*. Auckland University Press, Auckland, New Zealand.

Tuggle, H. D., and M. J. Tomonari-Tuggle
1980 Prehistoric agriculture in Kohala, Hawaii. *Journal of Field Archaeology* 7:297–312.

Valeri, V.
1985 *Kingship and Sacrifice: Ritual and Society in Ancient Hawaii*. University of Chicago Press, Chicago.

Vitousek, P. M., T. Ladefoged, A. Hartshorn, P. V. Kirch, M. Graves, S. Hotchkiss, S. Tuljapurkar, and O. Chadwick
2004 Soils, agriculture, and society in precontact Hawai'i. *Science* 304:1665–1669.

Wilder, G. P.
1928 *The Breadfruit of Tahiti*. Bernice P. Bishop Museum Bulletin 50, Honolulu.

Wittfogel, K. A.
1957 *Oriental Despotism: A Comparative Study of Total Power*. Yale University Press, New Haven, Connecticut.

Wozniak, J. A.
2001 Landscapes of food production on Easter Island: Successful subsistence strategies. In *Pacific 2000: Proceedings of the Fifth International Conference on Easter Island and the Pacific*, edited by C. M. Stevenson, G. Lee, and F. J. Morin, 91–102. Easter Island Foundation, Los Osos.

Yen, D. E.
1971 The development of agriculture in Oceania. In *Studies in Oceanic Culture History*, vol. 2, edited by R. C. Green and M. Kelly, 1–12. Pacific Anthropological Records 12. Bernice P. Bishop Museum, Honolulu.
1973 The origins of Oceanic agriculture. *Archaeology and Physical Anthropology in Oceania* 8:68–85.
1974a *The Sweet Potato and Oceania: An Essay in Ethnobotany*. Bernice P. Bishop Museum Bulletin 236, Honolulu.
1974b Arboriculture in the subsistence of Santa Cruz, Solomon Islands. *Economic Botany* 28:247–284.
1988 Easter Island agriculture in prehistory: The possibilities of reconstruction. In *First International Congress, Easter Island and East Polynesia, Vol. 1: Archaeology*, edited by C. Cristino, P. Vargas, R. Izaurieta, and R. Budd, 59–82. University of Chile, Santiago.

Yen, D. E., P. V. Kirch, P. Rosendahl, and T. Riley.
1972 Prehistoric agriculture in the upper Makaha Valley, Oahu. In *Makaha Valley Historical Project: Interim Report No. 3*, edited by D. E. Yen and E. Ladd, 59–94. Pacific Anthropological Records 18. Bernice P. Bishop Museum, Honolulu.

Mesoamerica

The Roles of Ritual and Technology in Mesoamerican Water Management

JOYCE MARCUS

Nearly 50 years ago we saw the publication of two influential books on irrigation, each utilizing an explicitly comparative approach. Those two works—Steward et al.'s (1955) *Irrigation Civilizations: A Comparative Study* and Wittfogel's (1957) *Oriental Despotism: A Comparative Study of Total Power*—are not cited as often today as they were in the 1960s, but their pioneering comparative approach continues to be fruitful (e.g., Doolittle 1990; Downing and Gibson 1974; Neely 2005; Scarborough 2003; Scarborough and Isaac 1993; Wilkinson 2003). Such an approach has, for example, led to the question, "Why do some civilizations invest in large-scale hydraulic systems while others do not?"

Judging by the recent flurry of publications on the subject, water management has once again become a hot topic. There are probably five principal reasons for this renewed interest. One obvious reason is that so much information on water management has now been amassed for geographic regions *not* represented in Steward's or Wittfogel's work. A second reason is new paleoenvironmental evidence of long-lasting droughts in various parts of the prehistoric world. This evidence suggests the likelihood that during drought *and* nondrought periods stored water would be managed carefully so that it lasted from one rainy season until the next (Curtis et al. 1996, 1998; Folan et al. 1983; Gill 2000; Gunn et al. 2002; Hodell et al. 1995, 2001; Scarborough 1994, 1998).

Third among the reasons is a growing interest in "rainfall" civilizations that stand in contrast to the more famous "hydraulic" civilizations of the Nile or Tigris-Euphrates valleys (see Adams, this volume). Fourth among the reasons is a rethinking of the notion that all civilizations in arid regions had grand-scale irrigation works administered by state personnel. It is now clear that some civilizations relied on a multiplicity of small-scale systems, each manageable by small groups or families.

The fifth reason is an outgrowth of our field's current focus on ritual of all kinds; we have developed an increasing awareness that ritual often played a role in the acquisition of water. Most studies of water management in the 1950s and 1960s were ecologically based and focused on technology. This focus on technology began to change during the following decade. In the mid-1970s Flannery and I argued that Zapotec water management could be better understood if we studied the dynamic interaction between ritual and technology (Flannery and Marcus 1976). Using ethnohistoric data, we suggested that the Zapotec used *technology* for variables they could control and *ritual* for variables they perceived to be controlled by the supernatural. Today we are in an even better position to combine technological and ritual data to forge a more holistic understanding of water management and document the uses of water, both agricultural and nonagricultural.

Let us begin the story of water management where many indigenous populations would begin: with ritual.

RITUAL, RECIPROCITY, AND PREDICTION

In most ancient societies all dealings with the sacred required ritual. Ritual was a form of communication between human beings and the supernatural, often enacted to ensure a desirable outcome. By making an offering and uttering a request, the individual sought to establish a set of reciprocal relations. A farmer said he would do his part, hoping that the supernatural forces would do theirs in a type of delayed reciprocity.

Ritual acts could also be performed to learn the future. The Shang of China inscribed bones with binary pairs of outcomes, such as "Auspicious: we will receive this harvest" versus "Inauspicious: we will not receive this harvest." Ritual specialists then applied heat to hollows bored into the bones (cattle scapulae or turtle plastrons), interpreting the resultant stress cracks either as "auspicious" or "inauspicious." The frequency with which scapulae and plastrons were collected, as well as the time spent on their cleaning, ritual consecration, and preparation for heating and cracking, indicates the degree to which forecasting the future was a central concern. It is known that the last nine Shang rulers were served by more than 120 diviners whose names are also known (Keightley 1999:236–237).

In ancient Mesoamerica similar acts of divination took place in households long before the state formed, and many of those ritual acts involved three important concerns: ancestors, life crises, and agriculture (Marcus 1998, 1999). As for water, its importance in Mesoamerican village life is obvious:

it was needed for drinking, cooking, washing, and bathing. Ancient villagers also needed water to make adobes, mud daub, and lime plaster and to irrigate thirsty crops to ensure a good harvest.

Mesoamerican peoples did not, however, think of water merely as a utilitarian resource. Water, rain, clouds, springs, caves, and waterholes were part of the cosmos and were considered alive, sacred, and worthy of reverence (Marcus 1978:181). Other living components of the Mesoamerican universe included the earth, the sun, the moon, mountains, lightning, thunder, and earthquakes. This animatistic view of the world affected the way Mesoamerican populations interacted with nature and its living elements (Marcus and Flannery 1978). From the perspective of indigenous peoples, one needed ritual to communicate with supernatural forces like Clouds and Lightning, the actual owners of water while it was in the sky. Once Clouds and Lightning had released that water, however, man's technology of ditches and canals could take over. In Mesoamerica ritual preceded technology.

The Nahua

According to Alan Sandstrom, an ethnographer working with Nahua speakers in the Huasteca of Mexico, the landscape surrounding each village is still alive with spirits. "Every hill, valley, spring, lake, stream, section of river, boulder, plain, grove, gorge, and cave has its proper name and associated spirit" (Sandstrom 1991:241).

This living landscape is approached with ritual today just as it was in the past. In olden times, the precolumbian calendar provided a structure for many major rituals. For example, each of the 18 months—*veintenas* of 20 days each—within the Aztec year featured rites, many of them relevant to the agricultural cycle. During the first month of their year, *Atl cahualo* (corresponding to early February), the Aztecs are said to have sacrificed children on hilltops to petition rain. If the children shed many tears before their sacrifice, the Aztecs considered this a favorable prognostication for a rainy year (Anderson and Dibble 1981:1–2). Archaeological evidence for such rites includes the remains of 42 sacrificed children found with 11 polychrome sculptures of the deity Tlaloc in a cache below Tenochtitlan's Main Temple (Matos 1987:Plates 15, 47). Sacrificing children at this temple, set on a pyramid thought to represent a sacred mountain, complemented the sacrifice of children on the actual mountains that surround the Basin of Mexico. Mountains expressed the concept of *altepetl*, or community, because they held the water that supported humanity (López Austin 1973:62–63).

Since the ultimate source of water was considered to be caves inside mountains, hilltops were an appropriate place to request rain or appease rain-bearing clouds. The mountains themselves were revered as living ancestors. To honor them, the Aztecs used corn dough to create effigy figures in the shape of mountains, with the top of each effigy taking the form of a human head. This rite took place during the month of *Tepeilhuitl*, corresponding to early October. Such an isomorphism, in which mountains = ancestors, also characterizes the Maya and other Mesoamerican groups (Broda 1970, 1971, 1987; Guiteras-Holmes 1961; Vogt 1969).

To request water and rain, the Aztec made offerings to the aforementioned deity Tlaloc. Significantly, the origin of Tlaloc has been traced to the adjective *tlallo*, meaning "made of earth" or "covered with earth," demonstrating the close link between earth, water, and rain (Sullivan 1974).

The Aztec cosmos was thought to have been created from the body of a female monster or earth goddess called Tlaltecutli:

> From one half they created earth and from the other, the heavens. To compensate her for the damage she had suffered, all the gods ordained that the sustenances of life issue from her. From her hair they made the trees and the flowers and plants; from her skin and its tiny hairs, the small plants and flowers; from her numerous eyes, the springs and fountains and caves; from her mouths, the rivers and great caverns; from her nose, the valleys and mountains. (Sullivan 1974:216)

In this cosmos Tlaloc and his helpers, the *tlaloque*, were depicted carrying vessels filled with water, lightning staffs, and maize. Rain would be released when the *tlaloque* struck these water-filled vessels with lightning or when the *tlaloque* tipped the vessels to pour water (Figure 9.1).

Figure 9.1. On this Aztec stone sculpture one of the *tlaloque* (assistants to the storm deity Tlaloc) is shown pouring water and maize from a vessel. Height: 23 cm (adapted from Nicholson 1971:Figure 14; and Baquedano 1984:Figure 59).

Tlaloc's *tlaloque* were assigned to the four great world quarters and the center, each direction associated with a different color (Nicholson 1971:414). Nahua-speaking populations petitioned Lightning to penetrate the clouds, called *ahauhqueh* or "owners of water" (Ruiz de Alarcón 1984:357), so that rain might be released to the ground, where it could be manipulated by human technology. Since rain was "owned" by the clouds, its release required appeals to Lightning to intervene on behalf of human beings. Such relations with clouds, lightning, and rain characterized other Mesoamerican groups as well.

Water was so fundamental a resource that the most common Nahua term for a town or province, *altepetl*, combines the words *water + mountain*. Many Nahua communities were, in fact, named after water, for example *Atlan* (Place of Water), *Atotonilco* (Place of Warm Waters), *Ameyalco* (Place of Springs), *Amaxac* (Where the Water Divides), *Atzacan* (Place Where Water Is Contained), *Atlhuelic* (Sweet Water), *Acocozpan* (Canal of the Yellow Water), *Atlatlauhcan* (Place of Red Water), and *Atlapulac* (Place Set in the Water). Cultivated or irrigated fields were also incorporated into the hieroglyphs for many Aztec place names, such as *Tecmilco* (Fields of the Ruler), *Tlaahuililpan* (Irrigated Orchard), and *Acamiltzinco* (Place of the Planted Fields) (see Figure 9.2; see also Peñafiel 1885 for more examples).

The Mixtec

Like their Nahua neighbors, the Mixtec of highland Oaxaca viewed the earth as a living being that could feel pain. Many contemporary Mixtec speakers still refer to the earth by using a series of body metaphors: the soil is its "flesh," rocks are its "bones," rivers are its "veins," and water is its "blood." The Mixtec say that when farmers poke their digging sticks into the ground, the soil "shouts in pain" from having its flesh punctured (Monaghan 1995:98). Each time people plant, break a rock, or excavate a house site, "they cause pain and should ask the place where this occurred for forgiveness" (ibid.).

The Mixtec term *ñu'un savi* can refer to rain clouds, to ponds that form on top of mountains, or to the supernatural controller of Lightning, whose offspring is maize. Through what rituals might rain be produced? In the Mixtec community of Nuyoo there are "rain people," whose job is to make rain fall. "They do this, one man explained, by 'calling' the Rain. They climb four prominent mountaintops (*shini yuku*), which contain Rain shrines, and smoke seven cigars. The cigars . . . cause clouds to form. . . . When [the rain people] become intoxicated by the tobacco, they begin to shout (*kanara*) and call the Rain. . . . We hear the booming shouts [of the rain people] as thunder" (Monaghan 1995:349).

When the farmers of Nuyoo talk about rain, they speak of it as a potent volatile substance (Monaghan 1995:105). They see rain as a vital element of renewal and

Figure 9.2. Aztec place names that incorporate the signs for water (*atl*), canal (*apantli*), or planted field (*milli*) (redrawn from Peñafiel 1885).

associate it with masculine sexuality. "Rain, people say, is the father of all plants, and they are its children (*se'ya savi*)" (ibid.:111). "Of all the things produced through the coming together of Earth and Rain, corn is by far the most important" (ibid.:112). If corn kernels are spilled or wasted, there are negative repercussions. "Disrespectful behavior toward any of the children of the Rain thus invites retaliation" (ibid.:114).

Like their sixteenth-century Aztec neighbors, the sixteenth-century Mixtec sacrificed children on hilltops to petition rain. During the Spanish Inquisition one Mixtec priest was asked how he had conducted this rite. He said that he had ascended the highest peak, removed a child's heart, and placed the heart in front of a stone image of Rain (*Dzahui* or *Savi*). After spending two days on the mountaintop, he burned the heart and placed the ashes with other items in a bundle (Terraciano 2001:266).

The Maya

In the world of the Maya it was *Hunab ku* who created sacred things like lightning (*chaak*), fire (*k'ak'*), the sun (*k'in*), the moon (*u'*), caves (*chen*), mountains (*witz*), and so on (Marcus 1978:181). The Tzotzil of the highlands say that there was a time when all Maya could see into the mountains and communicate directly with the ancestors who dwell there; today, however, only *h'ilol*, or "seers," possess this ability to communicate directly with those ancestors (Vogt 1969:416). Like the mountains, the "sacred earth," or *ch'ul balamil*, was also a living being. When a Maya "cuts down the forest to make his milpa, he apologizes to the earth for 'disfiguring' its face" (Thompson 1970:165).

The Mayas' relations with these living, sacred beings were characterized by delayed reciprocity. For example, since rain clouds were thought to emanate from caves or waterholes, offerings were made on mountaintops, in caves, and at cenotes in anticipation of the gift of water. If the rain was still withheld by the clouds, ritual acts were needed to bring it to earth. Ceremonies among the modern Tzotzil are dedicated to the Earth Lord, "that he should order the clouds to come out of the earth, clouds to rain on our corn, so our corn should not die" (Vogt 1969:459).

The Yucatec Maya beseeched Lightning (*hats' chaak*) to split the clouds (*muyal*) and release rain. These petitions to Lightning (*chaak*) and other guardians of the bush (*kuil kaaxob*) were witnessed by various ethnologists, including Robert Redfield (1941:117–122) and Alfonso Villa Rojas (1945). Although these Maya rites were affected by Catholicism (as was evident from the use of the cross and rosary), they still retained indigenous concepts, especially the idea that a pact or reciprocal obligation exists between farmers on the one hand and natural forces on the other. These forces were believed to affect the winds, the soil, the rains, and the harvest (Redfield and Villa Rojas 1934).

When a Yucatec farmer selects a piece of land to plant, he contracts his first obligation to the gods. On the land selected he must set up a cross and place before it three or five gourds of *zaca*, or maize gruel. "Then, in shouts loud and prolonged, so that they will be heard over the entire tract to be cultivated," the farmer invites "the yuntzilob [spirits] to come and receive the offering" (Villa Rojas 1945:111). The farmer is informing the guardians of the bush that the land is about to be used by a friend, and he asks these guardians to make sure that the trees he is about to cut will put up no resistance. The farmer is obligated to fell no more forest than he can plant; otherwise, the guardians of the bush will punish him.

The next rite occurs when the field has been cleared and is ready for burning. The farmer makes an offering of seven dishes of *zaca* in his field. The purpose of this rite is to ask the whirlwind, *kakal mozon ik*, to come fan the flames so that the whole field will burn. If the wind's arrival is delayed, the farmer whistles for it to come.

The planting is also accompanied by ritual acts performed in the field (Redfield and Villa Rojas 1934). Offerings are placed on a small altar erected on the east side of the field. Seven dishes of maize gruel sweetened with honey must be offered on each of the first seven days of the planting, while the farmer recites prayers. On the eighth day seven packages of 13 tortillas each and seven dishes of cooked meat are offered in addition to maize gruel. The offering of food is referred to as *u hanli chaakob*, "dinner of the lightnings," and one of Villa Rojas' informants told him that the offering is passed on to the *kunku chaak*, the supreme god of all the lightnings, who gives part of it to the subordinate *chaakob* with the recommendation that they frequently water the field of the supplicant (Villa Rojas 1945:112).

The most elaborate Yucatec Maya agricultural rite occurs once a year, during August or September, when the maize is in its most crucial growth phase. This rite is performed by the whole community, with every family contributing offerings. All the men cooperate in a hunt for wild animals to be used as meat offerings, while all the women prepare food. Although the principal purpose is to ensure sufficient rainfall, this ceremony (*okotbatam*, or "petition") is performed even in rainy years, since to fail to do so might incur punishment.

When a drought does occur, a still larger *okotbatam* involving all the villages in the area is held (Villa Rojas 1945:113). If the early maturing maize cobs develop in September, each Maya farmer offers them as the first ears of corn. On an altar he offers 13 pairs of corn on the cob and 13 dishes of gruel. The November harvest is followed by two rites, sometimes held months later. One rite is called *tupp k'ak'* "put out the fire" (telling the winds to stop creating heat in the fields) and the other *u hanli col* "dinner of the field" (thanking Lightning and rain spirits); both involve offerings of food and gratitude for the harvest. The Yucatec Maya farmer calls all the necessary beings—the spirits of the bush, the wind, lightning, and rain—to come to his fields and to bring water so he can practice rainfall agriculture.

Water for household consumption usually comes from waterholes, which include, for the Tzotzil Maya, limestone sinks, caves, and all openings in the earth's crust; these openings were also important because they were the ultimate sources of lightning bolts and clouds (Vogt 1969:386–387). The location of waterholes continues to affect Tzotzil settlement patterns and social organization, so much so that the three units that constitute the hamlet are (1) the *domestic group*, living in one or more buildings in a compound; (2) the *sna*, consisting of two or more domestic groups; and (3) the *waterhole group*, composed of two or more *snas*. These waterhole groups vary in size from 4 houses and fewer than 15 people to more than 40 houses with 150 people (Vogt 1969:141). Each waterhole group maintains a series of shrines; some are erected on hills to communicate with the ancestors, others in caves to communicate with the earth.

The Zapotec

Like the Maya, the Zapotec of highland Oaxaca used ritual performances to petition Lightning (*Cociyo*) to pierce the clouds (*Zaa*) to release rain (considered a form of *niça*, "water"). Once that rain reached the surface of the earth, it could be manipulated through technology (Flannery and Marcus 1976; Marcus 1989). Thus ritual was needed to liberate the water when it was inside the clouds, but after it reached the ground it became amenable to technology (Figure 9.3).

Clouds were particularly important because the ancient Zapotec believed they had descended from them; they referred to themselves as *peni zaa*, "People of the Clouds" or "Cloud People" (Marcus and Flannery 1996).

The powerful supernatural *Cociyo*, or Lightning, not only could split the clouds to release rain but also had the power to create thunder, called *Xoo cociyo* because it rumbled like *Xoo* "earthquake." When the Zapotec prayed and sacrificed to Lightning, they addressed him as *Pitào Cociyo*, "Great Spirit of Lightning" (Marcus 1983a). In addition the Zapotec ritual calendar was divided into four time units called *cociyo*, or "lightnings" (Marcus and Flannery 1978). Offerings made to these lightnings included food and drink; one's own blood; sacrificed quail, turkeys, or dogs; captives taken in war; or young children.

Not only was Lightning a major force in the ancient past, but it remained so in the twentieth century, and stories about it continued to be told:

> On the summit of a mountain, long before the dawn of the world, lived Cocijoguí, the Old Lightning of Fire. He was the lord of all the "lightnings," large and small. At the foot of his dazzling throne he had in his custody four immense clay jars. In one he kept the clouds shut up; in another water (rain); in the third, hail; and in the fourth, wind. In turn, each one of these jars was watched over by a lesser lightning in the form of a *chintete* or small lizard. (Cruz 1946:33–35)

Some Late Classic (ca. AD 500–800) ceramic vessels from Oaxaca seem to depict Cociyo with four attached jars for wind, clouds, rain, and hail (see Figure 9.4). Cruz's (1946) informants tell us something further about the celestial hierarchy. Under "Old Lightning of Fire" were four subordinate lightnings,

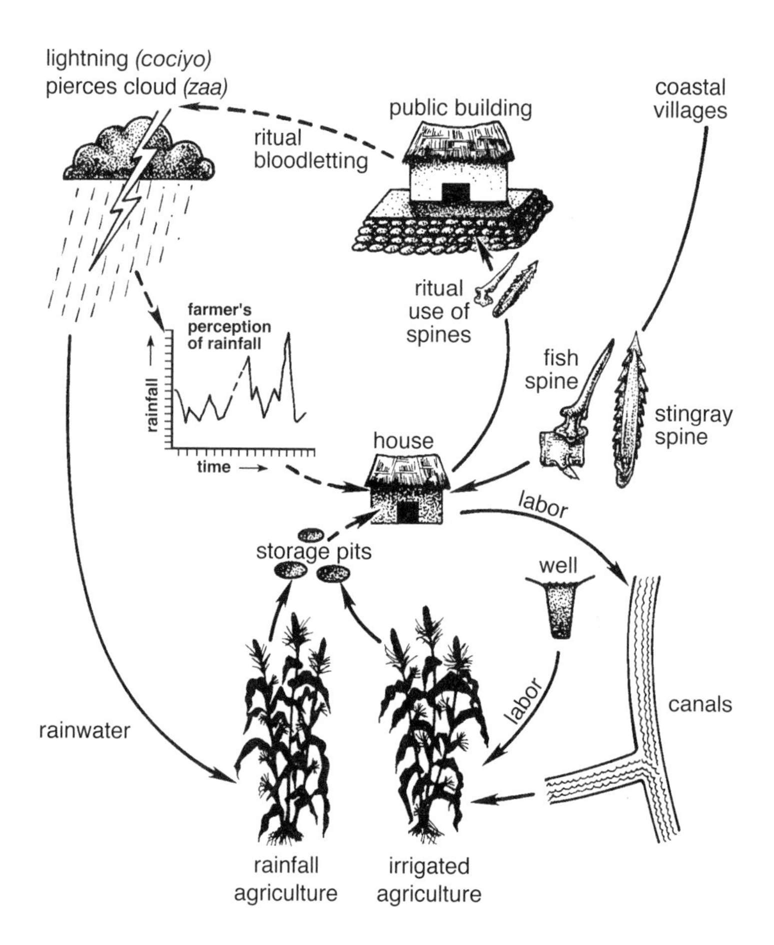

Figure 9.3. Simplified model for the matter-energy transactions (solid arrows) and information transactions (*dashed arrows*) connected with prehistoric Zapotec agriculture. A farmer makes predictions about the adequacy of the year's precipitation based on the volume of spring rainfall. If he predicts inadequacy, the farmer offers his own blood to Cociyo (Lightning) at a public building, asking Lightning to split the clouds to produce more rain, and he increases his planting on land irrigated by wells and canals. Predictions of adequate rainfall, however, encourage the farmer to gamble on cultivating more nonirrigated land (redrawn from Flannery and Marcus 1976:Figure 3).

called *cociyozaa* (in charge of clouds), *cociyoniza* (in charge of water), a third *cociyo* in charge of hail, and *cociyopi* (in charge of wind). When Cociyoguí would not respond to the petitions of his people, they prayed to the Great Spirit, to *Pitào*, who had given him life. We learn from Cruz that Lightning resided on the summit of a mountain and that each of the four lesser *cociyo* had specific obligations (Marcus 1983a). Given the power attributed to Lightning, it is not surprising that various sixteenth-century Zapotec rulers incorporated Cociyo into their names (for example, Cociyoeza and Cociyopii).

Figure 9.4. Zapotec effigy vessel from the Valley of Oaxaca depicting the face of *Cociyo* (Lightning) with four jars attached behind him, probably for hail, wind, clouds, and rain. Height of vessel: 15 cm (redrawn from Marcus 1983:Figure 9.13).

In the 1930s Elsie Clews Parsons (1936:211) recorded many acts of reverence for Lightning on the part of the people of Mitla. They made offerings to Lightning in the center of their agricultural fields, as well as in a cave in the mountains. According to Parsons, the Mitla Zapotec associated Lightning with two animals in particular: (1) the hummingbird, because it revives at "first lightning," and (2) the lizard (chintete), because it was considered "the embodiment of lightning." She says, "When Lightning stays on earth as a lizard he re-ascends into the sky through fire; the lizard is placed in a little gourd of water which is thrown into a fire" (Parsons 1936:212).

Among some Mesoamerican groups maize was regarded as the creation of Lightning or his offspring. For example, the Zapotec of Loxicha in southern Oaxaca regard maize as the offspring of Lightning (the father) and Earth (the mother) (Weitlaner 1965:561). In a Zapotec tale recorded in Mitla (Parsons 1936:211) Lightning is said to have created corn of different colors by dropping it in different sacks. For the Trique people of Oaxaca, Longacre (in Nader 1965:414) recorded the story of a woman who said she knew she was going to be killed by Lightning. When asked why, she said it was because Lightning was angry that she had spilled corn, his offspring, on the ground.

From the perspective of indigenous Mesoamerican populations, the rituals that are conducted to address Lightning—the petitioning, incantations, sacrifices, and offerings—precede any and all technological manipulations of the water sent to them.

THE ROLE OF TECHNOLOGY

When scholars mention "water control," they are often thinking of thousands of workmen building vast irrigation systems. Water management, however, had very humble beginnings. In its initial stages in Mesoamerica single families or small social units might construct ditches, canals, or drains. Such small work teams could accomplish much more than most scholars usually assume.

Beginning on an extended family scale, some water management evolved into more complex systems; other systems, however, stayed quite simple. Why did some systems expand, while others did not? We can illustrate this theme by comparing the Valley of Oaxaca with other Mesoamerican cases.

Early Water Management in the Valley of Oaxaca

By 8000 BC the occupants of the Valley of Oaxaca were growing gourds and squash in the humid barrancas. Prior to 1200 BC they are believed to have relied exclusively on a combination of rain and humid bottomland to grow maize. The amount of rainfall in a given year was, and still is, the major factor in determining whether a year is "good," "average," or "poor." Rainfall farming is still the most common type in the valley, even though it is the most risky, since 2 to 3 years in 10 the rainfall is below the 500 mm minimal level that will ensure a good maize harvest. The Indians' concern with rainfall was clearly expressed in the sixteenth-century Zapotec language; there were expressions for "to rain in one place," "to rain with large drops," "to rain hard," "to rain gently," "the first rains of the year," and so forth (Córdova 1942:251v [1578]). The Zapotec also divided their fields into two types: *quèela huizoa*, those that could be irrigated, and *quèela pichijta*, those that depended on rainfall alone (Figure 9.3).

By 1000 BC we see the first archaeological evidence for man-made wells of the type used for "pot irrigation" (Figure 9.5). Such *riego a brazo* or *riego a cántaro* ("irrigation by hand or jar") is labor intensive; each farmer draws water up in a 10-liter jar from a depth of c. 3 m and then pours the water on each individual plant (Marcus and Flannery 1996:108). The sixteenth-century

Figure 9.5 A Zapotec farmer, irrigating by hand, uses a jar to pour water directly on his plants at Abasolo in the Valley of Oaxaca, Mexico.

Zapotec phrase for this pot irrigation was *tiquilla niçaya*, "to irrigate with a cántaro" (Córdova 1942 [1578]). Formative sites where the water table was high enough to have permitted well irrigation include Abasolo, Mitla, Zaachila, San Lázaro Etla, and San José Mogote. So far, actual Formative wells have been found at Abasolo and Mitla (Flannery 1983), but as Kirkby (1973:119) has indicated, this "shows only that wells were used as a means of obtaining water, not necessarily that the water was used for irrigation." Some wells, of course, may have served both houses and gardens.

Contemporaneous with the first man-made wells were the first drainage ditches and cisterns. By 1000 BC at the village of San José Mogote, the earliest inhabitants of Area B[1] were deliberately leveling bedrock to create flat terraces by hacking the soft volcanic tuff. In addition to house terraces they carved out ditches, canals, and a large bedrock cistern using stone celts of metamorphic rock (Flannery and Marcus 1994). The short canals served to divert rain runoff away from houses, directing it to a large cistern that had an overflow canal in case it was filled to the brim (Figures 9.6, 9.7). During heavy rains the system of ditches would have directed water to the cistern, kept the house dry, and allowed the water to be stored for later use. A similar drainage ditch was found in bedrock just uphill from a house built around 1000–900 BC at the site of Tierras Largas (Flannery 1983:326), leading us to suspect that such features were widespread.

Figure 9.6. View of the bedrock in Area B, San José Mogote, Oaxaca, showing the runoff canal leading from the large cistern (Feature 58) that served to store water for House 16/17 at c. 1000 BC.

We can suggest that these simple ditches were excavated by individual households and that the original inspiration for them was the need to remove *excess* water, especially during the rainy season. Preventing water from entering houses built on slopes seems to have been the initial purpose. Later, canals were used in conjunction with brush dams to bring water from piedmont streams to dry but otherwise fertile lands below. Some of the sixteenth-century Zapotec terms for these canals were *piaça*, "canal excavated in earth"; *toçaaya niça piaça*, "to bring water by canal"; *xineçaniça*, "route of the water" or canal; and *yaga-pethoça*, "wooden canal" (possibly made of hollowed logs). Small-scale canal systems were built on the slopes of Monte Albán almost as soon as the city was founded, but they served only a small number of families living along the canal (O'Brien et al. 1982).

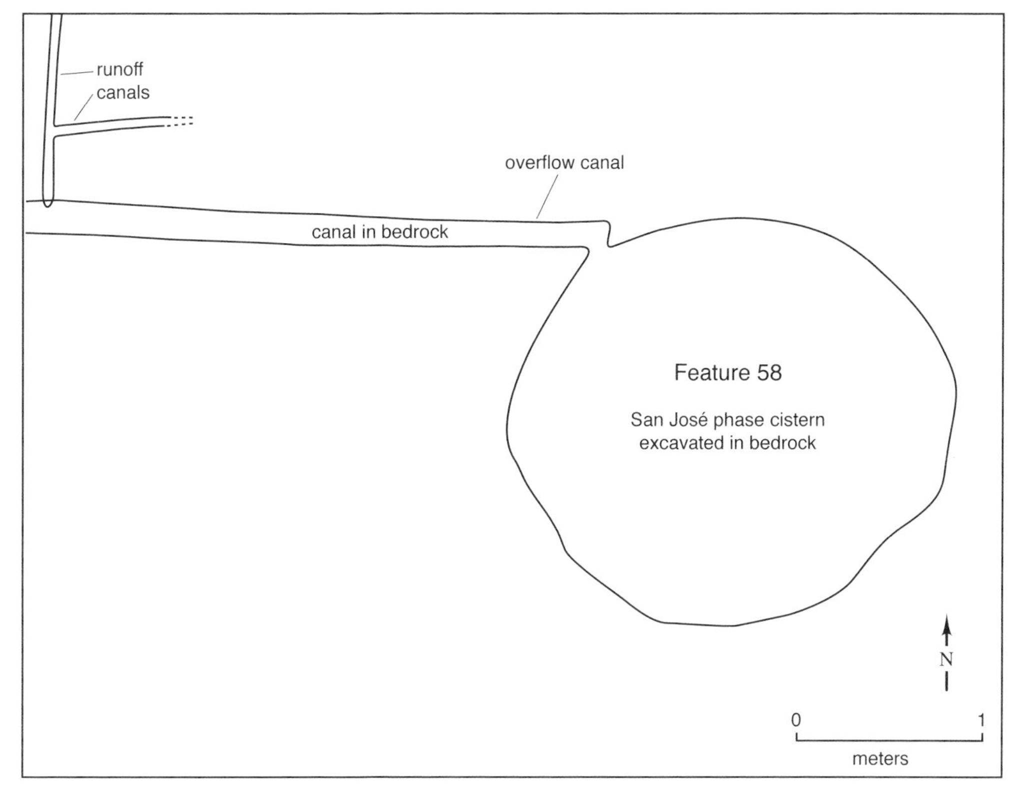

Figure 9.7. Cistern from Area B, San José Mogote, Oaxaca, Mexico (simplified version of Figure 15.4 in Flannery and Marcus 1994).

Now we will turn to the Tehuacán Valley, an arid region that reveals a very different story from that of the Valley of Oaxaca.

The Tehuacán Valley

Water control was very early in Tehuacán. James A. Neely (personal communication) has found preceramic-era water wells and Formative canal systems there.

Though far drier than the Valley of Oaxaca, the Tehuacán Valley had a far greater potential for large-scale irrigation, including aqueducts and canals that could run for kilometers and serve several communities. Some canal systems linked a prodigious set of perennial springs in the northern valley to a series of frost-free plains in the southern valley. In addition, a spectacular local system was developed during the Formative in the Arroyo Lencho Diego, a steep-sided canyon in the southern valley (Spencer 1979; Woodbury and Neely 1972). Here a Middle Formative dam was constructed 1 kilometer above the arroyo mouth, creating a reservoir with a capacity of 37,000 m^3 of water. This initial dam was 6 m wide, 2.8 m high, and 175 m long. Spencer (1993) has estimated that this dam could have been built in a single dry season by a team of eight energetic men, people drawn from a nearby village of perhaps 9 to 12 households.

By 600–500 BC the population of the arroyo had increased to c. 30 households living in two villages. They completed a second stage, giving the dam a width of 100 m, a height of 8 m, and a length of 400 m, creating a reservoir of 1,430,000 m^3. Spencer calculated that a labor force of 41 to 106 persons would have been necessary to construct a cofferdam and spillway during one dry season; then the dam itself could be built during three dry seasons. By this time, labor from several villages had to work together and were evidently under the control of elite families living upstream from the dam (Spencer 1979, 1993). At this writing, Neely (personal communication) is making new discoveries in the area; one of his earliest is an Archaic well 4.7 m deep, dating to 7900 BC, at San Marcos Necoxtla.

Comparing the Valley of Oaxaca and the Tehuacán Valley

The Valley of Oaxaca seems to have been an area where dozens of small streams could be diverted with brush dams to provide floodwater or canal irrigation. Multiple small-scale systems emerged during the period from 500 BC to AD 500, but none would have required the administration of a centralized government. Even after the state had formed by 300 BC (Spencer and Redmond 2001), there is no archaeological evidence for large-scale irrigation works or a coordinated valleywide strategy. The Tehuacán Valley offers a different story because the massive water sources in its northern region could not be maximized until *after the state had formed and sufficient labor was available.* "If there is a lesson to be learned here, it is that very powerful states [the Zapotec of the Valley of Oaxaca] can be supported by rather simple farming techniques, and that very tiny states [like that in the Tehuacán Valley] can produce some very impressive irrigation works" (Flannery 1983:339).

Now we turn to the Maya area, one of the most densely populated areas of Mesoamerica between AD 600 and 800.

Northern Maya Lowlands

The Yucatán Peninsula lacks rivers, but its porous bedrock has enabled rainwater to percolate down to form subterranean caverns. Where natural faults or weaknesses exist in bedrock, the karstic crust collapses, creating circular waterholes known as cenotes. These caverns and cenotes constituted critical sources of water for the ancient inhabitants, and not surprisingly, their distribution affected the location of human settlement. In many localities such natural sources of water were lacking or insufficient to meet all the needs of the large population centers. Thus, artificial means of water storage had to be developed.

In the Edzná Valley annual precipitation can range from 1000 to 1300 mm. Here the rainy season runs from June through December, and the dry season from January to May. Subterranean water at Edzná occurs at a depth of 20 m. Although excavations at Dzibilnocac, c. 50 km from Edzná, revealed wells that reached a depth of more than 13 m, Ray Matheny (1978) found no wells at Edzná. Instead, he found *aguadas*, natural depressions, and more.

Beginning at about 600 BC, settlers were probably attracted to Edzná's numerous natural depressions, which fill up with water during the rainy season. By c. 300 BC to AD 250 Edzná invested in a hydraulic system involving several canals and reservoirs. One 12-km-long canal was, in fact, connected to a moat. That long canal was capable of holding 900,000 m^3 of water and the moat another 200,000 m^3. Twelve other canals form a radial pattern, like the spokes of a wheel, around the center of ancient Edzná (see Figure 9.8). Matheny (1978:200–201) argues persuasively that the canals were not afterthoughts of a population under stress but "clearly part of a grand scheme of original city planning." Although all of these canals stored rainwater, Matheny suggests that the largest canal, which averages 50 m in width and 1.5 m in depth, could also have been used for canoe transportation.

Although population estimates at most Mesoamerican sites are calculated from the extent of construction and/or extent of the sherd scatter on the surface, Matheny's data provide us with still another method. We can now use "water storage capability" as a general index of how many people *could* have lived at Edzná. In the Edzná area today a Maya family of five uses about 100 liters of water per day during the dry season and about 50 liters per day during the rainy season. Each person would thus need c. 15 liters per day. Annual use would thus approximate 15 liters × 365 days = 5,475 liters per year per individual. Given that the Edzná water system could have stored 2,000,000 m^3, it was potentially capable of supporting 20,000 to 30,000 people (lower population figures would result if we take into account drier years, evapotranspiration, and other factors that led to the loss of water).

At several northern Yucatán sites annual rainfall rates were much lower than at Edzná, and other strategies for water storage had to be employed. Edward H. Thompson (1897) reports that water was stored in *chultuns*, artificial cisterns made by the ancient Maya. These cisterns were excavated into limestone bedrock, cracks were filled with rocks and mortar, and the interior was then coated with a layer of lime plaster. A very large *chultun* might hold 35,000 liters. At sites such as Labná there were stone surfaces and plastered pavements that directed rainwater into the *chultuns* (see Figures 9.9, 9.10). In the 1890s

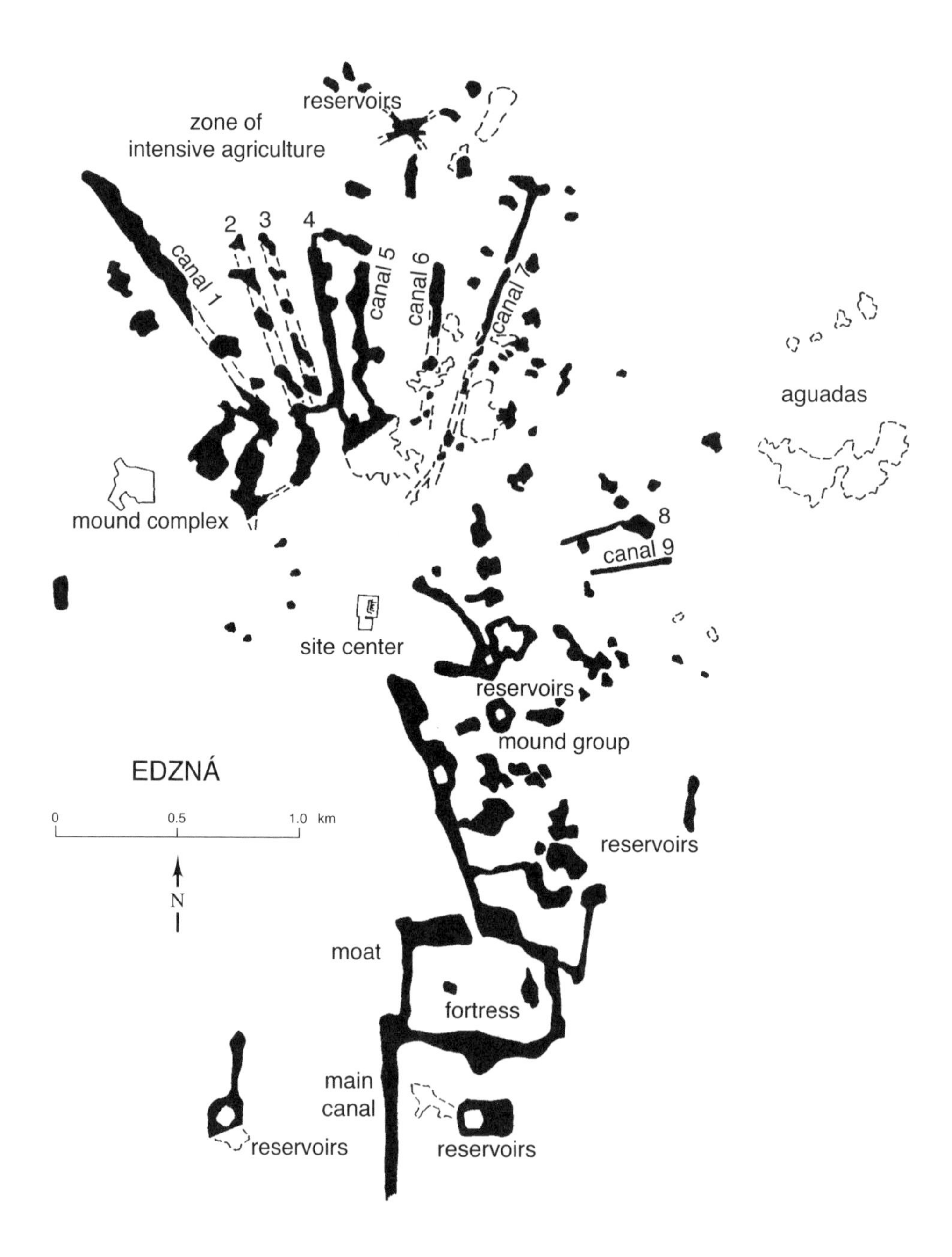

Figure 9.8. The Maya site of Edzná in Mexico features an impressive water-storage system of canals, reservoirs, and a moat (redrawn from Matheny 1978:Figure 10.8).

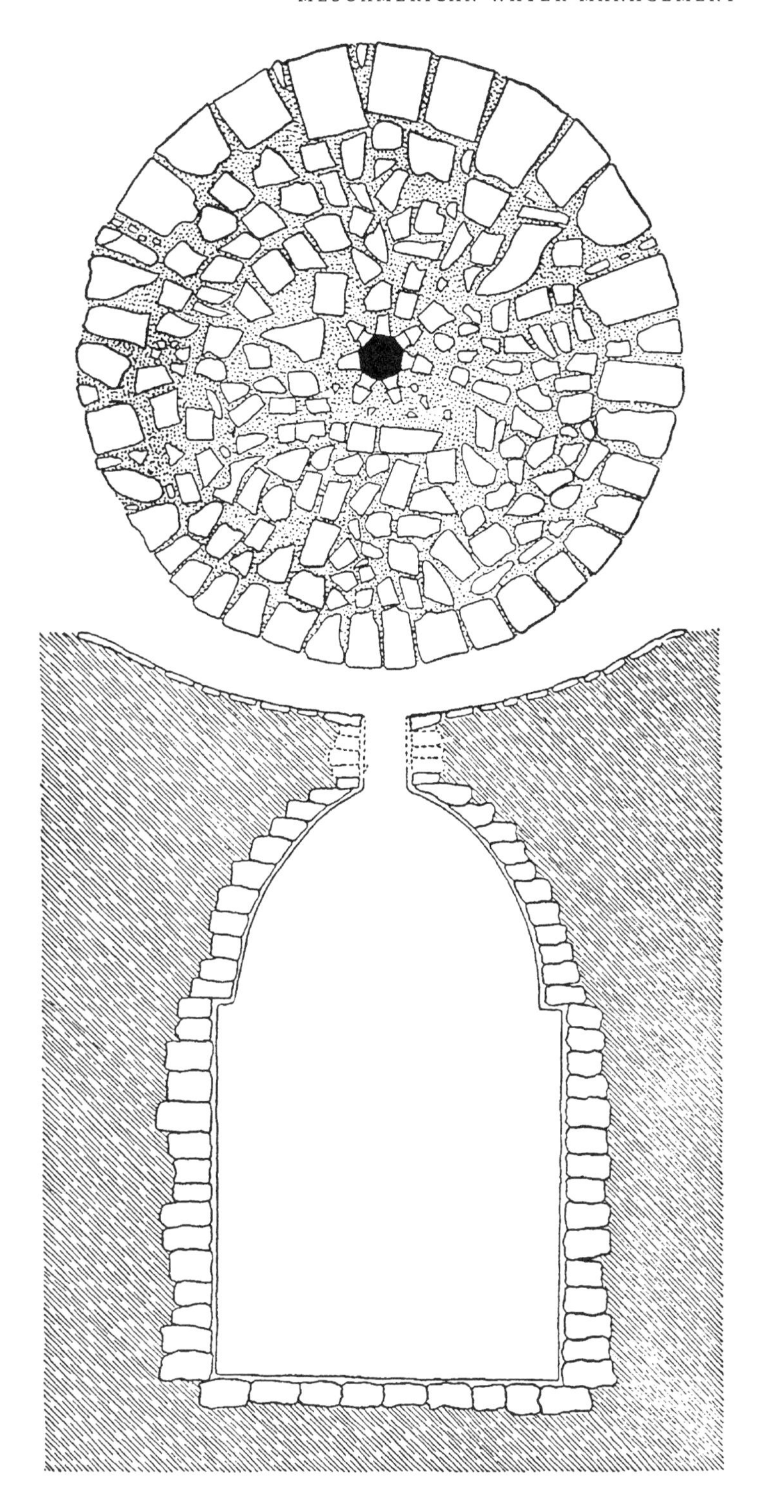

Figure 9.9. The circular paved area (*top*) led rainwater to the chultun opening (shown as a solid black dot). The cross section of that Labná chultun is given below (redrawn from E. H. Thompson 1897:Figure 2).

Figure 9.10. Twelve examples of chultuns, aptly called "giant chianti flasks" by J. Eric S. Thompson (1975:xv). Although water storage was the initial purpose of these Labná chultuns, several were later reused for burials or discarded rubbish (redrawn from E. H. Thompson 1897:Plates V, VIII, and III).

Thompson was among the first to point out that the "built environment"—the man-made pavements and plastered surfaces on plazas, patios, courts, roofs, and buildings—provided ideal "water catchment" areas that directed rainwater to *chultuns*. Pollock (1980:561) concludes, "There is little question in my mind that it was the exploitation of the chultun as a method of storing water during the prolonged annual dry season that permitted the large population indicated by the number of ruin sites in the Puuc area" (northwest Yucatán).

More quantification, however, is needed for most Maya sites. Calculating the total paved area at each site, together with mapping all the associated drains, cisterns, and storage tanks, could lead to the discovery of some general correspondences between (1) length of occupation and water storage capacity and/or (2) population estimates and potential water supply available during the dry season.

Southern Maya Lowlands

Tikal, in northern Guatemala, occupies an area that receives 1350 to 2000 mm of rainfall per year, quite a bit more than the Edzná area. In both areas, of course, four to six months of the year saw little or no rain.

Tikal's population has been estimated as 60,000 to 80,000 from AD 600 to 800. How Tikal's large population met its water needs during the dry season has been the subject of detailed studies by Scarborough (1994, 1998; Scarborough and Gallopin 1991). He divided the site of Tikal into six catchment areas, or watersheds, noting that the plastered plazas, platforms, and courts were the pavements that "provided a deliberate set of impervious catchment areas subtly designed to seal the underlying, naturally porous limestone and direct rainy season runoff into the sizable tanks within the central precinct of the site" (Scarborough 1998:141). The most completely altered catchment was at the monumental summit of Tikal, where more than 900,000 m^3 of water could be collected in a series of reservoirs in association with monumental public buildings. Scarborough suggests that Tikal, originally just a limestone hillock, had been converted into a "water mountain," a built environment that combined monumentality with efficient water capture and storage (Figure 9.11). Near the foot of the hillock were swampy terrain and storage tanks that had a combined water storage capacity of perhaps 50,000 to 175,000 m^3.

Thinking of Classic Maya cities as convex paved watersheds and reservoir systems is quite a change in our perspective (Marcus 1982). Such a focus on water capture and storage can begin to address long-standing questions about how Maya cities with thousands of occupants met their water needs during the dry season. Since the Tikal system was based exclusively on rainfall, it was precarious. If rainfall decreased dramatically over a period of several years, as

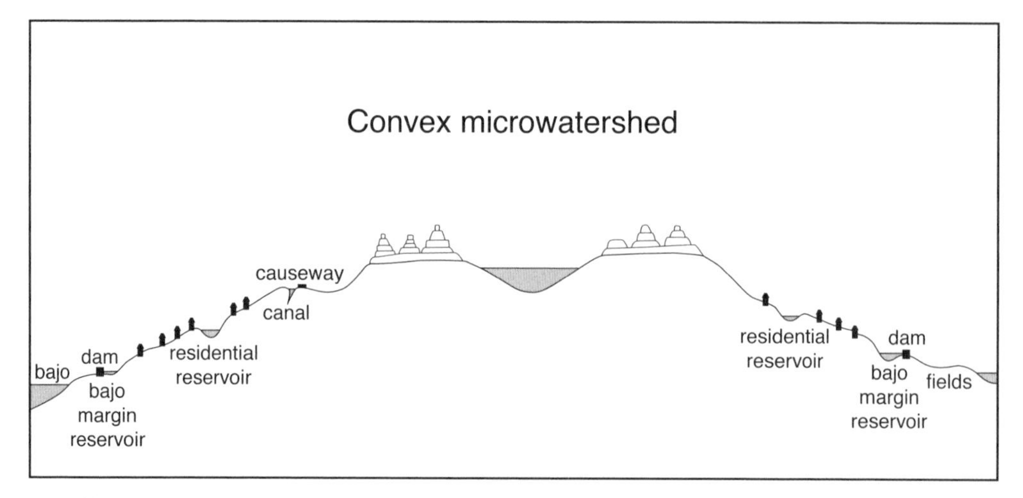

Figure 9.11. Cross-section of a convex microwatershed or "water mountain" (redrawn from Scarborough 1994:189).

some scholars have been suggesting, it could have contributed to site abandonments and local political collapse.

Maya sites most able to withstand multiyear droughts were those that had access to other sources of water, such as lakes, rivers, aguadas, and caves, and/or those that could disperse their populations over the landscape (Weiss-Krejci and Sabbas 2002). Empirical evidence for multiyear droughts affecting each Maya city has yet to be fully assembled, but some work along those lines has already been done (see Gill 2000).

The Postclassic Aftermath

After the collapse of major Petén cities around AD 900, significant populations continued to cluster around the lakes of northern Guatemala and Belize (Demarest et al. 2004; Marcus 1995:24; 2003:106). Elsewhere in Mesoamerica other lake systems supported truly huge Postclassic populations. The best-known example comes from the Basin of Mexico, the center of the Aztec Empire, which included five interconnected lakes: Zumpango, Xaltocan, Texcoco, Xochimilco, and Chalco. The island capital of Tenochtitlan, with perhaps 100,000 inhabitants, was connected to the mainland by three causeways, each 8 m wide (Figure 9.12). The northern causeway was 3 km long and ran to Tepeyac, west of Atzacualco; the western causeway was 3 km long and ran to Tlacopan (modern Tacuba) and Chapultepec; the southern causeway ran for 6 km to a fort where it divided into two branches, one running to Coyoacán and the other to Ixtapalapan. The causeways were interrupted at intervals to allow canoe traffic to pass. The western causeway was accompanied by an aqueduct bringing drinking water from springs at Chapultepec (it also had tubes piping water directly into the residences of the Aztec nobles). The bulk of the population purchased drinking water from traveling canoemen.

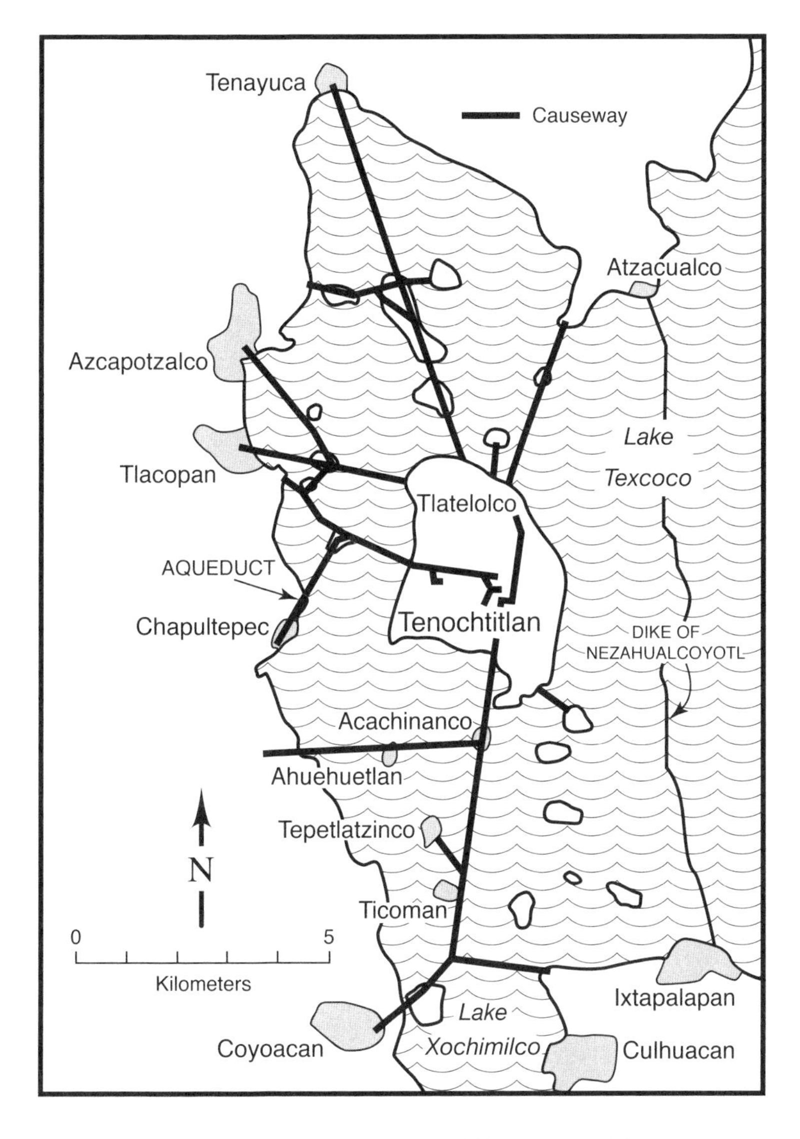

Figure 9.12. Map showing an Aztec aqueduct and dike and some of the communities on the mainland that were linked by causeways to Tenochtitlan and Tlatelolco on the island (redrawn from Doolittle 1990:Figure 5.8; and Marcus 1983b:Figure 10.13).

The sixteenth-century judge Alonso de Zorita describes the Basin of Mexico aqueducts in AD 1556 as follows:

> Along one of the causeways leading to the city, there are two cement aqueducts, each two paces broad, and an estado high, through one of which a volume of very good fresh water, as big around as a man's body, flows into the heart of the city and supplies all with sufficient water for drinking and other purposes. The other aqueduct, which is empty, is used when they wish to clean the pipe, and remains in use until the cleaning is finished. . . . Canoemen peddle the water through all the streets. (Keen 1963:160–161)

This description refers to the later aqueduct (Figure 9.13b), situated on higher islands, not the earlier aqueduct built in AD 1418 and destroyed by a flood in AD 1449. To support such aqueducts as they passed over the lake, the Aztec constructed a chain of artificial islands, using woven reed mats covered with mud and rocks (Bribiesca Castrejón 1958; Doolittle 1990:122–123). The Aztec placed hollowed-out tree trunks to carry the water across the troughs between islands, and next to those logs they placed wooden planks for pedestrian traffic across the lake (Figure 9.13a). The labor force that built the second aqueduct was so large and industrious that the workers were said to "look like ants" (Alvarado Tezozomoc 1944:381). These laborers came from Texcoco, Azcapotzalco, Tlacopan, Coyoacán, Xochimilco, and four other towns.

Chapultepec ("Grasshopper Hill") was a sacred "water mountain," not only providing fresh drinking water to the capital but also displaying the portraits of Aztec rulers in its living rock. Although it is very much damaged today, one can still see the outlines of a portrait of Motecuhzoma II, reportedly carved in AD 1519 by 12 sculptors working for 30 days (Nicholson 1961). Chapultepec may have been a sacred place—associated in sixteenth-century texts with the selection of the first Aztec ruler—but it also was a garden spot, with canals, terraces, flower gardens, palaces, temples, and life-size carvings of the Aztec rulers.

The Aztec also turned the hill of Tetzcotzingo, 7 km southeast of Texcoco, into a "water mountain," with gardens, temples, terraces, pools, and baths, by bringing water from Cerro Metecatl via an aqueduct (Niekler 1919; Parsons 1971). Alva de Ixtlilxochitl (1985:153–154) describes Tetzcotzingo's rock-cut temples, gardens, sweet-smelling flowers, and singing birds and speaks of serene pools and water that "fell like rain" from higher basins to lower ones.

Such water mountains are not unexpected creations when one considers how the Aztec transformed the lake system of the Basin of Mexico. They used drainage canals to reclaim swampy terrain; they built dikes to control floods and keep brackish water from mixing with fresh; they constructed *chinampas* served by traffic canals, built aqueducts, and irrigated terraces. On soggy

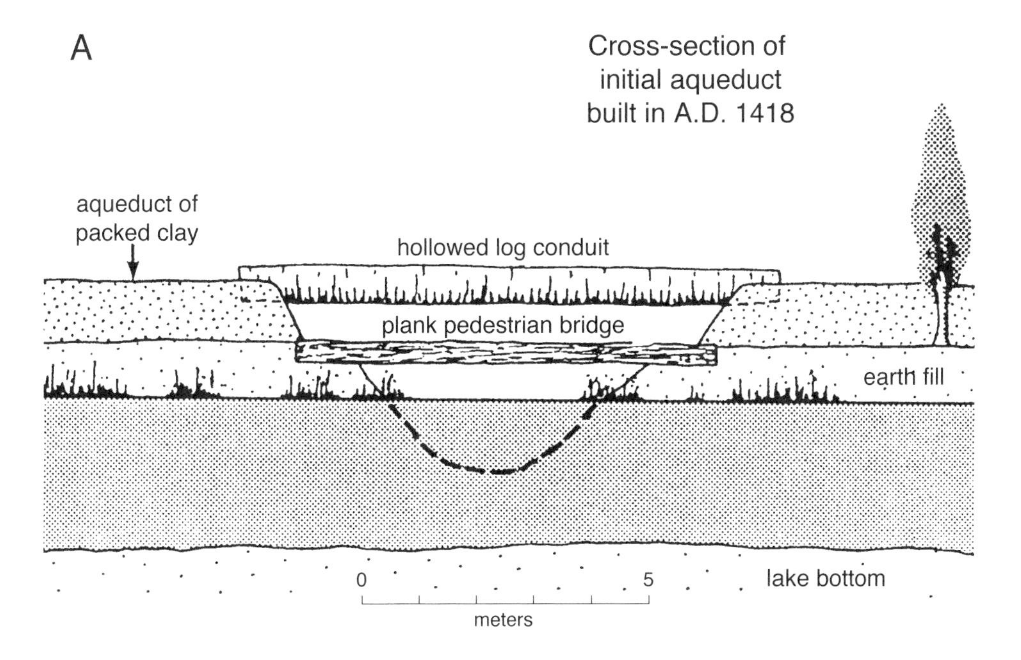

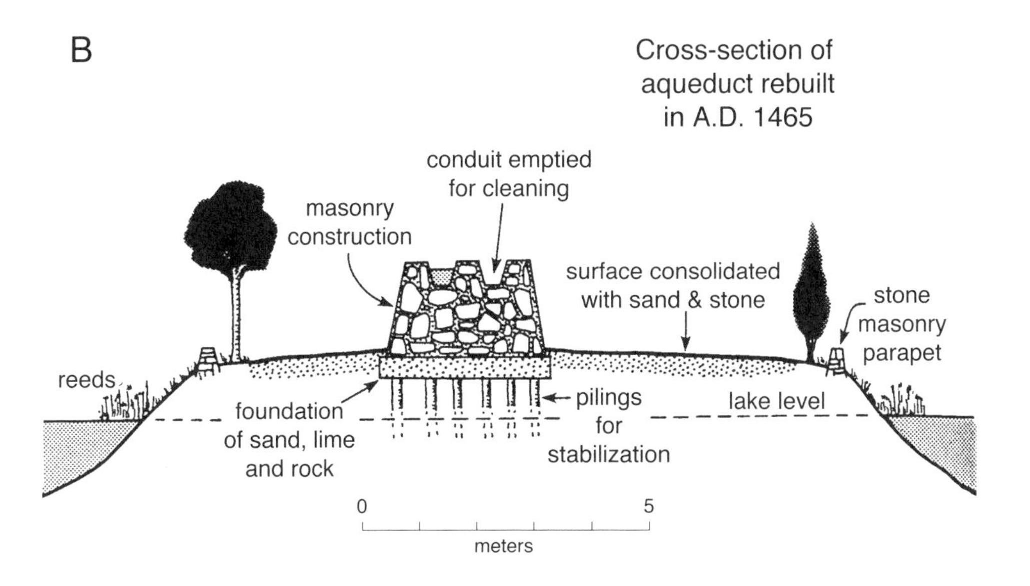

Figure 9.13. Both the first aqueduct (*top*) and second aqueduct (*bottom*) brought drinking water from Chapultepec to the Aztec capital, Tenochtitlan, in the Basin of Mexico (redrawn from Doolittle 1990: Figures 5.6, 5.7).

land (*chiauhtlalli*) no additional water was needed, but Sahagún (Dibble and Anderson 1963:252–254) shows how they diverted water to make dry land productive (*atlalli*) (Figure 9.14).

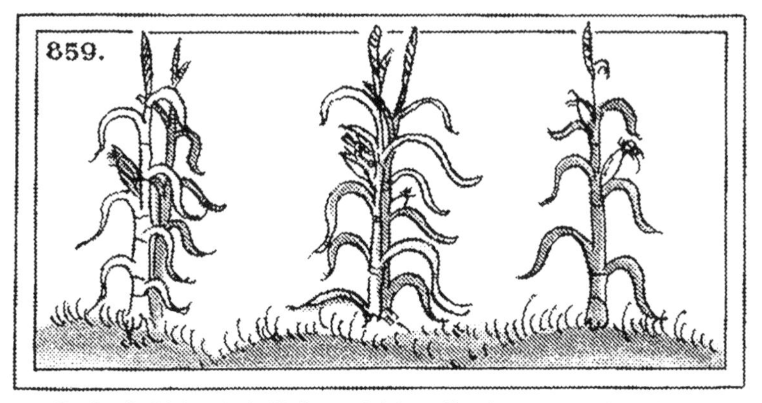

Called *Chiauhtlalli* (humid land), these wetlands did not require irrigation.

Called *Atlalli* (water + earth), these are the irrigated fields.

Figure 9.14. The sixteenth-century Aztec farmer distinguished "humid land that did not need irrigation" (*top*) from "land that required irrigation" (*bottom*) (redrawn from Dibble and Anderson 1963:252–254).

COMPARISONS AND CONTRASTS

From a comparative standpoint Mesoamerican water-control systems find parallels in the Far East. For example, agriculture in Shang dynasty China also involved performances by diviners or "ritual technicians." Along with their prayers for rain and good harvests the Shang offered millet ale, cattle, dogs, sheep, grains, and human victims to the ancestors in the manner of their Mesoamerican counterparts (Keightley 1999:258). "Every successful harvest would have produced a profound sense not just of economic security but of spiritual vindication" (Keightley 1999:279).

We have seen that among early Zapotec and Maya farmers actual irrigation may have been preceded and anticipated by the use of drainage ditches and cisterns to remove excess water (Lohse and Findlay 2000; Marcus and Flannery 1996:107). In much the same way Chinese texts "suggest that the Shang may have been engaged in the drainage of low-lying fields, but no archaeological evidence indicates that they constructed large-scale irrigation works" (Keightley 1999:279). In both regions, perhaps, ditches designed to get rid of excess water may have led eventually to ditches that carried water to drier areas. And some Mesoamerican cultures, like the Maya, rather than wasting excess water during the rainy season foresaw the utility of storage systems that kept water available during the dry season.

Scarborough's (2003) suggestion that Maya cities, such as Tikal or La Milpa, were "water mountains" is of great interest, since the Aztec called their cities *altepemeh*, "water mountains," and made hills with springs the venue for gardens, water basins, rock-cut terraces, canals, aqueducts, temples, and palaces.

Our brief survey showed us the diversity of water-control strategies implemented by Mesoamerican polities. Some areas, like the northern Maya lowlands, relied on water in deep natural wells because there were no rivers. Other areas, like the Valley of Oaxaca, had abundant streams and rivers but never developed any centralized system of control because no single source of water was large enough to warrant state intervention. Still other regions, like the Tehuacán Valley, had such a gap between the biggest springs and the best frost-free land that optimal irrigation was only possible after the rise of a state that could build long canals.

What might be the future of water management studies? We need to create a larger set of diachronic case studies from various parts of the world so that we can evaluate entire evolutionary sequences, from simple to maximal elaboration, documenting both the advances in technology and the changing size of labor forces. We also need to document the synchronic operation of each hydraulic system at a specific time and place, discussing a range of topics from (1) the degree of specialization displayed by the labor force to (2) which crops were grown, (3) the potential yields and number of harvests per year, and (4) who might have consumed the crops.

In the early studies of water management, technology was often the primary focus. Today we realize that that was our secular Western perspective at work. Ancient societies often relied as heavily on ritual as technology, persuading the supernatural masters of water to cooperate. At last we are in a position to integrate both the ritual and technological components of water management to construct a more holistic view of each society's relationship with water.

NOTES

1. Area B is one of the residential wards at an early Zapotec village now called San José Mogote. For more details on the cistern and rain runoff canals, see Flannery and Marcus (2005:307–310).

REFERENCES

Alva de Ixtlilxochitl, F. de

1985 *Historia de la Nación Chichimeca.* Edición de Germán Vázquez. RAYCAR, Madrid, Spain.

Alvarado Tezozomoc, H.

1944 *Crónica Mexicana. Escrita hacia el año de 1598.* Editorial Leyenda, México, DF, México.

Anderson, A. J. O., and C. E. Dibble (translators)

1981 *Florentine Codex. General History of the Things of New Spain,* by Fray Bernardino de Sahagún. *Book 2—The Ceremonies.* Monographs of the School of American Research 14, Part III. School of American Research, Santa Fe, New Mexico.

Baquedano, E.

1984 *Aztec Sculpture.* Trustees of the British Museum, London.

Bribiesca Castrejón, J.

1958 El agua potable en la República Mexicana: Los abastecimientos en la época prehispánica. *Ingeniería hidráulica en México* 12(2):69–82.

Broda, J.

1970 Tlacaxipeualiztli: A reconstruction of an Aztec calendar festival from 16th century sources. *Revista Española de Antropología Americana* 5:197–273.

1971 Las fiestas aztecas de los dioses de la lluvia: Una reconstrucción según las fuentes del siglo XVI. *Revista Española de Antropología Americana* 6:245–327.

1987 Templo Mayor as ritual space. In *The Great Temple of Tenochtitlan,* edited by J. Broda, D. Carrasco, and E. Matos Moctezuma, 61–123. University of California Press, Berkeley.

Córdova, J. de
1942 [1578] *Vocabulario en lengua zapoteca*. Pedro Charte y Antonio Ricardo, Mexico.

Cruz, W. C.
1946 *Oaxaca recóndita: Razas, idiomas, costumbres, leyendas y tradiciones del Estado de Oaxaca, México*. México, D.F.

Curtis, J. H., M. Brenner, D. A. Hodell, R. A. Balser, G. A. Islebe, and H. Hoogiemstra
1998 A multi-proxy study of Holocene environmental change in the Maya lowlands of Petén, Guatemala. *Journal of Paleolimnology* 19:139–159.

Curtis, J. H., D. A. Hodell, and M. Brenner
1996 Climate variability on the Yucatan Peninsula (Mexico) during the past 3500 years, and the implications for Maya cultural evolution. *Quaternary Research* 46:37–47.

Demarest, A. A., P. M. Rice, and D. S. Rice (editors)
2004 *The Terminal Classic in the Maya Lowlands*. University Press of Colorado, Boulder.

Dibble, C. E., and A. J. O. Anderson (translators)
1963 *Florentine Codex. General History of the Things of New Spain*, by Fray Bernardino de Sahagún. *Book 11—Earthly Things*. Monographs of the School of American Research 14, Part XII. School of American Research, Santa Fe, New Mexico.

Doolittle, W. E.
1990 *Canal Irrigation in Prehistoric Mexico: The Sequence of Technological Change*. University of Texas Press, Austin.

Downing, T. E., and M. Gibson (editors)
1974 *Irrigation's Impact on Society*. University of Arizona Press, Tucson.

Flannery, K. V.
1983 Precolumbian farming in the valleys of Oaxaca, Nochixtlán, Tehuacán, and Cuicatlán: A comparative study. In *The Cloud People*, edited by K. V. Flannery and J. Marcus, 323–339. Academic Press, San Diego.

Flannery, K. V., and J. Marcus
1976 Formative Oaxaca and the Zapotec cosmos. *American Scientist* 64(4):374–383.
1994 *Early Formative Pottery of the Valley of Oaxaca, Mexico*. University of Michigan Museum of Anthropology, Memoir 27. Ann Arbor.
2005 *Excavations at San José Mogote 1: The Household Archaeology*. University of Michigan Museum of Anthropology, Memoir 40. Ann Arbor.

Folan, W. J., J. Gunn, J. D. Eaton, and R. W. Patch
1983 Paleoclimatological patterning in southern Mesoamerica. *Journal of Field Archaeology* 10:453–468.

Gill, R. B.
2000 *The Great Maya Droughts: Water, Life, and Death*. University of New Mexico Press, Albuquerque.

Guiteras-Holmes, C.
1961 *Perils of the Soul: The World View of a Tzotzil Indian*. Free Press of Glencoe, New York.

Gunn, J. D., R. T. Matheny, and W. J. Folan
2002 Climate-change studies in the Maya area: A diachronic analysis. *Ancient Mesoamerica* 13(1):79–84.

Hodell, D. A., M. Brenner, J. H. Curtis, and T. Guilderson
2001 Solar forcing of drought frequency in the Maya lowlands. *Science* 292:1367–1370.

Hodell, D. A., J. H. Curtis, and M. Brenner
1995 Possible role of climate in the collapse of Classic Maya civilization. *Nature* 375:391–394.

Keen, B.
1963 *Life and Labor in Ancient Mexico: The Brief and Summary Relation of the Lords of New Spain by Alonso de Zorita*. Rutgers University Press, New Brunswick, New Jersey.

Keightley, D. N.
1999 The Shang: China's first historical dynasty. In *The Cambridge History of Ancient China: From the Origins of Civilization to 221 B.C.*, edited by M. Loewe and E. L. Shaughnessy, 232–291. Cambridge University Press, Cambridge, UK.

Kirkby, A. V. T.
1973 *The Use of Land and Water Resources in the Past and Present Valley of Oaxaca, Mexico*. University of Michigan Museum of Anthropology, Memoir 5. Ann Arbor.

Lohse, J. C., and P. N. Findlay
2000 A Classic Maya house-lot drainage system in northwestern Belize. *Latin American Antiquity* 11(2):175–185.

López Austin, A.
1973 *Hombre-dios: Religión y política en el mundo náhuatl*. Instituto de Investigaciones Históricas, Universidad Nacional Autónoma de México, Mexico.

Marcus, J.
1978 Archaeology and religion: A comparison of the Zapotec and Maya. *World Archaeology* 10(2):172–191.
1982 The plant world of the sixteenth- and seventeenth-century lowland Maya. In *Maya Subsistence: Studies in Memory of Dennis E. Puleston*, edited by K. V. Flannery, 239–273. Academic Press, New York.
1983a Zapotec religion. In *The Cloud People*, edited by K.V. Flannery and J. Marcus, 345–351. Academic Press, San Diego.
1983b On the nature of the Mesoamerican city. In *Prehistoric Settlement Patterns: Essays in Honor of Gordon R. Willey*, edited by E. Z. Vogt and R. M. Leventhal, 195–242. University of New Mexico Press, Albuquerque.

1989 Zapotec chiefdoms and the nature of Formative religions. In *Regional Perspectives on the Olmec*, edited by R. J. Sharer and D. C. Grove, 148–197. Cambridge University Press, Cambridge, UK.

1995 Where is lowland Maya archaeology headed? *Journal of Archaeological Research* 3:3–53.

1998 *Women's Ritual in Formative Oaxaca: Figurine-Making, Divination, Death and the Ancestors*. University of Michigan Museum of Anthropology, Memoir 33. Ann Arbor.

1999 Men's and women's ritual in Formative Oaxaca. In *Social Patterns in Pre-Classic Mesoamerica*, edited by D. C. Grove and R. A. Joyce, 67–96. Dumbarton Oaks, Washington, DC.

2003 Recent advances in Maya archaeology. *Journal of Archaeological Research* 11(2):71–148.

Marcus, J., and K. V. Flannery

1978 Ethnoscience of the sixteenth-century Valley Zapotec. In *The Nature and Status of Ethnobotany*, edited by R. I. Ford, 51–70. University of Michigan Museum of Anthropology, Ann Arbor.

1996 *Zapotec Civilization: How Urban Society Evolved in Mexico's Oaxaca Valley*. Thames and Hudson, London.

Matheny, R. T.

1978 Northern Maya lowland water-control systems. In *Pre-Hispanic Maya Agriculture*, edited by P. D. Harrison and B. L. Turner II, 185–210. University of New Mexico Press, Albuquerque.

Matos Moctezuma, E.

1987 The Templo Mayor of Tenochtitlan: History and interpretation. In *The Great Temple of Tenochtitlan*, edited by J. Broda, D. Carrasco, and E. Matos Moctezuma, 15–60. University of California Press, Berkeley.

Monaghan, J.

1995 *The Covenants with Earth and Rain: Exchange, Sacrifice, and Revelation in Mixtec Sociality*. University of Oklahoma Press, Norman.

Nader, L.

1965 The Trique of Oaxaca. In Ethnology, ed. E. Z. Vogt, 400–416. *Handbook of Middle American Indians*, Vol. 7, R. Wauchope, general editor, University of Texas Press, Austin.

Neely, J. A.

2005 Mesoamerican Formative period water management technology: An overview with insights on development and associated method and theory. In *New Perspectives on Formative Mesoamerican Cultures*, edited by Terry G. Powis. BAR International Series 1377, 127-146. Archaeopress, Oxford, UK.

Nicholson, H. B.

1961 The Chapultepec Cliff Sculpture of Motecuhzoma Xocoyotzin. *El México Antiguo* 9:379–444.

1971 Religion in pre-Hispanic central Mexico. In Archaeology of Northern Mesoamerica, edited by G. F. Ekholm and I. Bernal, 395–446. *Handbook of Middle American Indians*, Vol. 10, R. Wauchope, general editor, University of Texas Press, Austin.

Niekler, O.

1919 Texcotzinco. *El México Antiguo* 1:110–112.

O'Brien, M. J., R. D. Mason, D. E. Lewarch, and J. A. Neely

1982 *A Late Formative Irrigation Settlement below Monte Albán: Survey and Excavation on the Xoxocotlán Piedmont, Oaxaca, Mexico*. University of Texas Press, Austin.

Parsons, E. C.

1936 *Mitla: Town of the Souls*. University of Chicago Press, Chicago.

Parsons, J. R.

1971 *Prehistoric Settlement Patterns in the Texcoco Region, Mexico*. University of Michigan Museum of Anthropology, Memoir 3. Ann Arbor.

Peñafiel, A.

1885 *Nombres geográficos de México: Catálogo alfabético de los nombres de lugar pertenecientes al idioma "Náhuatl."* Secretaría de Fomento, Mexico.

Pollock, H. E. D.

1980 *The Puuc: An Architectural Survey of the Hill Country of Yucatan and Northern Campeche, Mexico*. Peabody Museum of Archaeology and Ethnology, Harvard University, Cambridge, Massachusetts.

Redfield, R.

1941 *The Folk Culture of Yucatan*. University of Chicago Press, Chicago.

Redfield, R., and A. Villa Rojas

1934 *Chan Kom, a Maya Village*. Carnegie Institution of Washington 448. Washington, DC.

Ruiz de Alarcón, H.

1984 *Treatise on the Heathen Superstitions and Customs that Today Live among the Indians Native to this New Spain, 1629*, edited by J. R. Andrews and R. Hassig. University of Oklahoma Press, Norman.

Sandstrom, A. R.

1991 *Corn Is Our Blood: Culture and Ethnic Identity in a Contemporary Aztec Indian Village*. University of Oklahoma Press, Norman.

Scarborough, V. L.

1994 Maya water management. *National Geographic Research & Exploration* 10:184–199.

1998 Ecology and ritual: Water management and the Maya. *Latin American Antiquity* 9:135–159.

2003 *The Flow of Power: Ancient Water Systems and Landscapes*. School of American Research Press, Santa Fe, New Mexico.

Scarborough, V. L., and G. G. Gallopin

1991 A water storage adaptation in the Maya lowlands. *Science* 251:658–662.

Scarborough, V. L., and B. L. Isaac (editors)

1993 Economic aspects of water management in the prehispanic New World. *Research in Economic Anthropology*, supp. 7. JAI Press, Greenwich, Connecticut.

Spencer, C. S.

1979 Irrigation, administration, and society in Formative Tehuacan. In *Prehistoric Social, Political, and Economic Development in the Area of the Tehuacan Valley*, edited by Robert D. Drennan, 13–109. University of Michigan Museum of Anthropology, Technical Report No. 11, Ann Arbor.

1993 Human agency, biased transmission, and the cultural evolution of chiefly authority. *Journal of Anthropological Archaeology* 12:41–74.

Spencer, C. S., and E. Redmond

2001 Multilevel selection and political evolution in the Valley of Oaxaca, 500–100 B.C. *Journal of Anthropological Archaeology* 20:195–229.

Steward, J. H., R. M. Adams, D. Collier, A. Palerm, K. A. Wittfogel, and R. L. Beals

1955 *Irrigation Civilizations: A Comparative Study*. Pan American Union, Washington, DC.

Sullivan, T. D.

1974 Tlaloc: A new etymological interpretation of the god's name and what it reveals of his essence and nature. In *Atti del XL Congresso Internazionale degli Americanisti*, 2:213–219. Tilgher, Roma-Genova, Italy.

Terraciano, K.

2001 *The Mixtecs of Colonial Oaxaca*. Stanford University Press, Stanford, California.

Thompson, E. H.

1897 *The Chultunes of Labná, Yucatan*. Peabody Museum of Archaeology and Ethnology, Harvard University, Cambridge, Massachusetts.

Thompson, J. E. S.

1970 *Maya History and Religion*. University of Oklahoma Press, Norman.

1975 Introduction. In *The Hill-Caves of Yucatan*, by Henry C. Mercer, vii–xliv. University of Oklahoma Press, Norman.

Villa Rojas, A.

1945 *The Maya of east central Quintana Roo*. Carnegie Institution of Washington, Publication 559. Washington, DC.

Vogt, E. Z.

1969 *Zinacantan: A Maya Community in the Highlands of Chiapas*. Harvard University Press, Cambridge, Massachusetts.

Weiss-Krejci, E., and T. Sabbas
2002 The potential role of small depressions as water storage features in the central Maya lowlands. *Latin American Antiquity* 13(3):343–357.

Weitlaner, R. J.
1965 Supervivencias de la religión y magia prehispánicas en Guerrero y Oaxaca. *Proceedings of the Thirty-Fifth International Congress of Americanists*, 557–563. Universidad Nacional Autónoma de México, Mexico City.

Wilkinson, T. J.
2003 *Archaeological Landscapes of the Near East.* University of Arizona Press, Tucson.

Wittfogel, K. A.
1957 *Oriental Despotism: A Comparative Study of Total Power.* Yale University Press, New Haven, Connecticut.

Woodbury, R. B., and J. A. Neely
1972 Water control systems of the Tehuacan Valley. In *Chronology and Irrigation: The Prehistory of the Tehuacan Valley*, vol. 4, edited by F. Johnson, 81–153. University of Texas Press, Austin.

The Economic Underpinnings of Prehispanic Zapotec Civilization

Small-Scale Production, Economic Interdependence, and Market Exchange

Gary M. Feinman

> Contrary to the popular belief that nature always remains the same—a belief that has led to static theories of environmentalism and to their equally static rejections—nature changes profoundly whenever man, in response to simple or complex historical causes, profoundly changes his technical equipment, his social organization, and his world outlook. Man never stops affecting his natural environment. He constantly transforms it.
>
> —Karl Wittfogel, *Oriental Despotism*

> Irrigation does not create the state, as some have argued; but the state can use canal irrigation to increase its centralization in ways that it cannot manipulate other land-use systems in Oaxaca.
>
> —Kent Flannery, Preface

The eloquent writings of Karl Wittfogel (1981) were seminal in the development of anthropological conceptions that irrigation farming was a central building block in the development and florescence of early civilization around the globe (e.g., Steward 1949). Yet as early as a half century ago the notion that large-scale canal irrigation necessarily preceded and prompted the emergence of early states in highland Mexico was met with skepticism and critique (e.g., Beals 1955; Price 1971; Wolf and Palerm 1955; see also Adams 1966). Despite a scarcity of archaeological information, especially before the later 1960s, these early debates were extremely provocative from a theoretical standpoint, posing questions regarding the relationship between household production and overarching suprahousehold management, as well as between humans, their technologies, and the surrounding environments. More specifically, did intensive farming systems, particularly those involving large-scale water control, require centralized management, or could individual householders cooperate without hierarchical infrastructure?

Following five decades of concerted fieldwork in Mexico's semiarid highland valleys, archaeologists still debate the nature of the causal linkages and temporal correspondences between the practice and scale of irrigation systems, on the one hand, and political development, on the other. For example, in the indigenous Basin of Mexico (e.g., Nichols 1987; Nichols and Frederick 1993; Parsons 1991), where canal systems have been found dating back to before the Common Era and the lakeside chinampa system of the Aztecs is well known (e.g., Armillas 1971), the link between irrigation and political change remain under discussion. Yet in the Valley of Oaxaca, also in the Mexican highlands, where one of Mesoamerica's earliest cities (Monte Albán) arose (Figure 10.1), few if any scholars would acknowledge that large-scale irrigation had any significant causal role in the rise and growth of the early Zapotec state. In fact, large-scale canal irrigation has been shown to have had little subsistence or managerial importance in either the past or present Valley of Oaxaca (Flannery et al. 1967; Kirkby 1973; Lees 1973).

As Joyce Marcus and Charles Stanish reiterate in this book's introduction, long-held, neo-Wittfogelian notions that all or most early civilizations were "irrigation societies" and that central governmental management of farming systems, especially water, prompted the rise (and supported the power) of states are still periodically touted both inside and outside of anthropology (e.g., Lucero 2002; see also Butzer 1996). Yet any effort to extend the propositions of Wittfogel too literally or to expect that his emphasis on large-scale irrigation would apply to the emergence of preindustrial civilizations in all global regions must come to grips with the complete absence of empirical support for these views in regard to the Valley of Oaxaca. For that reason it is worthwhile to review these findings here (see also Isaac 1993). Yet at the same time, this chapter endeavors to illustrate that if one takes a step back from some of the more specific propositions of Wittfogel (1955, 1972, 1981) and Julian Steward (1949, 1955a, 1955b), but retains their broader focus on the changing relations between labor, agrarian production strategies, and political organization (Blaikie and Brookfield 1987; Gilman 1981; Netting 1990), there remains much to be learned regarding the precolonial history of the Valley of Oaxaca and, perhaps, beyond.

We can build on these earlier important works and resulting debates to address a series of more specific questions. If large-scale canal irrigation was not a factor, how and why did the driest sector of the semiarid Valley of Oaxaca, the eastern or Tlacolula arm, become one of the most densely inhabited parts of this region during the Classic (AD 200–800) and Postclassic (AD 800–1520)

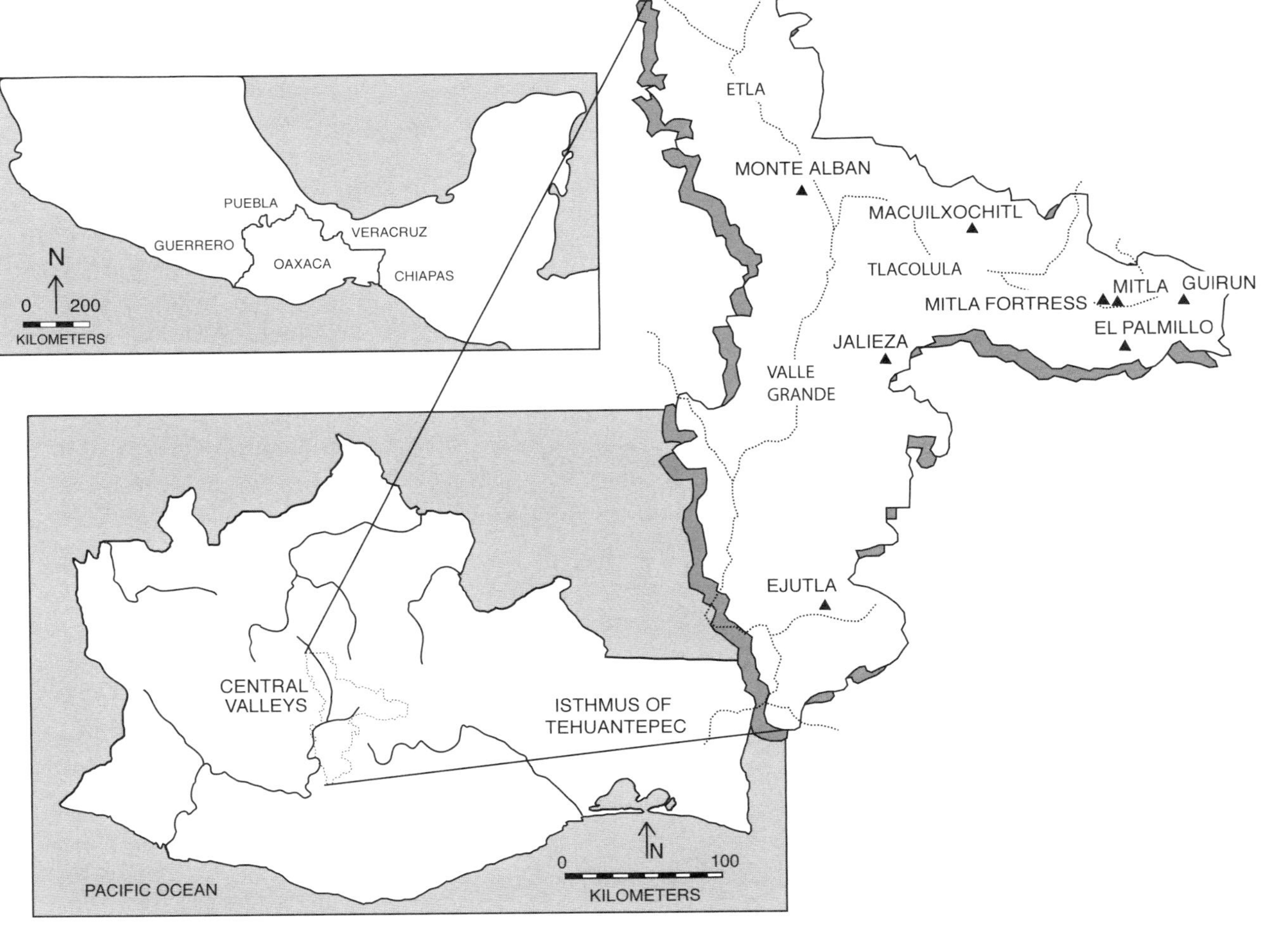

Figure 10.1. Map of Mexico, showing the location of the Valley of Oaxaca and key sites mentioned in the text.

periods (Kowalewski et al. 1989)? How can this subvalley's demographic importance be understood in the face of evidence that a dependable maize crop frequently cannot be grown in much of the Tlacolula arm. Finally, what were the apparent larger (valleywide) political and economic causes and ramifications of the later prehispanic demographic expansion in Tlacolula and the agrarian strategies that seem to have maintained them? By addressing these issues, we illustrate that many of the queries and relations between suites of variables explored by Wittfogel and Steward were indeed on the mark for constructing a basic understanding of archaic states; however, the nature of the relations between agricultural intensification and political formations were less deterministic and more synergistic than these important theorists first proposed a half century ago (see also Downing and Gibson 1974; Scarborough and Isaac 1993).

BACKGROUND TO THE VALLEY OF OAXACA: HISTORY AND FARMING STRATEGIES

To confirm Wittfogel's irrigation hypothesis, one might look to discover a chain of events in the past in which canal irrigation (or another system of intensive water control) led to central control or management, thereby culminating in the rise of a more hierarchical political formation, the state. These changes should occur in the specified sequence and without considerable time lag. As early as the 1970s a unified program of archaeological, geographic, and ethnographic studies (Prehistory and Human Ecology of the Valley of Oaxaca), under the overarching direction of Kent Flannery, illustrated that these expectations are not met in the prehispanic Valley of Oaxaca and that the specific hypotheses advanced by Wittfogel could not possibly have direct application there.

The semiarid, highland Valley of Oaxaca is *y*-shaped, divided into three arms. In much of the area average annual rainfall is adequate for a single maize harvest; however, there also is considerable spatial and temporal variation, which makes rainfall-dependent farming rather unpredictable. Guilá Naquitz cave, situated near Mitla in the eastern end of the Tlacolula arm of the valley, is the site of the earliest (roughly 10,000 years ago) domesticated plant (squash) in all of the Americas (Smith 1997). Nevertheless, sedentary life and villages were not established in the valley until millennia later, between 4,000 and 3,500 years ago (Marcus and Flannery 1996). It was not until sometime around 2,500 years ago that the earliest city was established at the hub of the valley's three arms. This centrally situated hilltop settlement at Monte Albán was the capital of the region's first state (Blanton 1978).

A key study by Anne Kirkby (1973) (part of the larger Prehistory and Human Ecology project) examined the relationship between agricultural land, water resources, settlement pattern, and political change in past and present Oaxaca.

Kirkby's research is still recognized as a landmark investigation, and several of its findings set the foundation for much of the rest of our discussion. Kirkby (1973:31) found that households were the primary units of agricultural labor in the Valley of Oaxaca; this seems to correspond to the principal prehispanic unit of production and consumption as well (e.g., Feinman et al. 2002; Marcus and Flannery 1996).

Through her examination of production practices and yields across the valley, Kirkby also found that the availability of water is the most critical factor in the determination of maize yields. Because water availability varies considerably from one part of the region to another, yields differ markedly from one area to another as well. The uneven distribution of perennial water sources, patchy rains, and rugged terrain all give a strong spatial component to these productivity differentials. In general, the eastern (Tlacolula) arm receives less rainfall (Kowalewski 1982) and has fewer opportunities for supplemental water through irrigation, so maize farming is somewhat riskier there than in the region's northern (Etla) or southern (Valle Grande) arms.

Overall, canal irrigation is feasible on less than 10 percent of the farmland in the Valley of Oaxaca (Figure 10.2) (Flannery et al. 1967; Kirkby 1973:50). Other sources of perennial water, primarily high water table, also are available on only a small proportion of the land, so well over 80 percent of the summer maize production across the valley depends on rainfall. Because of the high dependence on annual precipitation, which is highly variable, more recent research (F. Dilley 1993; M. Dilley 1997) illustrates that harvest failures due to drought occur roughly one year out of every four in the valley. In Tlacolula more than six out of ten years are excessively dry (Kowalewski 1982).

Based on archaeological findings (Flannery et al. 1967) from the Valley of Oaxaca, it is evident that some forms of water control and irrigation (namely the use of wells to exploit high water table land) were practiced in the region going back almost to the time of the earliest villages in Oaxaca. Such use of groundwater to supplement rainfall clearly enhances agricultural production. Yet this kind of well or pot irrigation is today entirely a household affair and likely entailed few if any suprahousehold managerial costs. Its advent in the Valley of Oaxaca precedes the rise of Monte Albán by at least 500 years. At the same time, there is no archaeological evidence from excavations (Flannery et al. 1967) or full-coverage valleywide surveys (Blanton et al. 1982; Kowalewski et al. 1989) that truly large-scale canal irrigation was ever practiced in Oaxaca, yet the region's Zapotec civilization has long been recognized as having been one of Mesoamerica's earliest and longest lived (Palerm and Wolf 1957). The only prehispanic canal irrigation systems that have been noted in the valley are extremely small and postdate the establishment of Monte Albán (e.g., O'Brien et al. 1982). In their totality such findings provided no empirical confirmation for the hydraulic hypothesis of Wittfogel and Steward.

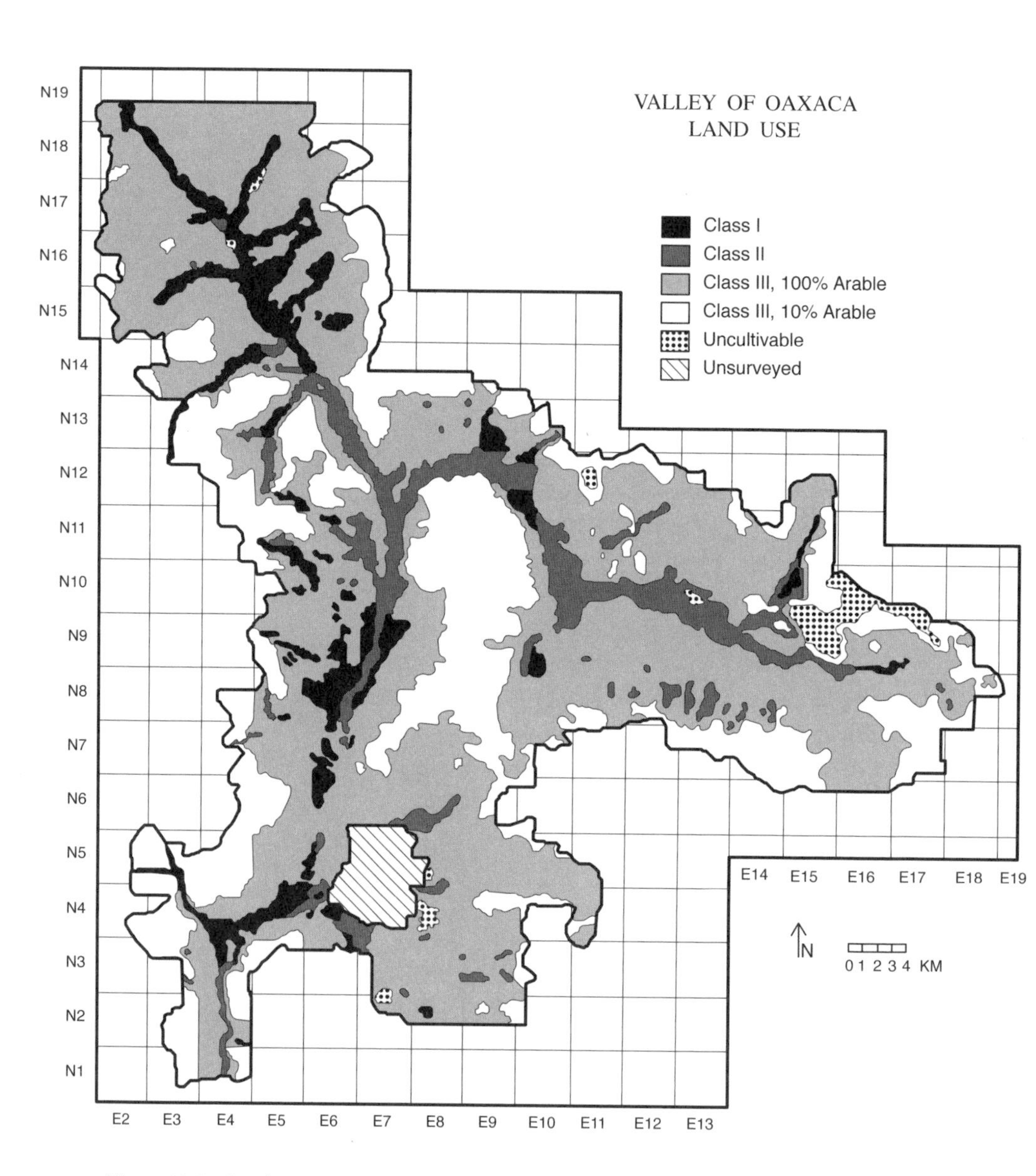

Figure 10.2. Land-use map for the Valley of Oaxaca. Class I land is mostly alluvial, including high water table and irrigated land. Class II land has access to occasional supplemental water. Class III land is rainfall dependent, largely in the foothills (100 percent arable) or mountains (10 percent arable).

LAND, WATER, AND POLITICAL ORGANIZATION: REVISING THE MODELS AND THE QUESTIONS

As Karl Butzer (1996:200) has noted, "the Wittfogel model, like Elvis[,] refuses to die." Although there is no rationale for its specific resuscitation, there are broader themes and research questions that arose from the works of Wittfogel (1981) and Steward (1955a, 1955b) that are useful and worth reviving and pursuing today. Most important, the approaches of Wittfogel and Steward, as well as their early critics, focused attention on the economic (specifically agrarian) underpinnings of early civilization—not just as a factor in the rise of these complex social formations but as related to the continuity and change of these formations over time (e.g., Palerm and Wolf 1957; Sanders 1976). More specifically, the hydraulic hypothesis concentrated attention on the relationships between agrarian producers and suprahousehold authorities. Questions such as, how did leaders support their positions? what permitted suprahousehold authorities to maintain their privileged roles over time? and why do householders often yield a proportion of their resources, labor, and autonomy to leaders? are not only highlighted in the works of Wittfogel and Steward but are at the forefront of the contemporary archaeological research agenda (e.g., Feinman and Nicholas 1987; Gilman 1981). In contrast, however, to the presumptions of the hydraulic hypothesis, large-scale irrigation or intensified agricultural systems cannot be considered proof presumptive of governmental control (e.g., Erickson 1993; Morrison 1994; Netting 1990).

Following general scholarly critique of the deterministic and unilineal aspects of the hydraulic hypothesis, a number of researchers have taken a broader and more synergistic perspective toward capital-intensive subsistence regimes or "landesque capital" (Blaikie and Brookfield 1987), that is, agricultural systems in which primary labor investment results in long-term or even permanent modification of the landscape. Antonio Gilman's (1981:5) study of Bronze Age Europe has been particularly significant in this regard as it illustrated how capital-intensive agricultural strategies, such as tree cropping, where households made continual labor investments in their land (often with delayed returns), have implications for the long-term relations between those agrarian producers and a ruling minority. For example, with productive systems that require heavy initial investments of work, the producers may be reluctant to relinquish the very resources or assets that they have created. These investments check other tendencies toward fission or having householders "vote with their feet," particularly in response to demands from above. As a consequence commitments to prior labor investments to the land provide opportunities for overarching political authorities even if those authorities do not directly control production, and, most important, such opportunities or implications are not restricted to canal irrigation or systems of water management.

EASTERN TLACOLULA

As discussed preliminarily above, issues concerning water, land, production, political organization, and population are central to understanding the later prehispanic occupation of the Tlacolula arm of the Valley of Oaxaca. At face value the rapid growth of population in this dry part of the valley, beginning in the Early Classic period and continuing up to the Spanish Conquest, requires examination and explanation. By the Late Postclassic period (AD 900–1520), the most densely settled part of the valley was the eastern part of the Tlacolula arm (Feinman and Nicholas 1990:221; Kowalewski et al. 1989). Yet in the Tlacolula arm canal-irrigable terrain makes up an even smaller proportion (only 3 percent) of the region's cultivable land than elsewhere (9 percent) in the valley (Kirkby 1973:50), and most of the irrigable land in Tlacolula sits in the western part of that arm, close to the valley's center. At the same time, the Tlacolula arm (and particularly eastern Tlacolula) is the driest part of the region, where crop losses due to inadequate water are most apt to occur. Much of the land yields only 0.2 to 0.4 metric tons of maize per hectare annually (Kirkby 1973:65), which is close to the threshold below which farmers did not find it worthwhile to plant maize in the 1970s (Kirkby 1973). In Tlacolula there is a 70 percent chance that less than 600 mm of rain will fall in a year, so yields from rainfall-dependent maize fields often fall below 0.4 metric tons per hectare (Kowalewski 1982). After the 1970s some of the land in eastern Tlacolula that had been regularly planted in maize (at least periodically) has been repeatedly left fallow or is now used solely for grazing.

Given the precarious nature of maize farming in eastern Tlacolula, it is not particularly surprising that the population in this part of the valley was relatively sparse (compared to the rest of the valley) from the dawn of sedentary village life through the end of the Formative period (c. 1600 BC–AD 200). The rise of the early city of Monte Albán at the center of one of Mesoamerica's first states had less apparent impact on settlement in eastern Tlacolula in comparison to areas closer to the valley hub. By the latter part of the Formative era the central parts of the valley had populations higher than eastern Tlacolula by an order of magnitude (Feinman and Nicholas 1990:221). This pattern began to change early in the Classic period, when a series of hilltop towns of hundreds (sometimes thousands) of people were established (or grew rapidly from earlier small communities). Although there were demographic ups and downs that followed, eastern Tlacolula tended to be occupied at densities more comparable to the rest of the region into the late prehispanic era, when it became the most densely settled part of the valley (Feinman and Nicholas 1990).

The remainder of this chapter explores the factors and consequences that underpinned the demographic expansion that occurred in eastern Tlacolula beginning during the first centuries AD. The complex, spatially variable patterns of population change that have been documented in this highland region over millennia point to the importance of understanding changes in the valley's political economy (e.g., Kowalewski 2003:214) in order to come to grips with the demographic growth in eastern Tlacolula despite the precarious nature of maize farming. What were the economic underpinnings for this expansion? How did the Classic and Postclassic inhabitants of Tlacolula sustain themselves, and how can this inform us regarding the nature of the regional economy? At present there is no indication that markedly new productive technologies were introduced early in the Classic period. And while political circumstances in the region must be taken into account, eastern Tlacolula's demographic takeoff began and was sustained during the height of Monte Albán's hegemony. Yet certain sites (e.g., Mitla, Macuilxochitl) in eastern Tlacolula achieved their greatest importance and this subvalley's population reached its greatest densities following the collapse of Monte Albán (Kowalewski et al. 1989). The Late Classic–Early Postclassic breakdown of the Monte Albán polity prompted a political balkanization that culminated in a more decentralized political landscape in which the valley was divided into a series of semiautonomous petty states (e.g., Marcus 1989). Two of the largest petty states were centered in eastern Tlacolula. So the economic and demographic factors associated with eastern Tlacolula's rise occurred in the context of an emerging and solidifying Monte Albán state. These factors and conditions were apparently, at least in part, sustainable after the Monte Albán state's collapse and perhaps even helped lay some of the seeds for it.

EL PALMILLO

More than eight years of research at the hilltop town of El Palmillo and its environs has provided clues to the economic and demographic expansion of eastern Tlacolula during the Classic period (Feinman et al. 2002). Like many settlements in this part of the valley, El Palmillo, situated in the community of Santiago Matatlán in the extreme eastern edge of the valley, grew rapidly early in the Classic period (Figure 10.3). El Palmillo became one of the most populous settlements (if not the largest) in the Tlacolula arm during the Classic period. Although the site's size diminished greatly during the Postclassic period, population continued to reside in its general vicinity (at lower elevations in the lands of contemporary Matatlán) during this latter era just before Spanish contact, when neighboring Mitla rose to regional prominence.

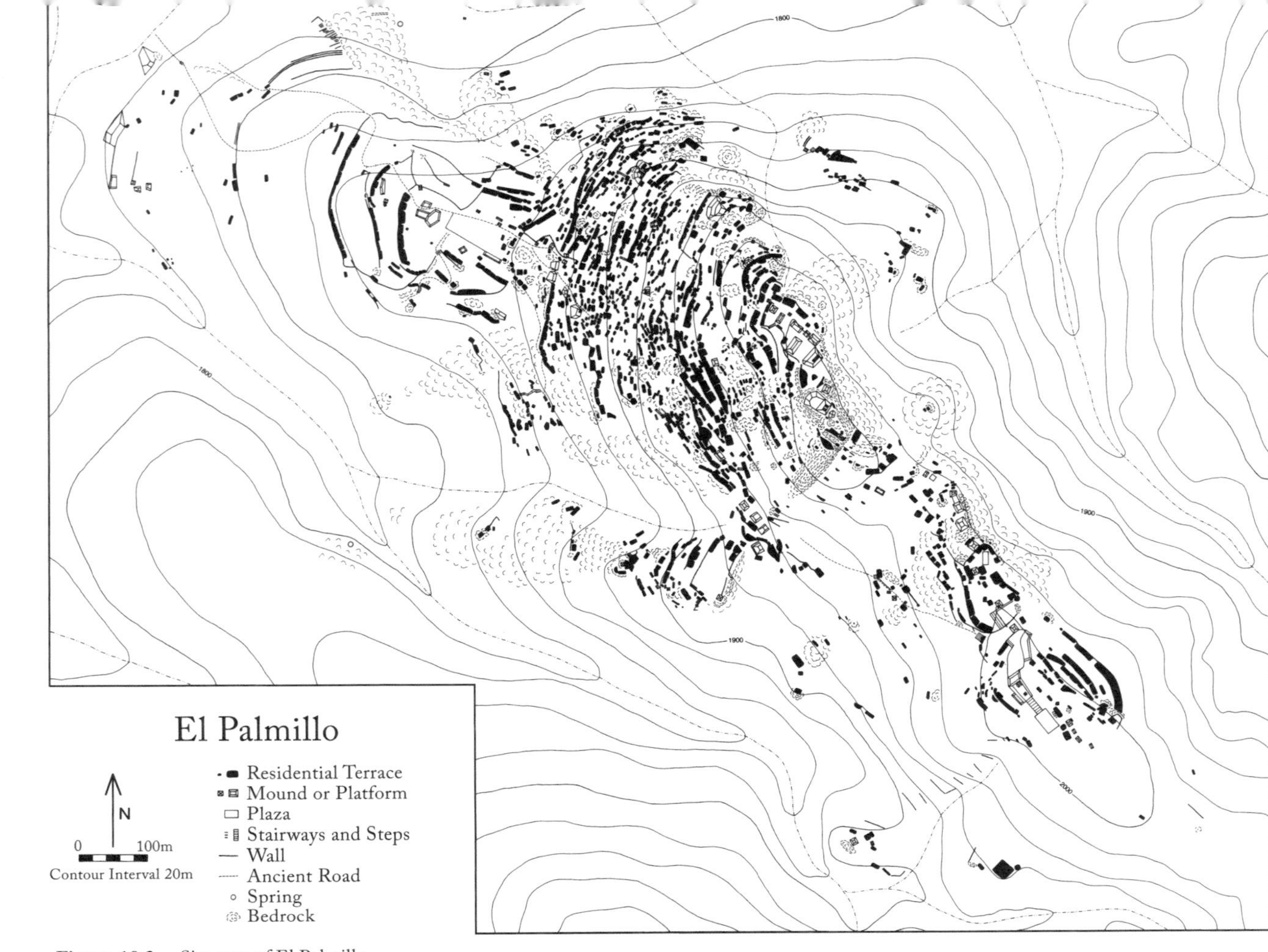

Figure 10.3. Site map of El Palmillo.

The site of El Palmillo, situated in the piedmont zone, was first located and mapped in 1980 as part of the valleywide settlement pattern survey (Blanton et al. 1982; Kowalewski et al. 1989). Following a more intensive survey and surface pickup of this site along with two neighboring hilltop terrace communities, the Mitla Fortress and Guirún (Feinman and Nicholas 2000, 2004a), five seasons of terrace excavations have been undertaken at El Palmillo (Feinman et al. 2002). As part of this latter study a botanical survey of the present vegetation at the site of El Palmillo was undertaken (Middleton et al. 2001). All of these studies provide the empirical foundation for the findings presented here.

During the El Palmillo excavations (1999–2004) seven terraces were examined almost in their entirety. One of the findings of these excavations is that the flattened surfaces of these terraces were completely residential. Almost all the flat space present on the terraces was covered by lime-plastered floors associated with rooms and patios (Figure 10.4). None of the space appeared agricultural in use. Of course, with more than 1,400 terraces mapped at the site, it is a bit risky to extrapolate too generally from the small sample of terraces that have been excavated. Yet in line with the excavated findings, domestic architecture and artifacts were observed consistently on the surfaces of other terraces during site surveys (Feinman and Nicholas 2000, 2004a; Kowalewski et al. 1989). Terrace excavations at Monte Albán also established the residential use of such flattened space there (e.g., Winter 1974).

THE IMPORTANCE OF XEROPHYTIC PLANTS

A diet reliant on the cultigens maize, beans, and squash has long been considered a core feature of prehispanic Mesoamerican civilization (e.g., Kirchhoff 1943). Nevertheless, the concentration of large communities and dense populations in areas that can be considered only marginal (or worse) for maize farming raises the issue of whether and to what extent maize was a staple crop for all of the inhabitants of the Valley of Oaxaca. Given the absence of a conjunction between reliable water sources and flat farmland near the base of El Palmillo, it is unlikely that maize farming could have sustained the thousands of people who inhabited the more than 1,400 terraces that were occupied at the site during the Classic period. If maize was not a key crop for the later prehispanic inhabitants of eastern Tlacolula, what did they do for a living? And why did apparent alternatives to maize farming emerge in eastern Tlacolula during the Early Classic period? As noted above, what are the implications of these economic options on broader regional political and economic relations?

Colonial period accounts from eastern Tlacolula (Horcasitas and George 1955:21) state that corn was not a staple for most people who lived in that part of the valley (near Mitla). "The provisions with which they sustained

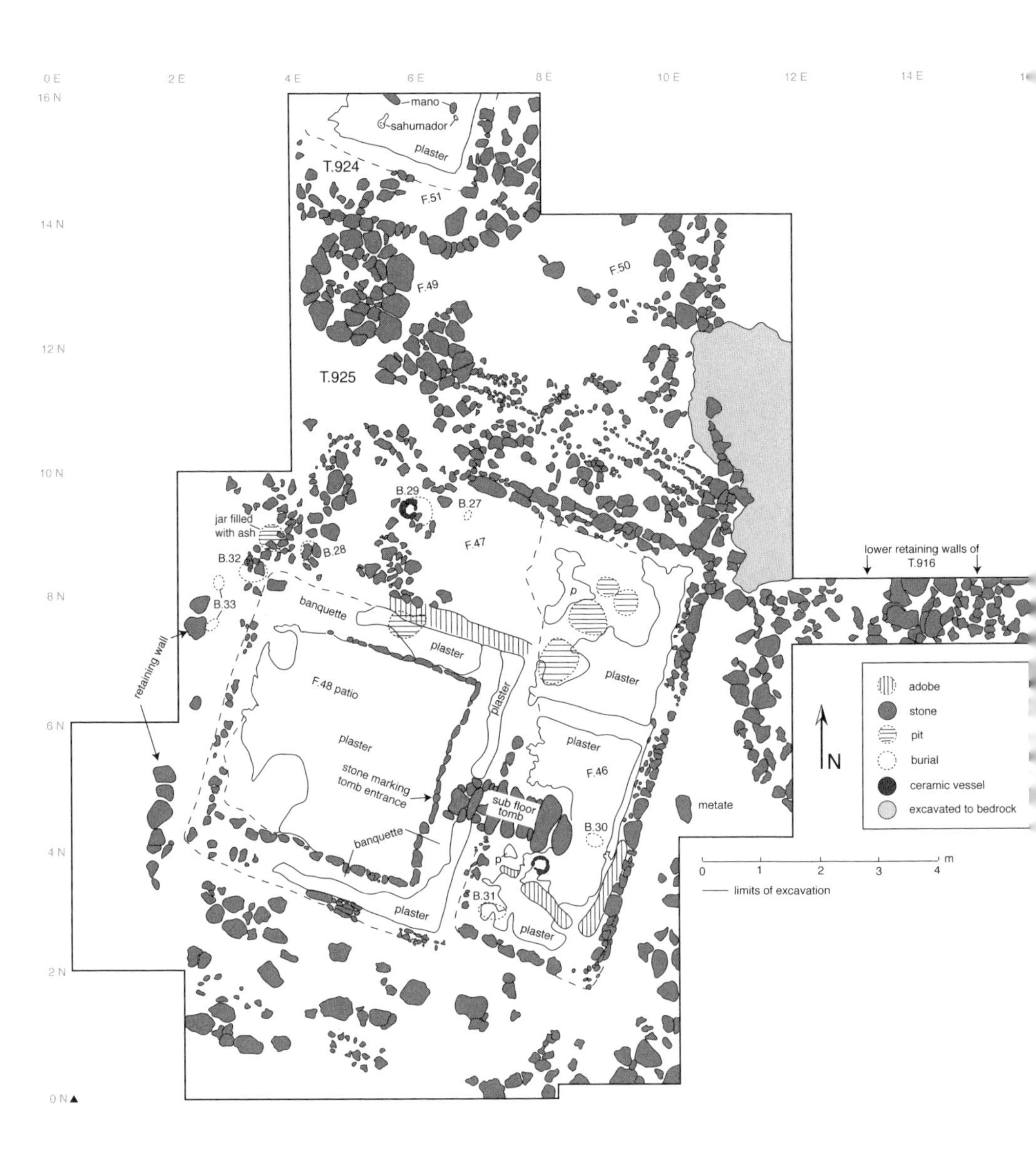

Figure 10.4. Plan of the final residential complex on Terrace 925 at El Palmillo.

themselves were some wild herbs that they did not have a name for except to know them as their food, and some wild prickly pears and sap of a tree that they call maguey. Some of the upper classes managed to eat rabbits. The peasants seldom ate corn as it did not exist." Although the principal occupation at El Palmillo preceded the writing of this account by centuries, it is worth considering the evidence from El Palmillo for economic activities that may have served as an alternative to maize production.

Most of the flattened space on the El Palmillo terraces was occupied by domestic architecture. Yet plants adapted to rocky terrain could have been fostered and cultivated along terrace walls and on the slopes in front of these features. Terrace walls at El Palmillo currently support dense xerophytic (drought tolerant) vegetation, particularly during the wet season. If planted on walls and slopes around domestic space, xerophytic plants such as cacti and maguey (*Agave* spp.) could easily have provided privacy (Parsons 1936:9) while helping reinforce retaining walls and preventing erosion (Ortiz de Montellano 1990:97; Patrick 1985; Wilken 1987:105). Most such sloping and rocky spaces are less amenable to growing seed crops, but they seem to provide little obstacle for cacti and succulents.

Drought-resistant plants such as maguey, prickly pear, and organ cactus have a long history of use in Oaxaca (Smith 1986), extending back to the Archaic period (9000–2000 BC). These plants, which collectively are used today for food, fiber, medicine, and to make intoxicating beverages (e.g., Druijven and Kruithof 1992; Parsons 1936), have previously been noted in relative abundance on archaeological sites in the Valley of Oaxaca (Martínez y Ojeda 2004), particularly in Tlacolula (Martínez y Ojeda 1996). More recently, a botanical survey (Feinman et al. 2002; Middleton et al. 2001) also confirmed the diversity and abundance of these cacti and succulents at El Palmillo, and we have proposed that the modern, uncultivated plants on-site represent to a considerable degree a relict or vestige of a prehispanic plant community that likely was fostered (see also Folan et al. 1979).

Several archaeological findings from El Palmillo provide support for the proposed economic importance of xerophytic plants at the site. Maguey roasting ovens were uncovered on two terraces, and large spindle whorls of the kind used for spinning *ixtle* or maguey fiber were abundant at the site (Feinman et al. 2002). Bone implements likely used for spinning and weaving also were found in significant numbers (Middleton et al. 2002). In addition to the use of maguey and perhaps other xerophytic plant resources for craft activities, chipped-stone tool manufacture also was important at El Palmillo. During the prehispanic period specialized craft productive activities seem to have supplemented subsistence pursuits, and some of these craft products are likely to have been exchanged for maize and other foods produced elsewhere in the valley (Feinman and Nicholas 2004b). This picture of the past, in which households in the Tlacolula arm of the valley relied heavily on craft manufacture and xerophytic resources, is similar to what is described for contemporary Tlacolula, with craft activities having a key economic role in domestic sustenance (Druijven and Kruithof 1992).

THE ECONOMIC ROLE OF MAGUEY

The importance of maguey as a source of food, fiber, rope, liquid, combustible substance, even construction material has been documented for past and present highland Mexico (Figure 10.5) (e.g., Evans 1990; Hough 1908; Parsons and Parsons 1990). A sixteenth-century account by Hernández (1959 [1577]), for example, highlights the importance of this resource for the indigenous inhabitants of central Mexico:

> As a whole it [maguey] can be used as a fuel or to fence fields. Its shoots can be used as wood and its leaves as roofing materials, as plates or platters, to make paper, to make cord with which they make shoes, cloth, and all kinds of clothes. . . . From the sap . . . they make wines, honey, vinegar and sugar. . . . From the root, they also make very strong ropes which are useful for many things. The thicker part of the leaves as well as the trunk, cooked underground . . . are good to eat and taste like cidra (a citruslike fruit) with sugar. . . . This plant, by itself, could easily furnish all that is needed for a simple frugal life since it is not harmed by storms, the rigors of the weather, nor does it wither in drought. There is nothing which gives a higher return. (Hernández (1959[1577] 1:348–349)

Hernández's observations have been supported by more recent research. Maguey has a long maturation period of 7 to 25 years. As a consequence, in a semiarid environment maguey is less susceptible than annuals, like maize, to a year or two of harsh drought. The plant thrives on rocky sediments and is not seasonally limited so that different plants can be exploited year-round.

Maguey sap, or *aguamiel*, is one of the most versatile dietary components of agave (Evans 1990; Hough 1908; Parsons and Darling 2000; Parsons and Parsons 1990). When fresh, aguamiel is a thin, watery liquid that can be drunk raw and in arid areas may serve as a substitute for water (Evans 1990:125). Aguamiel is not only refreshing but also nutritious; the sap contains protein, carbohydrates, and a variety of vitamins and minerals including calcium, iron, and vitamin C.

Aguamiel can be fermented into a nondistilled, mildly alcoholic beverage called *pulque*. One liter of this nutritious drink contains roughly 600 calories and supplies important minerals, amino acids, and vitamins (Parsons and Darling 2000:83). Studies conducted in the 1940s among Otomí villagers in

Figure 10.5. Uncultivated maguey on the slopes of El Palmillo.

north Mexico found that pulque contributed as much as 12 percent of the people's total calories (second only to tortillas) and 48 percent of their total vitamin C (Anderson et al. 1946:Table 2). Taken in moderation, pulque's nutritional value makes it a reasonable substitute for meat (Gentry 1982:11).

A fermented beverage (most likely pulque) was consumed in Oaxaca in the sixteenth century (del Paso y Troncoso 1905:145 [1580]), and the Postclassic consumption of pulque is documented in the *Codex Vindobonensis* (a Mixtec document from Oaxaca) (Furst 1978:201–203). Similar images also are present in the *Codex Nuttall* (Nuttall 1975:Plate 82). The drink could have been an important dietary staple for earlier prehispanic people as well (Evans 1986:305; Martín del Campo 1938). In Oaxaca today pulque has largely been replaced by *mezcal*, which is produced by distillation that was introduced by the Spaniards, although small-scale pulque production is still practiced in some Oaxacan villages, including Matatlán.

Aguamiel also can be boiled down into a sugar that can be formed into cakes for portability; these cakes are high in calories and can be stored for long periods of time (Evans 1990:126–127; Parsons and Darling 2000:83). The heart and leaves can be roasted and eaten. The highly nutritious flesh contains roughly 350 calories and 4.5 grams of protein per 100 grams (about a 3.5 ounce serving) (Fish et al. 1985:112). Importantly, the productivity per hectare of maguey (Parsons and Parsons 1990:336–338) tops maize (Kirkby

1973:127) in a semiarid setting. Whereas certain varieties of maguey yield 5,000 to 9,000 liters per hectare annually (2.87–5.17 million calories), maize returns of 0.4 metric tons of maize per hectare provide not quite a million calories of food. In arid parts of Mexico during the Mexican Revolution, early in the twentieth century, when access to markets was cut off, rural people relied virtually exclusively for long periods on maguey and nopal (Parsons and Parsons 1990:11). If the greater seasonal and interannual risk factors associated with maize also are taken into account, then one can imagine the prehispanic inhabitants of El Palmillo relying heavily on maguey and other xerophytic plants, especially since the craft products associated with certain fiber-producing maguey varieties (along with pulque and local chipped-stone materials) could have been exchanged for maize and other foods. Perhaps the so-called Mesoamerican triumvirate of maize, beans, and squash was not the staple diet for every prehispanic household.

RELIANCE ON XEROPHYTIC PLANTS: IMPLICATIONS FOR ECONOMY AND SETTLEMENT

If the residents of El Palmillo and those of neighboring hilltop terrace towns in the Tlacolula arm were indeed heavily reliant on xerophytic plants, then a suite of queries and issues are worth pursuing. Why did the population of eastern Tlacolula grow so rapidly during the Early Classic period, and why is so much of that settlement clustered in reasonably large hilltop, terraced communities, like El Palmillo, that are off the valley floor? Parallel to canal irrigation systems and Mediterranean polyculture, including vines and olives (Gilman 1981; Renfrew 1972), maguey cultivation represents a capital-intensive subsistence regime (e.g., Blaikie and Brookfield 1987; Gilman 1981) in that there is a significant investment in the agricultural setting before the farmer reaps a return. Following the initial planting of maguey, there is a wait of 5 to 10 (or more) years for the plant's maturation. At terrace sites there also would have been a significant and continuing investment in the domestic context itself. Terrace residents would have had to build and repair the stone retaining walls that support the flat terrace spaces that were used for domestic edifices and maintain drainage channels that removed excess water that periodically fell (Kowalewski et al. 2004). At El Palmillo almost all domestic room and patio floors, as well as structure walls, were lime plastered, even for the most basic dwellings (Feinman et al. 2002), requiring frequent maintenance. Although in one sense terrace sites like El Palmillo can be viewed as a network of modularized residences on terraces, as a totality they represent an anthropogenic sculpted landscape. Like small-scale canal irrigation systems, the maintenance of terrace sites would have required the coordinated labor from a set of adjacent households during construction and rebuilding episodes (Kowalewski et al. 2004), but the nature and the scale of the work would not have necessitated the involvement of a central authority.

Although terrace sites have a long history in prehispanic Oaxaca, they are particularly prevalent in the Tlacolula arm (Kowalewski et al. 1989), where they were established in the latter half of the first millennium BC (Feinman and Nicholas 1999; Kowalewski et al. 1989). The dramatic growth of these hilltop communities at the fringes of the valley early in the Classic period may stem in part from macroregional factors that prompted the shrinking of Monte Albán's territorial reach outside the valley and its consolidation of its core physiographic heartland (Feinman 1998; Feinman and Nicholas 1999; Kowalewski et al. 1989). At the same time, raised off the valley floor, generally away from immediate irrigation opportunities, many terrace settlements, particularly in Tlacolula, lie on lands where xerophytic plants may have fared as well or better than maize. At these locations the capital investment in the shaping and maintenance of the anthropogenic landscape may have at times required interhousehold cooperation. The long maturation period of maguey would have entailed that new arrivals to the community would have had to depend on their neighbors to a degree, just as part-time craft producers would have had to do the same through interhousehold (and intersettlement) modes of exchange. Intracommunity ties and integration mechanisms likely also would have been important, as once maguey was in the ground the risks and costs of social fissioning or "voting with one's feet" would have been steep for any domestic unit. At the same time, the nucleation and high densities of these hilltop towns may have provided a degree of security (in numbers) for the capital investments that the occupants originally made and continued to have in their fields and their settlement.

Nevertheless, even if macroregional factors encouraged Oaxaca householders to move from outside (and congregate inside) the Valley of Oaxaca early in the first millennium AD (Feinman and Nicholas 1999), were there other factors that could have contributed to the feasibility of an economic lifeway heavily reliant on xerophytic plants, and why do we not see this earlier? The use of maguey fiber (*ixtle*) to produce cloth and clothing for exchange may have become economically viable for the residents of a sizable number of households only at the outset of the Classic period, at which point there appears to have been a "clothing revolution" (Stark et al. 1998). Although cloth itself has a longer history in Mesoamerica, its intensity of manufacture and role in status and dress shifted dramatically during the Classic period (Stark et al. 1998; see also Hall 1997; Mahler 1965:582). In the Valley of Oaxaca, as in neighboring regions (Mahler 1965; Stark et a1. 1998), figurines prior to the Classic period often are shown without explicit garments (e.g., Marcus 1998), while those of the Classic period are generally dressed, in some instances with elaborate clothing. The growing importance of cloth in Oaxaca and beyond may have contributed to the viability of this economic regime.

At the same time, the production for exchange of goods like cloth, rope, pulque, chipped-stone artifacts, and possibly cochineal at sites like El Palmillo could have been possible only if householders at terrace sites were able to exchange their products with their neighbors, occupants of other nearby communities, and even (perhaps indirectly) with people in other arms of the valley. Certainly, there is little question that with the comparatively high population levels of Oaxaca's Classic period, exchange across valley arms would have lessened the risk and impact of local droughts and crop failures (Feinman and Nicholas 1992). By the Classic period the inhabitants of the Tlacolula arm appear to have been far more integrated into valleywide political and economic networks than they had been earlier (Feinman 1997, 1998). The marked growth of Classic period population in this dry arm likely could not have been sustained without greater levels of intravalley house-to-house economic integration (see Murphy et al. 1997 for a contemporary parallel). Since there are no indications that production for exchange was enacted at any scale above the domestic unit or house, and no massive storage facilities or storehouses have been found anywhere in the Valley of Oaxaca (despite decades of concerted excavation at Monte Albán), much of this domestic exchange across sites and regions may have occurred in markets (Feinman and Nicholas 2004b; see also Appel 1982 and Spores 1965 for information on sixteenth-century Oaxaca markets). The economic webs that tied together communities and houses in Tlacolula may have contributed to the continued demographic growth and expanded political importance of eastern Tlacolula settlements even following the decline of Monte Albán's hegemony (c. AD 800). At that time in the Postclassic period important political and economic links were made outside the valley.

CONCLUDING THOUGHTS

More than five decades ago Pedro Armillas (1951:21–22) perceptively recognized the "contrast between [the] relatively low technology and [the] highly developed socio-political structure and intellectual life" in prehispanic Mesoamerica. Shortly thereafter, Palerm (1955:39) categorized this relationship as a contrast, but not a contradiction, and proposed that human labor and its organization held the key to understanding it. Broadly following Wittfogel (1957) and Steward (1955b), Armillas (1951) and Palerm (1955) looked to irrigation and the coercive patterns of management that were thought to result from it as the explanatory bridge. As elaborated above, decades of archaeological, ecological, and related fieldwork have shown unequivocally that these early postulations and causal links cannot lie behind the rise or the hegemony of Monte Albán or Zapotec civilization in the Valley of Oaxaca.

Yet these foundational scholars did point the way toward understanding prehispanic Oaxaca and Mesoamerica through careful studies of agriculture, economy, and their synergistic relations with other facets of social, ideological, and political life. Prehispanic Oaxaca was a land in which most production (agrarian and otherwise) was situated domestically, yet states could be monumental and powerful. As Flannery (1983:339) has noted: "If there is a lesson to be learned here, it is that very powerful states can be supported by rather simple farming techniques."

In line with the hydraulic hypothesis the specific patterns of domestic labor and the economic interdependency between householders provide keys for understanding the building of these larger political formations and cultural traditions. But rather than canal irrigation and heavily centralized, despotic relations, the Classic period polity, centered at Monte Albán, appears to have featured rather diverse domestic production strategies, household interdependency, regional markets, and perhaps even a less-than-uniform reliance on maize. With these new building blocks and the guiding intellectual direction of our forebears, it is my hope that the next 50 years are even more productive than the last in defining the nature and the underpinnings of the early Zapotec state.

REFERENCES

Adams, R. McC.

1966 *The Evolution of Urban Society*. Aldine, Chicago.

Anderson, R. K., J. Calvo, G. Serrano, and G. C. Payne

1946 A study of the nutritional status and food habits of Otomí Indians in the Mezquital Valley of Mexico. *American Journal of Public Health and the Nation's Health* 36(8):883–903.

Appel, J.

1982 A summary of the ethnohistorical information relevant to the interpretation of Late Postclassic settlement pattern data, the central and Valle Grande survey zones. In *Monte Albán's Hinterland, Part I: The Prehispanic Settlement Patterns of the Central and Southern Parts of the Valley of Oaxaca, Mexico*, by R. E. Blanton, S. Kowalewski, G. Feinman, and J. Appel, 139–148. University of Michigan, Museum of Anthropology, Memoir 15. Ann Arbor.

Armillas, P.

1951 Tecnología, formaciones socio-económicas, y religión en Mesoamérica. In *The Civilizations of Ancient America, Twenty-Ninth International Congress of Americanists*, edited by Sol Tax, 19–30. University of Chicago Press, Chicago.

1971 Gardens on swamps. *Science* 174:653–661.

Beals, R. L.

1955 Discussion: Symposium on irrigation civilizations. In *Irrigation Civilizations: A Comparative Study*, edited by J. H. Steward, 53–57. Pan American Union, Social Science Monographs 1, Washington, DC.

Blaikie, P., and H. Brookfield

1987 Defining and debating the problem. In *Land Degradation and Society*, edited by P. Blaikie and H. Brookfield, 149–157. Methuen, London.

Blanton, R. E.

1978 *Monte Albán: Settlement Patterns at the Ancient Zapotec Capital.* Academic Press, New York.

Blanton, R. E., S. Kowalewski, G. Feinman, and J. Appel

1982 *Monte Albán's Hinterland, Part I: The Prehispanic Settlement Patterns of the Central and Southern Parts of the Valley of Oaxaca, Mexico.* University of Michigan, Museum of Anthropology, Memoir 15. Ann Arbor.

Butzer, K. W.

1996 Irrigation, raised fields, and state management: Wittfogel redux. *Antiquity* 70:200–204.

del Paso y Troncoso, F.

1905 [1579–1581] *Papeles de Nueva España: Segunda serie, geografía y estadística*, Vol. 4. Sucesores de Rivadeneyra, Madrid.

Dilley, F. B.

1993 *Climate Change and Agricultural Transformation in the Oaxaca Valley, Mexico.* PhD dissertation, Pennsylvania State University.

Dilley, M.

1997 Climatic factors affecting annual maize yields in the Valley of Oaxaca, Mexico. *International Journal of Climatology* 17:1549–1557.

Downing, T. E., and M. Gibson (editors)

1974 *Irrigation's Impact on Society*. University of Arizona Press, Tucson.

Druijven, P., and A. Kruithof

1992 Tlacolula Valley, Oaxaca Mexico. In *Coping with Semiaridity: How the Rural Poor Survive in Dry-Season Environments*, edited by H. Reitsma, T. Dietz, and L. de Haan, 132–152. Netherlands Geographical Studies 146, Amsterdam.

Erickson, C. L.

1993 The social organization of prehispanic raised field agriculture in the Lake Titicaca Basin. In *Economic Aspects of Water Management in the Prehispanic New World*, edited by V. L. Scarborough and B. L. Isaac, 369–426. JAI Press, Greenwich, Connecticut.

Evans, S. T.

1986 Analysis of the surface sample ceramics. In *The Toltec Period Occupation of the Valley Part 1 – Excavations and Ceramics. The Teotihuacan Valley Project Final*

Report – Volume 4, edited by W. T. Sanders, pp. 283–366. Occasional Papers in Anthropology No. 13. Pennsylvania State University, University Park.

1990 The productivity of maguey terrace agriculture in Central Mexico during the Aztec period. *Latin American Antiquity* 1:117–132.

Feinman, G. M.

1997 Macro-scale perspectives on settlement and production in ancient Oaxaca. In *Economic Analysis beyond the Local System*, edited by R. E. Blanton, P. N. Peregrine, D. Winslow, and T. D. Hall, 13–42. University Press of America, Lanham, Maryland.

1998 Scale and social organization: Perspectives on the archaic state. In *Archaic States*, edited by G. M. Feinman and J. Marcus, 95–133. School of American Research Press, Santa Fe, New Mexico.

Feinman, G. M., and L. M. Nicholas

1987 Labor, surplus, and production: A regional analysis of Formative Oaxacan socio-economic change. In *Coasts, Plains, and Deserts: Essays in Honor of Reynold J. Ruppé*, edited by S. W. Gaines, 27–50. Anthropological Research Papers 38. Arizona State University, Tempe.

1990 At the margins of the Monte Albán state: Settlement patterns in the Ejutla Valley, Oaxaca, Mexico. *Latin American Antiquity* 1:216–246.

1992 Human-land relations from an archaeological perspective: The case of ancient Oaxaca. In *Understanding Economic Process*, edited by S. Ortiz and S. Lees, 155–178. University Press of America, Lanham, Maryland.

1999 Reflections on regional survey: Perspectives on the Guirún area, Oaxaca, Mexico. In *Settlement Pattern Studies in the Americas: Fifty Years since Virú*, edited by B. R. Billman and G. M. Feinman, 172–190. Smithsonian Institution Press, Washington, DC.

2000 Intensive survey of hilltop terrace sites in Oaxaca, Mexico. *Antiquity* 74:21–22.

2004a *Hilltop Terrace Sites of Oaxaca, Mexico: Intensive Surface Survey at Guirún, El Palmillo, and the Mitla Fortress*. Fieldiana: Anthropology, New Series No. 37. The Field Museum, Chicago.

2004b Unraveling the prehispanic Mesoamerican economy: Production, exchange, and consumption in the Classic-period Valley of Oaxaca. In *Archaeological Perspectives on Political Economies*, edited by G. Feinman and L. Nicholas, 167–188. University of Utah Press, Salt Lake City.

Feinman, G. M., L. M. Nicholas, and H. R. Haines

2002 Houses on a hill: Classic period life at El Palmillo, Oaxaca, Mexico. *Latin American Antiquity* 13(3):251–277.

Fish, S. K., P. R. Fish, C. Miksicek, and J. Madsden

1985 Prehistoric agave cultivation in southern Arizona. *Desert Plants* 7:107–112.

Flannery, K. V.

1973 Preface. In *Socio-Political Aspects of Canal Irrigation in the Valley of Oaxaca, Mexico*, by S. H. Lees, v–vi. University of Michigan, Museum of Anthropology, Memoir 6. Ann Arbor.

1983 Precolumbian farming in the valleys of Oaxaca, Nochixtlán, Tehuacán, and Cuicatlán: A comparative study. In *The Cloud People: Divergent Evolution of the Zapotec and Mixtec Civilizations*, edited by K. V. Flannery and J. Marcus, 323–339. Academic Press, New York.

Flannery, K. V., A. V. T. Kirkby, M. J. Kirkby, and A. W. Williams Jr.

1967 Farming systems and political growth in ancient Oaxaca. *Science* 158:445–453.

Folan, W. J., L. A. Fletcher, and E. R. Kintz

1979 Fruit, fiber, bark, and resin: Social organization of a Maya urban center. *Science* 204:697–701.

Furst, J. L.

1978 *Codex Vindobonensis Mexicanus I: A Commentary*. Institute for Mesoamerican Studies, State University of New York, Albany.

Gentry, H. S.

1982 *Agaves of Continental North America*. University of Arizona Press, Tucson.

Gilman, A.

1981 The development of social stratification in Bronze Age Europe. *Current Anthropology* 22:1–23.

Hall, B. A.

1997 Spindle whorls and cotton production in Middle Classic Matacapan and in the Gulf lowlands. In *Olmec to Aztec: Settlement Patterns in the Ancient Gulf Lowlands*, edited by B. J. Stark and P. J. Arnold III, 115–135. University of Arizona Press, Tucson.

Hernández, F.

1959 [1577] *Historia natural de la Nueva España*. 2 vols. Universidad Autónoma de México, Mexico.

Horcasitas, F., and R. George

1955 The Relación de Tlacolula y Mitla. *Mesoamerican Notes* 4:13–25.

Hough, W.

1908 The pulque of Mexico. *Proceedings of the United States National Museum (Smithsonian)* 33(1579):577–592.

Isaac, B. L.

1993 AMP, HH, and OD: Some Comments. In *Economic Aspects of Water Management in the Prehispanic New World*, edited by V. L. Scarborough and B. L. Isaac, 429–471. JAI Press, Greenwich, Connecticut.

Kirchhoff, P.

1943 Mesoamérica, sus límites geográficos, composición étnica y carácteres culturales. *Acta Americana* 1:92–107.

Kirkby, A. V. T.
1973 *The Use of Land and Water Resources in the Past and Present Valley of Oaxaca.* University of Michigan, Museum of Anthropology, Memoir 5. Ann Arbor.
Kowalewski, S. A.
1982 Population and agricultural potential: Early I through V. In *Monte Albán's Hinterland, Part I: The Prehispanic Settlement Patterns of the Central and Southern Parts of the Valley of Oaxaca, Mexico*, by R. E. Blanton, S. Kowalewski, G. Feinman, and J. Appel, 149–180. University of Michigan, Museum of Anthropology, Memoir 15. Ann Arbor.
2003 Scale and the explanation of demographic change: 3,500 years in the Valley of Oaxaca. *American Anthropologist* 105(2):313–325.
Kowalewski, S. A., G. M. Feinman, L. Finsten, R. E. Blanton, and L. M. Nicholas
1989 *Monte Albán's Hinterland, Part II: Prehispanic Settlement Patterns in Tlacolula, Etla, and Ocotlán, the Valley of Oaxaca, Mexico.* University of Michigan, Museum of Anthropology, Memoir 23. Ann Arbor.
Kowalewski, S. A., G. M. Feinman, L. M. Nicholas, and V. Heredia
2004 Hilltowns and valley fields: Great transformations, labor, and long-term history in ancient Oaxaca. In *Labor in Anthropology*, edited by E. P. Durrenberger and J. E. Marti. AltaMira Press, Walnut Creek, California.
Lees, S. H.
1973 *Socio-Political Aspects of Canal Irrigation in the Valley of Oaxaca, Mexico.* University of Michigan, Museum of Anthropology, Memoir 6. Ann Arbor.
Lucero, L. J.
2002 The collapse of the Classic Maya: A case for the role of water control. *American Anthropologist* 104:814–826.
Mahler, J.
1965 Garments and textiles of the Maya lowlands. In *Archaeology of Southern Mesoamerica*, edited by Gordon R. Willey, 581–593. *Handbook of Middle American Indians*, Vol. 3, R. Wauchope, general editor, University of Texas Press, Austin.
Marcus, J.
1989 From centralized systems to city-states: Possible models for the Epiclassic. In *Mesoamerica after the Decline of Teotihuacan, A.D. 700–900*, edited by R. A. Diehl and J. C. Berlo, 201–208. Dumbarton Oaks Research Library and Collection, Washington, DC.
1998 *Women's Ritual in Formative Oaxaca: Figurine-Making, Divination, Death and the Ancestors.* University of Michigan, Museum of Anthropology, Memoir 33. Ann Arbor.
Marcus, J., and K. V. Flannery
1996 *Zapotec Civilization: How Urban Civilization Evolved in Mexico's Oaxaca Valley.* Thames and Hudson, London.

Martín del Campo, R.
1938 El pulque en el México precortesiano. *Anales del Instituto de Biología* 9:5–23.
Martínez y Ojeda, E.
1996 Guía ilustrada de las plantas de Yagul. *Proyecto Yagul 96: Conservación de los recursos ecológicos*. Centro INAH, Oaxaca.
2004 *Recursos naturales de la zona arqueológica de Monte Albán*. Conaculta-INAH, Mexico City.
Middleton, W. D., G. M. Feinman, and L. M. Nicholas
2001 *An Investigation of the Use of Xerophytic Plant Resources in the Economy and Subsistence of El Palmillo, Oaxaca, Mexico*. Project report submitted to the Heinz Family Foundation, Pittsburgh.
2002 Domestic faunal assemblages from the Classic period Valley of Oaxaca, Mexico: A perspective on the subsistence and craft economies. *Journal of Archaeological Science* 29:233–249.
Morrison, K. D.
1994 The intensification of production: Archaeological approaches. *Journal of Archaeological Method and Theory* 1:111–159.
Murphy, A. D., M. Winter, and E. W. Morris
1997 Household adaptations in a regional urban system: The central valleys of Oaxaca, Mexico. In *Economic Analysis beyond the Local System*, edited by R. E. Blanton, P. N. Peregrine, D. Winslow, and T. D. Hall, 235–254. University Press of America, Lanham, Maryland.
Netting, R. M.
1990 Population, permanent agriculture, and polities: Unpacking the evolutionary portmanteau. In *The Evolution of Political Systems: Sociopolitics in Small-Scale Sedentary Societies*, edited by S. Upham, 21–61. Cambridge University Press, Cambridge, UK.
Nichols, D. L.
1987 Risk and agricultural intensification during the Formative period in the Northern Basin of Mexico. *American Anthropologist* 89(3):596–616.
Nichols, D. L., and C. D. Frederick
1993 Irrigation canals and chinampas: Recent research in the northern Basin of Mexico. In *Economic Aspects of Water Management in the Prehispanic New World*, edited by V. L. Scarborough and B. L. Isaac, 123–150. JAI Press, Greenwich, Connecticut.
Nuttall, Z.
1975 *The Codex Nuttall: A Picture Manuscript from Ancient Mexico*. Dover Publications, New York.
O'Brien, M. J., R. D. Mason, D. E. Lewarch, and J. A. Neely
1982 *A Late Formative Irrigation Settlement below Monte Albán: Survey and Excavation on the Xoxocotlán Piedmont, Oaxaca, Mexico*. University of Texas Press, Austin.

Ortiz de Montellano, B. R.
1990 *Aztec Medicine, Health, and Nutrition.* Rutgers University Press, New Brunswick, New Jersey.

Palerm, A.
1955 The agricultural basis of urban civilization in Mesoamerica. In *Irrigation Civilizations: A Comparative Study*, edited by J. H. Steward, 28–42. Pan American Union, Social Science Monographs 1, Washington, DC.

Palerm, A., and E. R. Wolf
1957 Ecological potential and cultural development in Mesoamerica. *Pan American Union Social Science Monograph* 3:1–37, Washington, DC.

Parsons, E. C.
1936 *Mitla: Town of Souls.* University of Chicago Press, Chicago.

Parsons, J. R.
1991 Political implications of prehispanic chinampa agriculture in the Valley of Mexico. In *Land and Politics in the Valley of Mexico: A Two-Thousand Year Perspective*, edited by H. R. Harvey, 17–42. University of New Mexico Press, Albuquerque.

Parsons, J. R., and J. A. Darling
2000 Maguey (*Agave* spp.) utilization in Mesoamerican civilization: A case for precolumbian "pastoralism." *Boletín de la Sociedad Botánica de México* 66:81–91.

Parsons, J. R., and M. H. Parsons
1990 *Maguey Utilization in Highland Central Mexico: An Archaeological Ethnography.* University of Michigan, Museum of Anthropology, Anthropological Papers 82. Ann Arbor.

Patrick, L. L.
1985 Agave and Zea in highland central Mexico: The ecology and history of the Metepantli. In *Prehistoric Intensive Agriculture in the Tropics*, edited by Ian S. Farrington, 2:539–546. British Archaeological Reports, Oxford.

Price, B.
1971 Prehispanic irrigation agriculture in nuclear America. *Latin American Research Review* 6(3):3–60.

Renfrew, C.
1972 *The Emergence of Civilisation: The Cyclades and the Aegean in the Third Millennium B.C.* Methuen, London.

Sanders, W. T.
1976 The agricultural history of the Basin of Mexico. In *The Valley of Mexico*, edited by E. R. Wolf, 161–178. University of New Mexico Press, Albuquerque.

Scarborough, V. L., and B. L. Isaac (editors)
1993 *Economic Aspects of Water Management in the Prehispanic New World.* JAI Press, Greenwich, Connecticut.

Smith, B. D.

1997 The initial domestication of *Cucurbita pepo* in the Americas 10,000 years ago. *Science* 276:932–933.

Smith, C. E.

1986 Preceramic plant remains from Guilá Naquitz. In *Guilá Naquitz: Archaic Foraging and Early Agriculture in Oaxaca, Mexico*, edited by K. V. Flannery, 265–274. Academic Press, New York.

Spores, R.

1965 The Zapotec and Mixtec at Spanish Contact. In *Archaeology of Southern Mesoamerica*, edited by Gordon R. Willey, 962–987. *Handbook of Middle American Indians*, Vol. 3, R. Wauchope, general editor, University of Texas Press, Austin.

Stark, B. L., L. Heller, and M. A. Ohnersogen

1998 People with cloth: Mesoamerican economic change from the perspective of cotton in south-central Veracruz. *Latin American Antiquity* 9:7–36.

Steward, J. H.

1949 Cultural causality and law: A trial formulation of the development of early civilization. *American Anthropologist* 51:1–27.

1955a Introduction. The irrigation civilizations: A symposium on method and result in cross-cultural regularities. In *Irrigation Civilizations: A Comparative Study*, edited by J. H. Steward, 1–5. Pan American Union, Social Science Monographs 1, Washington, DC.

1955b Some implications of the symposium. In *Irrigation Civilizations: A Comparative Study*, edited by J. H. Steward, 58–78. Pan American Union, Social Science Monographs 1, Washington, DC.

Wilken, G. C.

1987 *Good Farmers: Traditional Agricultural Resource Management in Mexico and Central America*. University of California Press, Berkeley.

Winter, M. C.

1974 Residential patterns at Monte Albán, Oaxaca, Mexico. *Science* 186:981–987.

Wittfogel, K. A.

1955 Developmental Aspects of Hydraulic Societies. In *Irrigation Civilizations: A Comparative Study*, edited by J. H. Steward, 43–52. Pan American Union, Social Science Monographs 1, Washington, DC.

1972 The hydraulic approach to Pre-Spanish Mesoamerica. In *The Prehistory of the Tehuacan Valley, Vol. 4: Chronology and Irrigation*, edited by F. Johnson, 59–80. University of Texas Press, Austin.

1981 *Oriental Despotism: A Comparative Study of Total Power*. Reprinted. Vintage Books, New York. Originally published 1957.

Wolf, E. R., and A. Palerm

1955 Irrigation in the Old Acolhua Domain, Mexico. *Southwestern Journal of Anthropology* 11:265–281.

Agricultural Intensification, Water, and Political Power in the Southern Maya Lowlands

Lisa J. Lucero

Ancient tropical societies, such as the Maya, are often relegated to the unknown or mysterious or, worse yet, are seen as a result of outside influences because of the traditional bias in anthropology of largely focusing on civilizations in temperate areas. Wittfogel, for example, in his classic *Oriental Despotism* (1957), argued that complex societies are underwritten by an agricultural base supported by irrigation. Because the Maya are not known to have built large-scale irrigation systems, he claimed that the Maya were a "marginal agrarian" (Wittfogel 1957:182) despotic society since the karstic topography of the Maya lowlands is "unsuitable for irrigation agriculture" (Wittfogel 1957:184). In another classic piece, Meggers (1954) presented a model of environmental determinism in which she categorized tropical zones as not suitable to support civilizations because of poor agricultural soils. Meggers based this assumption on her studies in the Amazon jungle, where many areas of tropical forests are poorly suited for intensive agriculture (though recent scholarship indicates otherwise; see Erickson 2003). Consequently, Maya civilization arose as the result of outside influences and "did not last" but witnessed "700 years of decline" (Meggers 1954:819) because conditions were not suitable to support a complex society. As recently as 1998, Meggers was still "struck by the seeming contradiction between the complexity of Maya culture and the relatively low agricultural potential of their environment" (Meggers 1998: xii). In fact, the southern Maya lowlands have a high percentage of mollisols, which are "considered by agronomists to be among the world's most important, naturally productive soils, with yields unsurpassed by other unirrigated areas" (Fedick 1988:106). Only 1 percent of the world's tropical soils are mollisols, yet they are the dominant soil type in the southern Maya lowlands, albeit in a relatively dispersed manner. The distribution of good agricultural land, in turn, affected settlement decisions and political histories.

There is no doubt, however, that tropical settings present unique conditions when compared with temperate regions, especially when attempting to account for sociopolitical complexity. "Limitations" in tropical areas include some or all of the following: endemic diseases (e.g., hepatic schistosomiasis), seasonal water quality problems, and the lack of beasts of burden and roads. Other parasites and bacteria also pose problems, as well as parasite-ridden flies and malaria-bearing mosquitoes that "flourish in environments disturbed by humans" (Miksic 1999:173). Standing, stagnant water provides ideal conditions for parasites, water-borne diseases (e.g., diarrhea and cholera), and pests to proliferate. Temperature variation throughout the year in the tropics is not extreme enough to kill off many pests. As a matter of fact, the more densely settled a tropical area is by humans, the more insects and other diseases spread and multiply. The increased reliance on intensive agriculture exacerbates this situation because land clearing creates conditions (open areas) for stagnant water to collect. Water quality issues are somewhat ironic in the humid tropics because people can lose 8 to 10 liters per day sweating and as a consequence need to drink more water than people living in temperate zones (Bacus and Lucero 1999). Because of these and other factors, Miksic goes so far as to suggest that "the unique problems connected with maintaining healthy drinking water supplies in a tropical environment help to explain why early urbanization in Indonesia, and in the tropical areas of the world in general, is relatively rare" (Miksic 1999:171).

Despite these challenging conditions, or because of them, two factors stand out when attempting to illuminate the development of sociopolitical complexity in tropical settings: (1) seasonal water issues are critical in tropical areas like the southern Maya lowlands; and (2) as with almost anywhere else that complex societies emerge, plentiful agricultural land is necessary to feed people and fund the political economy. How did these factors articulate in Classic Maya political histories (c. AD 250–950)? I attempt to address these issues in the remainder of this chapter. A major assumption with which I begin is that although there are numerous material and social conditions worldwide, there exist only a limited number of political responses or choices (Friedman and Rowlands 1978:241). In tandem with increasing political complexity is the need for surplus to fund an expanding political economy and feed people. As the papers in this volume indicate, in many cases, if not most, agricultural intensification provides the means for surplus production, which in turn provides the material basis for wealth differentiation. Political power—the ability to exact tribute—is another matter entirely. Agricultural intensification in and of itself is inadequate to account for the emergence of political centralization (see Erickson, this volume), at least in the Maya case. Other factors need to be taken into consideration, a few of which I focus on here: the distribution of people across the landscape, subsistence technology (e.g., reservoirs), means of transporting staples, and storage. I explore these factors in the southern Maya lowlands and discuss the relationship of agricultural intensification, water issues, and Classic Maya political power.

CLASSIC MAYA FARMING LIFE

In brief, Classic (c. AD 250–950) Maya farmers were rainfall dependent, dispersed across the landscape near centers and in hinterland areas, and largely economically self-sufficient at the community level (Lucero 2001). The Maya relied on scattered and small-scale subsistence systems because resources were diverse and dispersed due to variability in soil and rainfall. They did not have beasts of burden and lacked extensive road systems. As Shaw (2001) demonstrates, nearly all *sacbeob* (raised roads) in the southern Maya lowlands are 1 km or less in extent and are located within centers to connect architectural complexes. Folan, Marcus, and Miller (1995), however, note what may appear to be sacbeob between the regional center of Calakmul, Mexico, and outliers. Sacbeob extend c. 10 km from both El Mirador and Calakmul toward each other (c. 38 km distant), though the remainder of the connecting sacbe, if it exists, is not readily apparent. A sacbe also may connect El Mirador and Nakbe, located c. 13 km apart. Beyond these two cases, Caracol is the only southern Maya site with a noticeable road system; sacbeob up to c. 8 km link the site core with outlying building complexes (Chase and Chase 1994). Basically, the Maya built sacbeob within centers and seldom otherwise. Consequently, the transportation of bulky staples and goods over long distances was not feasible in areas lacking rivers (e.g., Tikal, Caracol, Calakmul, and others), especially since Maya bearers could not travel more than a few days without eating all the food they were carrying (Drennan 1984). The transport of small luxury items, however, was feasible (e.g., manufactured jade and obsidian items). People living in centers near rivers faced problems as well; raging rivers in the rainy season and low and murky waters in the dry season undoubtedly affected canoe travel.

SEASONAL LANDSCAPE AND SETTLEMENT

The southern Maya lowlands encompass a mosaic of resources and political histories. The distribution of people across the landscape is important because where people live affects the ability of political leaders to communicate with potential contributors and to organize the building of public works (Roscoe 1993). Densely settled and nucleated farmers are much easier to reach

(Carneiro 1970; Gilman 1981) than dispersed people. Where people live has to do, of course, with the location and nature of resources crucial for daily survival—land and water. Before detailing the significance of land and water, I want to briefly address prehispanic population size. Several scholars estimate high population numbers and densities, especially in the Late Classic period (c. AD 550–850) (see Rice and Culbert 1990). Scarborough et al. (2003:xvi), based on this assumption, "suggest that absolute village autonomy during the Classic period was not an option." As I have discussed and summarized elsewhere (Lucero 1999a), however, evaluating prehispanic population size and density is a challenge for several reasons (Ford 1991b). First, there is the lack of precise chronologies. We also need to take into account seasonal mobility and migration (discussed below) and structure function—that is, percentage of domiciles versus specialized buildings including workshops, kitchens, storage facilities, religious structures, sweat houses, field houses, and administrative buildings. Further complicating matters are "invisible" mounds, which are difficult to identify in the archaeological record (Johnston 2004; Pyburn 1997). Whatever the actual numbers of prehispanic Maya, there is little doubt that the Late Classic period witnessed the highest population size and density (Rice and Culbert 1990) and that the largest number of Maya lived near fertile land (Fedick and Ford 1990).

The best agricultural soils are found throughout the southern lowlands, but in a mosaic pattern (Fedick 1988, 1996; Fedick and Ford 1990; Ford 1986, 1991a, 1996; Sanders 1977). Soil variability relates to soil parent material, soil fertility, workability, root zone, drainage, slope, erosion factors, and other features (Fedick 1995). Based on these factors, Fedick (1988, 1996) has developed a hand-cultivation land-capability classification system consisting of five soil classes where "I" is the best and "V" is the worst. He found the highest residential density on Class II soils, followed by Class I alluvium along rivers. Interestingly, alluvial soils (Class I) are not the most densely settled, as they typically are in temperate regions. Because of the varied distribution of water and land, many Maya farmers lived dispersed across the hinterland (de Montmollin 1995; Fedick 1988, 1995, 1996; Ford 1986; Lucero 2001; Puleston 1983). Consequently, hinterland settlement consists largely of relatively scattered solitary mounds and patio groups, or farmsteads (Ashmore 1995; Gonlin 1994). Settlement maps also show homesteads clustered near centers (Fedick 1988; Ford 1986; Puleston 1983). Drennan (1988) and Killion (1990) suggest that this settlement pattern indicates intensive agriculture where farmers lived dispersed and separated by intensively utilized fields (Robin 2002).

Rainfall also varies throughout the southern lowlands and can range anywhere from 1,300 to 3,700 mm per year. Further, the beginning of the rainy season can vary up to five months in any given area, and rainfall patterns vary year to year (Gunn et al. 2002), both of which affect agricultural schedules. For example, if the rainy season starts later than usual, the seeds that have been planted in anticipation of timely rain will rot. If rains come sooner than predicted, seeds will not germinate. Also, if the rains start even earlier, farmers would not have burned their *milpas* (fields), which they cannot do if brush is wet. The annual wet-dry seasonal round also affected agricultural practices and settlement patterns, especially if dry season water needs resulted in farmers moving, at least temporarily, near water (e.g., artificial reservoirs in centers) (Lucero 1999b). During the annual six- to seven-month rainy season there was plenty of water for everyone, which was just as well since many agricultural tasks occur during this time (Atran 1993). The dry season was another matter. For all intents and purposes many parts of the southern Maya lowlands are transformed into a green desert during the dry season; temperatures and humidity are high, and it does not rain for four months. Access to water at such times was critical, even though it was the agricultural downtime. Many Maya congregated at water sources, both natural and artificial, at centers. This situation provided just the means the politically ambitious needed to acquire power over others, as I have described elsewhere (Lucero 1999b, 2006) and will briefly summarize below. The point is that the location of agricultural land indeed was critical, but potable water was as or even more critical, at least seasonally, and influenced where people lived and to what degree and when they interacted with each other.

Subsistence practices, along with seasonal water issues and soil variability, also impacted residential and seasonal mobility (de Montmollin 1995:189; Ford 1996; Lucero 1999b; Webster 1997; Webster and Kirker 1995). For example, Reina (1967), based on his ethnographic work among Maya farmers at Lake Petén Itzá, Guatemala, describes how farmers abandon their residences every three to four years to find new land. Farmers also may have used field houses during the height of the agricultural season, when they needed to weed, protect their fields from pests, harvest their crops, and store their foodstuffs (Atran 1993). Faust (1998:56–57) found this to be the case in her account of Maya farmers of Pich, in southeastern Yucatán, Mexico, where each family has rights to several *rancherías*. Prehispanic Maya also probably practiced residential and seasonal mobility (Webster 1997). Finally, several Maya archaeologists, based on household and community research, suggest that ancient Maya farmsteads were basically economically self-sufficient (Freter 1994; Hayden 1994; Lucero 2001), a factor that needs to be taken into account when discussing political systems. In other words, families were likely mobile and had all the resources necessary to survive without outside (elite or royal) interference—land, water, and raw materials for manufacturing farming tools, ceramic vessels, and other items. Something, then, had to bring people to centers where kings could tap their surplus labor and goods, which I argue was both water and ritual. I have detailed the role of ritual elsewhere (Lucero 2003), as have others (Marcus, this volume; Scarborough 1998, 2003). Here I focus on water and other material factors.

Seasonal water was the linchpin in the development and success of Maya kings (Lucero 1999b, 2003, 2006). Consequently, even if rulers owned or controlled productive land, they likely only could monitor land in the vicinity of centers. Outside the immediate area it would have been difficult to control access to land and extract payment for its use, whether or not they owned it. For land they owned or controlled beyond center environs, they probably had to grant laborers a portion of the crop as payment for working their land rather than with corvée labor. Hinterland farmers were also out of reach of the political net and could not easily be coerced to contribute to political coffers unless there was a way to bring farmers to royal centers.

SUBSISTENCE TECHNOLOGY

Different environmental settings are suitable for varied kinds of subsistence technology, most of which is handled at the household or community level (Johnson and Earle 2000). However, even if this is the case, if people are tied to the land because of subsistence technology, they are less likely to leave if faced with the choice of leaving or paying tribute—and they often have no choice but to do the latter (Gilman 1981). Consequently, households and/or communities can be relatively socially and economically self-sufficient; but if it is difficult to leave behind their means of support (e.g., plowed fields, fish ponds, canals, dams, terraces, and storage facilities), they are more accessible and more susceptible to political machinations. Political leaders, in other words, can interact with subjects since people are tied to the land; farmers would lose all if they chose independence and left their holdings rather than remain and pay tribute.

In the Maya area, agricultural and water systems are scattered, mirroring the distribution of resources and homesteads, and are largely small-scale with some exceptions. Maya farmers used several subsistence strategies including intensive house gardening, short-fallow infield (intensive) and long-fallow outfield (extensive) farming, terraces, dams, canals, raised fields, and drainage systems (Dunning et al. 1998; Fedick 1994; Flannery 1982; Harrison and Turner 1978; Killion 1990). Some intensive agricultural strategies might not have left obvious evidence. For example, to date archaeologists have not found any evidence for subsistence systems (terraces, dams, and so forth) near Tikal, the largest Maya center (Harrison 1993). At least two crops a year provided for families year-round. Average farmers either used storage bins near their fields or collected corn every few days from their fields when needed, as do the Lacandon farmers of Chiapas at present (McGee 1990:36). They turn the ears of corn downward to prevent damage from rain and pests such as birds. The most common and obvious large-scale subsistence technology was for water storage during the annual drought (4–6 months) (Lucero 1999b, 2002; Scarborough 1993, 1996). The Maya built well-engineered reservoirs with channels and drainage ditches (Scarborough 2003; Scarborough and Gallopin 1991).

Finally, the Maya also had access to diverse flora and fauna. Unlike their counterparts in temperate zones, the lack of large nucleated settlements meant that uninhabited jungles interspersed among farmsteads and centers were not necessarily denuded of wild animals, nuts, berries, medicinal flora, and fruits (Bacus and Lucero 1999).

MEANS OF TRANSPORTING STAPLES AND STORAGE

The means of transporting staples is another critical factor to take into account when attempting to understand political centralization. Surplus goods are only useful if they can reach political coffers, be stored (D'Altroy and Earle 1985), and be distributed. The same goes for labor in that the political elite have to be able to reach people to appropriate their surplus labor and goods.

Without beasts of burden, extensive road systems, and wheeled carts in the southern Maya lowlands, staples (maize, beans, and squash) had to be transported by human bearers. As mentioned earlier, this meant that Maya bearers could not travel more than a few days without eating all the food they were carrying (Drennan 1984). Maya in all probability, however, participated in exchange at *local* markets (Scarborough and Valdez 2003). In areas without lakes and rivers, travelers had to know the location of *aguadas* (rain-fed natural sinkholes), though water in most aguadas would have evaporated during the course of the dry season (Scarborough 1996). And travel is really only logistically feasible during the dry season. In the rainy season foot travel is difficult (Lucero in press): sloping terrain can be slippery and dangerous because of saturated soils, clay, and wet ground cover (leaves and debris); low-lying areas tend to be inundated and swamplike. Some goods could have been transported on rivers via canoes. However, only people near rivers would benefit and not during the height of the rainy season (turbulent) or dry season (low waters). For example, Houston (1998) describes the Usumacinta River at Piedras Negras, Guatemala, during the rainy season as dangerous and during the dry season as low, murky, and disease-ridden. Taken together, these conditions suggest that staples were largely grown and eaten in the vicinity of the fields in which they were grown. Prestige goods (cacao beans, jade, obsidian, feathers, cotton mantas, and others), though, were exchanged farther afield.

In addition, there is a lack of large-scale centralized storage facilities at Maya centers. Small-scale storage existed (*chultuns*), but it clustered at elite and royal residences and likely provided food only for household members. Chultuns typically are shoe-shaped chambers dug into the relatively soft and porous limestone bedrock. Puleston (1971) conducted an experiment to determine if they could store foods; they are too porous for water storage in the southern Maya lowlands unless sufficiently lined to prevent seepage, and most are not. He found that they would not store foods long because of the heat and humidity. However, dried or smoked maize and other staples can be stored longer (Reina and Hill 1980). Dahlin and Litzinger (1986) suggest that chultuns instead were used for fermenting alcoholic beverages or for pickling. They also note that the inconsistent distribution of chultuns indicates a specialized function rather than a purely economic and domestic one (e.g., for feasts and ceremonies). For example, Ford (1991b:Table 1) has noted that the highest density of chultuns is near or at centers, especially at elite residences. The lack of large-scale storage facilities indicates that rulers would not have been much help in a crisis caused by, for example, crop failures and food shortages. However, since crops were grown dispersed throughout the landscape, so, too, were the risks.

While it is likely that Classic Maya rulers could not amass large amounts of food to feed people for any length of time, they could still integrate them through sponsoring key ceremonies (e.g., water rites) and feasts, at which time they could distribute exotic wealth items to favored underlings, such as obsidian items, jade objects, and polychrome ceramics, most of which were acquired via long-distance exchange (Rice 1987). And even if Maya kings relied on wealth finance, they still had to acquire and distribute goods, which they accomplished at the local level. There is no clear evidence for markets "in which the tribute goods used as state payment are sold for subsistence goods" (D'Altroy and Earle 1985:188), though as mentioned earlier, local markets probably were used. Transportation issues, however, would still have limited the type and amount of items exchanged. Consequently, while intensive agriculture provided a surplus to fund the political economy, only the fruits of land in the immediate vicinity of centers were readily available. Agricultural surplus, together with seasonal water issues (royal water rites and centrally located reservoirs), however, provided the key means to support Maya rulership.

POLITICAL POWER IN THE SOUTHERN MAYA LOWLANDS

In this section I focus on the role that various resources and concomitant intensive agricultural strategies played in Classic Maya politics. Of course other factors were critical, such as religious and social mechanisms; here I focus on material features.

Table 11.1. Late Classic Maya centers (c. AD 550–850).

Scale	Minor Center	Secondary Center	River Regional Center	Nonriver Regional Center
Distribution of resources	River, extensive alluvium	River, uplands with small pockets of dispersed agricultural soils	River, concentrated alluvium	No rivers or lakes, uplands with large tracts of dispersed agricultural land and reservoirs
Political economy	No tribute	Some tribute	Tribute	Tribute
Annual rainfall (mm)[a]	2,160	2,200–2,850	Copán: 1,300 Palenque: 3,700	1,670–2,100
Seasonal water issues	Annual inundation and recession	Varies	Noticeable	Noticeable
Subsistence systems	None	Small-scale	Large-scale	Large-scale
Settlement patterns and density	Relatively dispersed and low settlement density;100–150[b] str/sq km	Slightly higher center density than hinterlands; 275[c] vs. up to 145 str/sq km	High center density vs. hinterlands; 1,449[d] vs. 28–99 str/sq km	High center and hinterland density; 235–557[e] vs. 39–313 str/sq km
Means of transportation	River and human bearers	River and human bearers	River and human bearers	Human bearers
Means of distribution	Interpersonal, local ritual events, local markets	Interpersonal, secondary royal events, local markets	Interpersonal, primary royal events, local markets	Interpersonal, primary royal events, local markets
Storage	None	Small-scale	Small-scale	Small-scale
Examples	Saturday Creek, Barton Ramie	Lamanai, Yalbac, Piedras Negras, Quiriguá, Seibal, Yaxchilan, Xunantunich, Cuello, Bonampak	Copán, Palenque	Tikal, Caracol, Calakmul

[a]Neiman 1997:Table 15.1.
[b]Lucero et al. 2004; Rice and Culbert 1990:Table 1.1.
[c]Ashmore 1990; Loten 1985; Rice and Culbert 1990:Table 1.1; Tourtellot 1990.
[d]Rice and Culbert 1990:Table 1.1; Webster and Freter 1990.
[e]Culbert et al. 1990; Folan, Marcus, Pincemin et al. 1995. Bajo (seasonal swamps) settlement likely accounts for lower densities.

The concept of environmental heterogeneity has been bandied about for decades (Sanders 1977). It is only relatively recently, however, that archaeologists have considered this diversity when attempting to explain the rise of complexity in the Maya lowlands (Dunning 1995; Dunning et al. 1998; Fedick and Ford 1990; Lucero 2003). This variability is reflected in the varied political systems and histories. I particularly focus on the Late Classic period (c. AD 550–850) because this was the period when Maya rulers attained their height of power. As Table 11.1 illustrates, the scale and degree of political power vary and are influenced by resource and settlement distribution and seasonal water issues and needs. I have defined these center types in greater detail elsewhere (Lucero 2002, 2003, 2006). I briefly summarize the key factors of each center type—minor, secondary, and regional—and add additional features that relate more directly to the topic of this chapter (means of transportation, distribution of goods, and storage). Not all centers fit neatly, of course, but in most cases when they do not, their political stories can be explained by local and historical circumstances (e.g., Dos Pilas, Uaxactun, El Mirador, and others).

Minor Centers

Barton Ramie and Saturday Creek are located on the Belize River on its relatively broad alluvium (1–2 km wide) on the eastern periphery of the southern Maya lowlands (Figure 11.1). Annual rainfall at Saturday Creek is 2,160 mm, similar to that of Barton Ramie (Figure 11.2). The rich river terrace soils are excellently suited for cash crops like cacao and cotton, not to mention maize, beans, and squash. Cacao and cotton require specific conditions and cannot grow in rocky upland soils but require saturated or moist soils (Gómez-Pompa et al. 1990). The Maya would have had plenty of water to last throughout the year, especially since they benefited from annual runoff from Guatemala and western Belize. The downside is that runoff also resulted in flooding and the deposition of clayey soils, which can remain saturated for most of the year. These communities comprised relatively low densities of dispersed farmsteads (e.g., 100–151 structures per sq. km).[1] Elites, or wealthy landowners, were unable to collect much tribute, if any, because they could not restrict extensive alluvium and politically integrate dispersed farmers. Farmers did not build conspicuous subsistence systems because water was plentiful but instead relied on the annual flooding and subsiding of rivers for agriculture (recession agriculture). Elites sponsored local small-scale public rituals and feasts at small temples and plazas and organized the construction of public works to promote solidarity (Arie 2001; Lucero et al. 2004).

The Maya in these areas could travel the rivers, except during the height of the rainy season, when water was too turbulent. Because the Belize River is

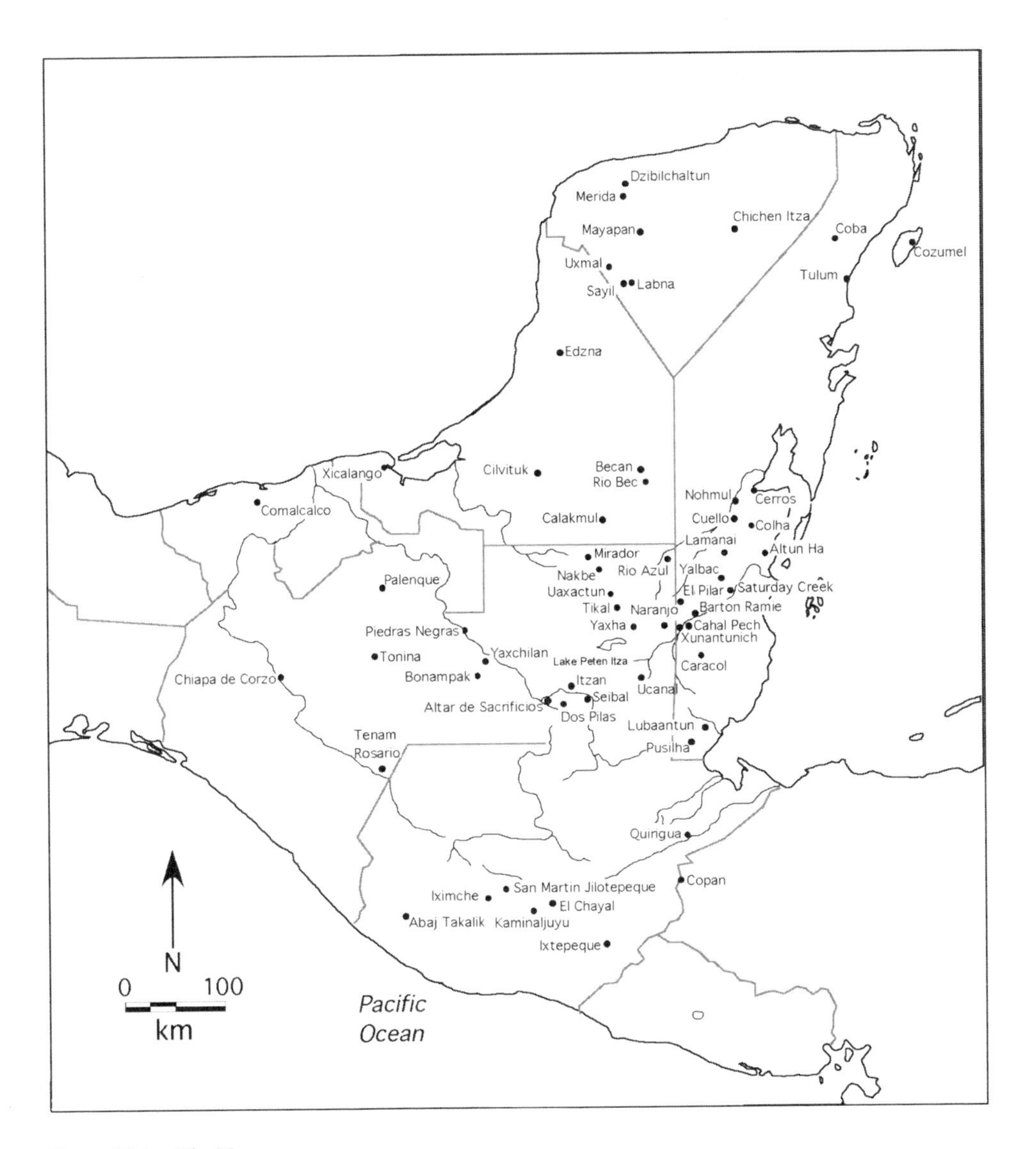

Figure 11.1. The Maya area.

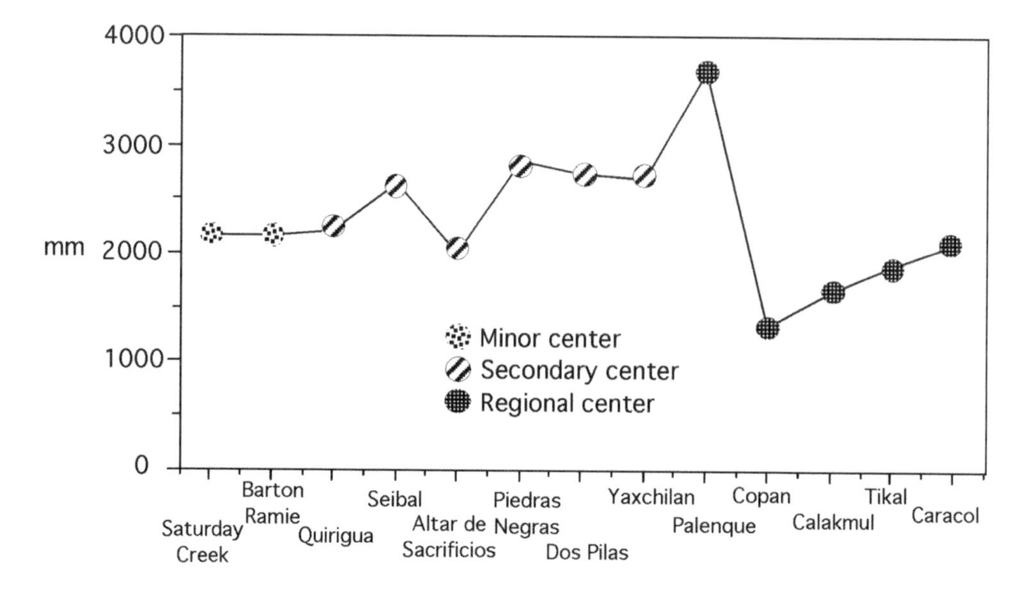

Figure 11.2. Annual rainfall in the southern Maya lowlands.

a major water catchment, the Maya could travel downriver toward the coast during the dry season. Elites could afford to obtain prestige items through their participation in the elite interaction sphere (long-distance exchange), some of which they gave as gifts to non-elites at local ceremonies for services rendered, such as working their land. Storage facilities are nonexistent; the limestone bedrock is buried deep under silt deposits, making chultun construction impractical. The rich alluvium, however, likely allowed farmers to plant crops throughout the year, as Mennonite farmers do at present, to offset the lack of storage facilities.[2] This also meant that farmers were more self-sufficient than their counterparts elsewhere because they did not rely on royal stores of water or capital to maintain subsistence systems since they had access to food and water the entire year.

Secondary Centers

Centers including Lamanai, Yalbac, Seibal, Yaxchilán, Quiriguá, Bonampak, Cuello, Piedras Negras, Xunantunich, and others are found along rivers in upland areas with limited or dispersed pockets of agricultural land.[3] Annual rainfall ranges from 2,220 to more than 2,800 mm. At most secondary centers, residents lived above rivers on ridges and hills, which meant that during the rainy season saturated hillsides influenced building decisions and agricultural practices (Turner 1974). Because of the dispersed pockets of agricultural land, Maya farmers relied on scattered small-scale subsistence systems including aguadas, dams, canals, and drainage ditches (Dunning et al. 1997; Fedick 1994). Settlement is typically dense near centers (e.g., up to 275 structures/km^2) and less so in hinterland areas (e.g., 145 structures/km^2). The uneven distribution of subsistence systems suggests their lesser political role. Maya kings, however, likely monopolized river trade routes (e.g., by controlling access to harbors and canoes) and acquired and distributed exotics such as obsidian and jade. Rulers could not exact as much tribute from subjects as primary rulers, because they could not completely control access to relatively small and dispersed pockets of agricultural land and small-scale water systems, not to mention relatively scattered farmers. Kings, however, may have provided the means to repair subsistence features damaged by rain in the immediate center environs and conducted integrative events (e.g., ceremonies and feasts) to acquire what tribute they could. They also participated in the royal interactions sphere (e.g., intercenter alliances, marriage, visitations, and warfare, and the adoption of writing, royal dynastic rites, and in some cases, the use of emblem glyphs) (Marcus 2003).

Similar to minor centers, the Maya in these areas could travel the rivers, except during the height of the dry and rainy seasons (e.g. Houston 1998). In addition to elites handing out exotics in exchange for services rendered by commoners, kings performed royal rites and showered gifts on lower-level elites and commoners. Storage facilities (chultuns) are small-scale and typically found only at elite and royal residences. Their size and distribution suggest that they could not supply much more beyond the immediate household.

River Regional Centers

The river centers of Palenque and Copán are located along rivers with concentrated alluvium, more in line with archaic states in temperate areas (e.g., China, Egypt, Mesopotamia, and western Andean South America) (Marcus 1998). For example, alluvial soils surrounding Copán are found within a 24-km^2 area (Webster 1999). Rulers collected tribute from densely settled farmers because of their ability to monopolize nearby concentrated resources. Kings also likely dominated trade with highland areas for jade and obsidian (Fash 1991). Annual rainfall is both above and below that found at nonriver regional centers—1,300 mm at Copán and 3,700 mm at Palenque. Copán's occupants, because of low annual rainfall, built reservoirs, which were managed and controlled by rulers (Fash 2005; Fash and Davis-Salazar in press). Inhabitants of Palenque lived in a hilly area overlooking fertile plains to the north. They built aqueducts and canals to drain water away from the center—a not-too-surprising fact given the high annual rainfall and the presence of more than 50 springs and streams in the site core (French 2002). Farmers also constructed terraces on the hillsides and possibly irrigation canals on the plains (Barnhart 2001:101). Settlement is typically dense around centers (1,449 structures/km^2 at Copán and 643 structures/km^2 at Palenque) and noticeably less dense in areas beyond the alluvium (28–99 structures/km^2 in rural Copán) (Barnhart 2001:Table 3.1; Rice and Culbert 1990:Table 1). At both centers the reliance on larger-scale subsistence systems suggests a greater need for capital—royal funds—to maintain them or at least rebuild them when they were damaged by heavy rains, flooding, or hurricanes (Lucero 2006).

Similar to secondary centers, Maya in these areas could travel the rivers, except during the height of the dry and rainy seasons. Again, exotics were given as gifts in exchange for services rendered by commoners and lower-level elites in large-scale royal ceremonies at open plazas at the foot of temples and palaces as part of the political legitimation process. As at secondary centers, storage facilities were small-scale and largely concentrated at elite and royal residences.

Nonriver Regional Centers

Regional centers such as Tikal, Calakmul, and Caracol are similar in scale to regional river centers but differ in several significant ways. They are located in upland areas with large pockets of dispersed fertile land but without permanent water sources (lakes and rivers). Figure 11.3 illustrates the fact that some of the largest Maya centers are not located near lakes or rivers. The rich land, however, supported many people, as well as political institutions. To deal with the lack of water, rulers organized the construction of large-scale and sophisticated reservoirs that collected enough water during the rainy season to last through the four-to-six-month dry season (Scarborough and Gallopin 1991). Reservoirs are located next to temples and palaces, making it easier to conduct large and public water rites to attract supporters, control their access, and highlight the prosperity of the king. The need for an adequate water supply is particularly crucial in these areas since rainfall is typically less than at secondary and regional river centers. For example, annual rainfall at secondary centers ranges from 2,200 to more than 2,800 mm; at Tikal it is just under 1,900 mm, at Calakmul just under 1,700 mm, and at Caracol 2,100 mm (see Figure 11.2). Rulers collected tribute from thirsty farmers in exchange for access to artificial reservoirs during the dry season, particularly from January through April or May (Folan, Marcus, Pincemin et al. 1995; Ford 1996; Lucero 1999b; Scarborough 1993, 1996; Scarborough and Gallopin 1991). Maya kings were responsible for providing enough potable water to last through the dry season, not only through organizing the continual maintenance to keep the reservoirs clean (Ford 1996) but also for performing rites necessary to propitiate key ancestors and deities such as Chac, the rain god (Scarborough 1998). Consequently, rulers were able to draw in farmers from hinterland areas (Lucero 1999b). This strategy was critical for political survival since in the immediate areas beyond center environs, people lived scattered throughout the landscape, making it challenging for rulers to organize work parties, feasts, and ceremonial events and extract surplus. Settlement, however, is relatively dense both around centers (e.g., 235–557 structures/km^2) and hinterland areas (e.g., up to 313 structures/km^2) (Rice and Culbert 1990:Table 1), likely reflecting a more mobile dry-season settlement pattern whereby farmers had more than one residence—near both hinterland fields and centers.[4]

Figure 11.3. The location of southern lowland Maya centers in relation to lakes and rivers. Palenque and Tonina are located along major tributaries of the Usamacinta River, which is not shown on the map.

The lack of major rivers resulted in a reliance on human bearers to bring in exotics; plentiful fertile land, though, provided for inhabitants. Rulers, however, funded large-scale rituals in public plazas and temples to attract and integrate farmers from the immediate area and beyond, at which they distributed exotics to underlings and supporters. Storage facilities are small-scale and cluster at elite and royal residences.

DISCUSSION AND CONCLUDING REMARKS

D'Altroy and Earle (1985) discuss the crucial significance of central storage in political centralization. This was the case for the Maya. Rather than the storage of agricultural surplus or other goods, however, some of the most powerful Maya rulers stored water, particularly during annual drought—or, in the case of Palenque, provided "stored" capital to repair damaged subsistence systems and agricultural fields. Agricultural surplus was still important, however, to support royals and their retinue of family members, retainers, military and religious specialists, and artisans, all who lived in centers surrounded by fertile land. The stored food, though, likely came from land they owned in the vicinity of their grand homes. Kings in areas with dispersed agricultural land (e.g., Tikal) attracted additional people during the dry season when farmers needed water for everyday purposes—not to mention the opportunity to gather for feasts, ceremonies, ball games, social interaction, and exchange. In return Maya commoners contributed tribute in the form of labor necessary to build the royal temples, palaces, and ball courts; to transport exotics long distances; and to manufacture utilitarian and craft items.

We do not find evidence for political power at minor centers because water and land were plentiful. As a result, farmers were largely self-sufficient year-round rather than only seasonally as in other areas (i.e., those with kings). As members of a larger society, however, they participated in integrative social and religious events at centers sponsored and organized by wealthier landowners. As noted earlier, alluvial soils are less densely settled than those in temperate areas of the world. It is possible that these areas served as large, private, and/or royal estates (de Montmollin 1989:203). However, the question of whether or not Maya farmers had the option to leave has to be addressed. Unlike in many temperate zones where farmers had little option but to stay (Gilman 1981), I think some Maya did have choices. They could flee into foothill and upland areas away from rivers near small aguadas, springs, and streams. This fact might account for low settlement densities in alluvial areas rather than the estate model; in other words, they comprised independent and self-sufficient communities. Of course, eventually in many areas options became increasingly limited as time passed and the agricultural mosaic landscape filled up to the point that small-scale subsistence systems became inadequate to supply the growing numbers of people. They then began to rely more on large-scale systems (e.g., artificial reservoirs).

At secondary centers the relatively scattered subsistence systems and food storage facilities were small-scale, as was their political power. Kings at regional centers centralized labor and goods because of what they provided in return—dry-season water or capital to offset weather damage. Clearly, it was only in areas with the greatest seasonal water issues and with the largest amounts of fertile land (dispersed or concentrated) that the most powerful polities emerged.

In response to those who have claimed that the Maya area is unsuitable to support political and social complexity because of the lack of major irrigation systems (Wittfogel) and fertile land (Meggers), I think the evidence speaks for itself. The Maya, like everyone else, adopted subsistence practices and technologies that best suited their needs in a tropical setting.

To conclude, I again pose the question -- How did water and agricultural surplus factor in Classic Maya political histories (c. AD 250–950)? Agricultural intensification provided a surplus. To be used for political purposes, however, it has to be acquired from people and harnessed to expand the political economy. Surplus is moot if its providers—farmers—are difficult to reach. In this situation means are necessary to bring people to rulers. The tropics, with all their unique traits, offered a way—seasonal water needs. Providing water during annual drought and conducting key water ceremonies furnished the means for the majority of rulers to draw farmers into centers, not to mention into their political folds. In return rulers made sure everyone had enough water when jungles became, for a time, green deserts.

NOTES

1. Because of the contentious issue of estimating population size and density, I use structure densities, a relative measure. Structure densities are also easier to compare among sites.

2. To quote a modernized Mennonite in Creole (in 1988), "Da Maya, de smarter dan we" (The Maya, they are smarter than we). Mennonite farmers only buy or lease land with a noticeable presence of Maya mounds because they know the ancient Maya lived on or near the best agricultural soils.

3. Yaxchilán, Piedras Negras, and perhaps other secondary centers, may in the future be defined as regional centers.

4. I do not claim that all farmers nucleated at centers during the dry season per se, but that many did. Some hinterland aguadas may have had enough water to last through the dry spell but, in most cases, not for many people. As mentioned, water in smaller aguadas likely evaporated long before the end of the dry season (Scarborough 1996).

REFERENCES

Arie, J. C.

2001 *Sun Kings and Hierophants: Geocosmic Orientation and the Classic Maya.* Unpublished master's thesis, Department of Sociology and Anthropology, New Mexico State University.

Ashmore, W.

1990 Ode to a dragline: Demographic reconstruction at Classic Quirigua. In *Precolumbian Population History in the Maya Lowlands*, edited by T. P. Culbert and D. S. Rice, 63–82. University of New Mexico Press, Albuquerque.

1995 Settlement archaeology at Xunantunich. In *Xunantunich Archaeological Report: The 1995 Field Season*, edited by R. M. Leventhal, 11–25. Report on file at the Department of Anthropology, University of California, Los Angeles; and at the Belize Institute of Archaeology, Belmopan.

Atran, S.

1993 Itza Maya tropical agro-forestry. *Current Anthropology* 34:633–700.

Bacus, E. A., and L. J. Lucero

1999 Introduction: Issues in the archaeology of tropical polities. In *Complex Polities in the Ancient Tropical World*, edited by E. A. Bacus and L. J. Lucero, 1–11. Archeological Papers of the American Anthropological Association Number No. 9. American Anthropological Association, Arlington, Virginia.

Barnhart, E. E.

2001 Palenque Mapping Project: Settlement and Urbanism at an Ancient Maya City. PhD dissertation, University of Texas, Austin.

Carneiro, R. L.

1970 A theory of the origin of the state. *Science* 169:733–738.

Chase, A. F., and D. Z. Chase

1994 Details in the archaeology of Caracol, Belize: An introduction. In *Studies in the Archaeology of Caracol, Belize*, edited by D. Z. Chase and A. F. Chase, 1–11. Monograph 7. Pre-Columbian Art Research Institute, San Francisco.

Culbert, T. P., L. J. Kosakowsky, R. E. Fry, and W. A. Haviland

1990 The population of Tikal, Guatemala. In *Precolumbian Population History in the Maya Lowlands*, edited by T. P. Culbert and D. S. Rice, 103–121. University of New Mexico Press, Albuquerque.

Dahlin, B. H., and W. J. Litzinger

1986 Old bottle, new wine: The function of chultuns in the Maya lowlands. *American Antiquity* 51:721–736.

D'Altroy, T. N., and T. K. Earle

1985 Staple finance, wealth finance, and storage in the Inka political economy. *Current Anthropology* 26:187–206.

de Montmollin, O.

1989 *The Archaeology of Political Structure: Settlement Analysis in a Classic Maya Polity*. Cambridge University Press, Cambridge, UK.

1995 *Settlement and Politics in Three Classic Maya Polities*. Prehistory Press, Madison, Wisconsin.

Drennan, R. D.

1984 Long-distance movement of goods in the Mesoamerican Formative and Classic. *American Antiquity* 49:27–43.

1988 Household location and compact versus dispersed settlement in prehispanic Mesoamerica. In *Household and Community in the Mesoamerican Past*, edited by R. R. Wilk and W. Ashmore, 273–293. University of New Mexico Press, Albuquerque.

Dunning, N. P.

1995 Coming together at the Temple Mountain: Environment, subsistence, and the emergence of lowland Maya segmentary states. In *The Emergence of Classic Maya Civilization*, edited by N. Grube, 61–69. Acta Mesoamericana 8. Verlag von Flemming, Möckmühl.

Dunning, N. P., T. Beach, P. Farrell, and S. Luzzadder-Beach

1998 Prehispanic agrosystems and adaptive regions in the Maya lowlands. *Culture and Agriculture* 20:87–101.

Dunning, N. P., T. Beach, and D. Rue

1997 The paleoecology and ancient settlement of the Petexbatun region, Guatemala. *Ancient Mesoamerica* 8:255–266.

Erickson, C.

2003 Historical ecology and future explorations. In *Amazonian Dark Earths: Origin, Properties, Management*, edited by J. Lehmann, D. C. Kern, B. Glaser, and W. I. Woods, 455–500. Kluwer Academic Publishers, Dordrecht, Netherlands.

Fash, B. W.

2005 Iconographic evidence for water management and social organization at Copán. In *Copán: The Rise and Fall of a Classic Maya Kingdom*, edited by E. W. Andrews and W. L. Fash, 103–138. School of American Research Press, Santa Fe.

Fash, B. W., and K. L. Davis-Salazar

in press Copan water ritual and management: Imagery and sacred place. In *Precolumbian Water Management: Ideology, Ritual, and Politics*, edited by L. J. Lucero and B. Fash. University of Arizona Press, Tucson.

Fash, W. L.

1991 *Scribes, Warriors, and Kings: The City of Copán and the Ancient Maya*. Thames and Hudson, London.

Faust, B. B.

1998 *Mexican Rural Development and the Plumed Serpent: Technology and Maya*

Cosmology in the Tropical Forest of Campeche, Mexico. Bergin and Garvey, Westport, Connecticut.

Fedick, S. L.

1988 *Prehistoric Maya Settlement and Land Use Patterns in the Upper Belize River Area, Belize, Central America*. PhD dissertation, Arizona State University, Tempe.

1994 Ancient Maya agricultural terracing in the upper Belize River area: Computer-aided modeling and the results of initial field investigations. *Ancient Mesoamerica* 5:107–127.

1995 Land evaluation and ancient Maya land use in the upper Belize River area, Belize, Central America. *Latin American Antiquity* 6:16–34.

1996 An interpretive kaleidoscope: Alternative perspectives on ancient agricultural landscapes of the Maya lowlands. In *The Managed Mosaic: Ancient Maya Agriculture and Resource Use*, edited by S. L. Fedick, 107–131. University of Utah Press, Salt Lake City.

Fedick, S. L., and A. Ford

1990 The prehistoric agricultural landscape of the central Maya lowlands: An examination of local variability in a regional context. *World Archaeology* 22:18–33.

Flannery, K. V. (editor)

1982 *Maya Subsistence: Studies in Memory of Dennis E. Puleston*. Academic Press, New York.

Folan, W. J., J. Marcus, and W. F. Miller

1995 Verification of a Maya settlement model through remote sensing. *Cambridge Archaeological Journal* 5:277–283.

Folan, W. J., J. Marcus, S. Pincemin, M. Domínguez Carrasco, L. Fletcher, and A. Morales López

1995 Calakmul: New data from an ancient Maya capital in Campeche, Mexico. *Latin American Antiquity* 6:310–334.

Ford, A.

1986 *Population Growth and Social Complexity: An Examination of Settlement and Environment in the Central Maya Lowlands*. Anthropological Papers No. 35. Arizona State University, Tempe.

1991a Economic variation of ancient Maya residential settlement in the upper Belize River area. *Ancient Mesoamerica* 2:35–46.

1991b Problems with evaluation of population from settlement data: Examination of ancient Maya residential patterns in the Tikal-Yaxhá intersite area. *Estudios de Cultura Maya* 18:157–186.

1996 Critical resource control and the rise of the Classic period Maya. In *The Managed Mosaic: Ancient Maya Agriculture and Resource Use*, edited by S. L. Fedick, 297–303. University of Utah Press, Salt Lake City.

French, K. D.
2002 *Creating Space through Water Management at the Classic Maya Site of Palenque, Chiapas, Mexico*. Unpublished master's thesis, University of Cincinnati, Cincinnati, Ohio.
Freter, A.
1994 The Classic Maya collapse at Copán, Honduras: An analysis of Maya rural settlement trends. In *Archaeological Views from the Countryside: Village Communities in Complex Society*, edited by G. M. Schwartz and S. E. Falconer, 160–176. Smithsonian Institution Press, Washington, DC.
Friedman, J., and M. J. Rowlands
1978 Notes toward an epigenetic model of the evolution of "civilization." In *The Evolution of Social Systems*, edited by J. Friedman and M. J. Rowlands, 201–267. Duckworth, London.
Gilman, A.
1981 The development of social stratification in Bronze Age Europe. *Current Anthropology* 22:1–23.
Gómez-Pompa, A., J. Salvador Flores, and M. Aliphat Fernández
1990 The sacred cacao groves of the Maya. *Latin American Antiquity* 1:247–257.
Gonlin, N.
1994 Rural household diversity in Late Classic Copán, Honduras. In *Archaeological Views from the Countryside: Village Communities in Complex Society*, edited by G. M. Schwartz and S. E. Falconer, 177–197. Smithsonian Institution Press, Washington, DC.
Gunn, J., R. T. Matheny, and W. J. Folan
2002 Climate-change studies in the Maya area. *Ancient Mesoamerica* 13:79–84.
Harrison, P. D.
1993 Aspects of water management in the southern Maya lowlands. *Research in Economic Anthropology* 7:71–119.
Harrison, P. D., and B. L. Turner II (editors)
1978 *Pre-Hispanic Maya Agriculture*. University of New Mexico Press, Albuquerque.
Hayden, B.
1994 Village approaches to complex society. In *Archaeological Views from the Countryside: Village Communities in Complex Society*, edited by G. M. Schwartz and S. E. Falconer, pp. 198–206. Smithsonian Institution Press, Washington, DC.
Houston, S. D.
1998 On the river of ruins: Explorations at Piedras Negras, Guatemala, 1997. *Mexicon* 20:16–22.
Johnson, A. W., and T. Earle
2000 *The Evolution of Human Societies: From Foraging Group to Agrarian State*. 2nd ed. Stanford University Press, Stanford, California.

Johnston, K. J.

2004 Lowland Maya water management practices: The household exploitation of rural wells. *Geoarchaeology* 19:265–292.

Killion, T. W.

1990 Cultivation intensity and residential site structure: An ethnoarchaeological examination of peasant agriculture in the Sierra de los Tuxtlas, Veracruz, Mexico. *Latin American Antiquity* 1:191–215.

Loten, H. S.

1985 Lamanai Postclassic. In *The Lowland Maya Postclassic*, edited by A. F. Chase and P. M. Rice, 85–90. University of Texas Press, Austin.

Lucero, L. J.

1999a Classic lowland Maya political organization: A review. *Journal of World Prehistory* 13:211–263.

1999b Water control and Maya politics in the southern Maya lowlands. In *Complex Polities in the Ancient Tropical World*, edited by E. A. Bacus and L. J. Lucero, 34–49. Archeological Papers of the American Anthropological Association Number 9. American Anthropological Association, Arlington, Virginia.

2001 *Social Integration in the Ancient Maya Hinterlands: Ceramic Variability in the Belize River Area*. Anthropological Research Paper No. 53. Arizona State University, Tempe.

2002 The collapse of the Classic Maya: A case for the role of water control. *American Anthropologist* 104:814–826.

2003 The politics of ritual: The emergence of Classic Maya rulers. *Current Anthropology* 44:523–558.

2006 *Water and Ritual: The Rise and Fall of Classic Maya Rulers*. University of Texas Press, Austin.

in press The power of water in ancient Maya politics. In *Precolumbian Water Management: Ideology, Ritual, and Politics*, edited by L. J. Lucero and B. Fash. University of Arizona Press, Tucson.

Lucero, L. J., S. L. Fedick, A. Kinkella, and S. M. Graebner

2004 Ancient Maya settlement in the Valley of Peace area, Belize. In *Archaeology of the Upper Belize River Valley: Half a Century of Maya Research*, edited by J. F. Garber, 86–102. University Press of Florida, Gainesville.

Marcus, J.

1998 The peaks and valleys of ancient states: An extension of the dynamic model. In *Archaic States*, edited by G. M. Feinman and J. Marcus, 59–94. School of American Research Press, Santa Fe.

2003 Recent advances in Maya Archaeology. *Journal of Archaeological Research* 11:71–148.

McGee, R. J.
1990 *Life, Ritual, and Religion among the Lacandon Maya*. Wadsworth, Belmont, California.

Meggers, B.
1954 Environmental limitation on the development of culture. *American Anthropologist* 56:801–824.
1998 Foreword. In *Mexican Rural Development and the Plumed Serpent: Technology and Maya Cosmology in the Tropical Forest of Campeche, Mexico*, by B. B. Faust, xi–xiii. Bergin and Garvey, Westport, Connecticut.

Miksic, J. N.
1999 Water, urbanization, and disease in ancient Indonesia. In *Complex Polities in the Ancient Tropical World*, edited by E. A. Bacus and L. J. Lucero, 167–184. Archeological Papers of the American Anthropological Association Number 9. American Anthropological Association, Arlington, Virginia.

Neiman, F. D.
1997 Conspicuous consumption as wasteful advertising: A Darwinian perspective on spatial patterns in Classic Maya terminal monument dates. In *Rediscovering Darwin: Evolutionary Theory and Archeological Explanation*, edited by C. Michael Barton and Geoffrey A. Clark, 267–290. Archeological Papers of the American Anthropological Association Number 7. American Anthropological Association, Arlington, Virginia.

Puleston, D. E.
1971 An experimental approach to the function of Maya chultuns. *American Antiquity* 36:322–335.
1983 *Tikal Report No. 13: The Settlement Survey of Tikal*. University Museum Monograph No. 48. University Museum, University of Pennsylvania, Philadelphia.

Pyburn, K. A.
1997 The archaeological signature of complexity in the Maya lowlands. In *The Archaeology of City-States: Cross-Cultural Approaches*, edited by D. L. Nichols and T. H. Charlton, 155–168. Smithsonian Institution Press, Washington, DC.

Reina, R. E.
1967 Milpas and Milperos: Implications for prehistoric times. *American Anthropologist* 69:1–20.

Reina, R. E., and R. Hill III
1980 Lowland Maya subsistence: Note from ethnohistory and ethnography. *American Antiquity* 45:74–79.

Rice, D. S., and T. P. Culbert
1990 Historical contexts for population reconstruction in the Maya lowlands. In *Precolumbian Population History in the Maya Lowlands*, edited by T. P. Culbert and D. S. Rice, 1–36. University of New Mexico Press, Albuquerque.

Rice, P. M.

1987 Economic change in the lowland Maya Late Classic period. In *Specialization, Exchange, and Complex Societies*, edited by E. M. Brumfiel and T. K. Earle, 76–85. Cambridge University Press, Cambridge, UK.

Robin, C.

2002 Outside of houses: The practices of everyday life at Chan Noohol, Belize. *Journal of Social Archaeology* 2:245–268.

Roscoe, P. B.

1993 Practice and political centralisation: A new approach to political evolution. *Current Anthropology* 34:111–140.

Sanders, W. T.

1977 Environmental heterogeneity and the evolution of lowland Maya civilization. In *The Origins of Maya Civilization*, edited by R. E. W. Adams, 287–297. University of New Mexico Press, Albuquerque.

Scarborough, V. L.

1993 Water management in the southern Maya lowlands: An accretive model for the engineered landscape. *Research in Economic Anthropology* 7:17–69.

1996 Reservoirs and watersheds in the central Maya lowlands. In *The Managed Mosaic: Ancient Maya Agriculture and Resource Use*, edited by S. L. Fedick, 304–314. University of Utah Press, Salt Lake City.

1998 Ecology and ritual: Water management and the Maya. *Latin American Antiquity* 9:135–159.

2003 *The Flow of Power: Ancient Water Systems and Landscapes*. School of American Research Press, Santa Fe.

Scarborough, V. L., and G. C. Gallopin

1991 A water storage adaptation in the Maya lowlands. *Science* 251:658–662.

Scarborough, V. L., and F. Valdez Jr.

2003 The engineered environment and political economy of the Three Rivers region. In *Heterarchy, Political Economy, and the Ancient Maya: The Three Rivers Region of the East-Central Yucatán Peninsula*, edited by V. L. Scarborough, F. Valdez Jr., and N. Dunning, 3–13. University of Arizona Press, Tucson.

Scarborough, V. L., F. Valdez Jr., and N. Dunning

2003 Introduction. In *Heterarchy, Political Economy, and the Ancient Maya: The Three Rivers Region of the East-Central Yucatán Peninsula*, edited by V. L. Scarborough, F. Valdez Jr., and N. Dunning, xiii–xx. University of Arizona Press, Tucson.

Shaw, J. M.

2001 Maya *Sacbeob*: Form and function. *Ancient Mesoamerica* 12:261–272.

Tourtellot, G.

1990 Population estimates for Preclassic and Classic Seibal, Peten. In *Precolumbian Population History in the Maya Lowlands*, edited by T. P. Culbert and D. S. Rice, 83–102. University of New Mexico Press, Albuquerque.

Turner, B. L., II
1974 Prehistoric intensive agriculture in the Mayan lowlands. *Science* 185:118–124.
Webster, D. L.
1997 City-states of the Maya. In *The Archaeology of City-States: Cross-Cultural Approaches*, edited by D. L. Nichols and T. H. Charlton, 135–154. Smithsonian Institution Press, Washington, DC.
1999 The archaeology of Copán, Honduras. *Journal of Archaeological Research* 7:1–53.
Webster, D. L., and A. Freter
1990 The demography of Late Classic Copan. In *Precolumbian Population History in the Maya Lowlands*, edited by T. P. Culbert and D. S. Rice, 37–61. University of New Mexico Press, Albuquerque.
Webster, D. L., and J. Kirker
1995 Too many Maya, too few buildings: Investigating construction potential at Copán, Honduras. *Journal of Anthropological Research* 51:363–387.
Wittfogel, K.
1957 *Oriental Despotism: A Comparative Study of Total Power*. Yale University Press, New Haven, Connecticut.

South America

Agricultural Innovation, Intensification, and Sociopolitical Development

The Case of Highland Irrigation Agriculture on the Pacific Andean Watersheds

Patrick Ryan Williams

In the Andes the relationship between irrigation and political development has been discussed in a number of forums. Among those most prominently debated has been the relationship to the intensification of raised field agriculture and the origins and collapse of the Tiwanaku state (Erickson 1987, 1993; Graffam 1992; Kolata 1986; Ortloff and Kolata 1993; Stanish 1994) and the nature of coastal irrigation in the Moche and Chimu states (Farrington and Park 1978; Moseley and Deeds 1982; Netherly 1984; Ortloff et al. 1982). In this chapter I address the nature of the relationship between irrigation agriculture and sociopolitical development in the highland irrigation systems in the Andean yungas and quichwa zones (c. 1,500–3,500 masl) of southern Peru (Figure 12.1).

Stanish (2001) encapsulates the view of most scholars that Wari's expansion just prior to AD 600 was one of three major first-generation state emergences occurring in the mid-first millennium AD. While previous studies have focused extensively on agrarian production in the Tiwanaku and Moche states (ibid.), Wari has received less attention, even though it was one of three pristine state origins dependent on agricultural production in the Andes. Both the Wari and Inca state systems relied heavily on agricultural production in the highland ecozones, and this study complements the excellent and extensive work carried out on the north coast and in the Andean altiplano. As scholars have argued for both the Moche and Tiwanaku polities, I concur that only by incorporating climatic variation and elite agency in resource production can we come to understand the development of Wari and Inca agrarian strategies.

In many irrigation societies the expansion of elite power is tied to increasing and extending control over resource production. This intensification is often

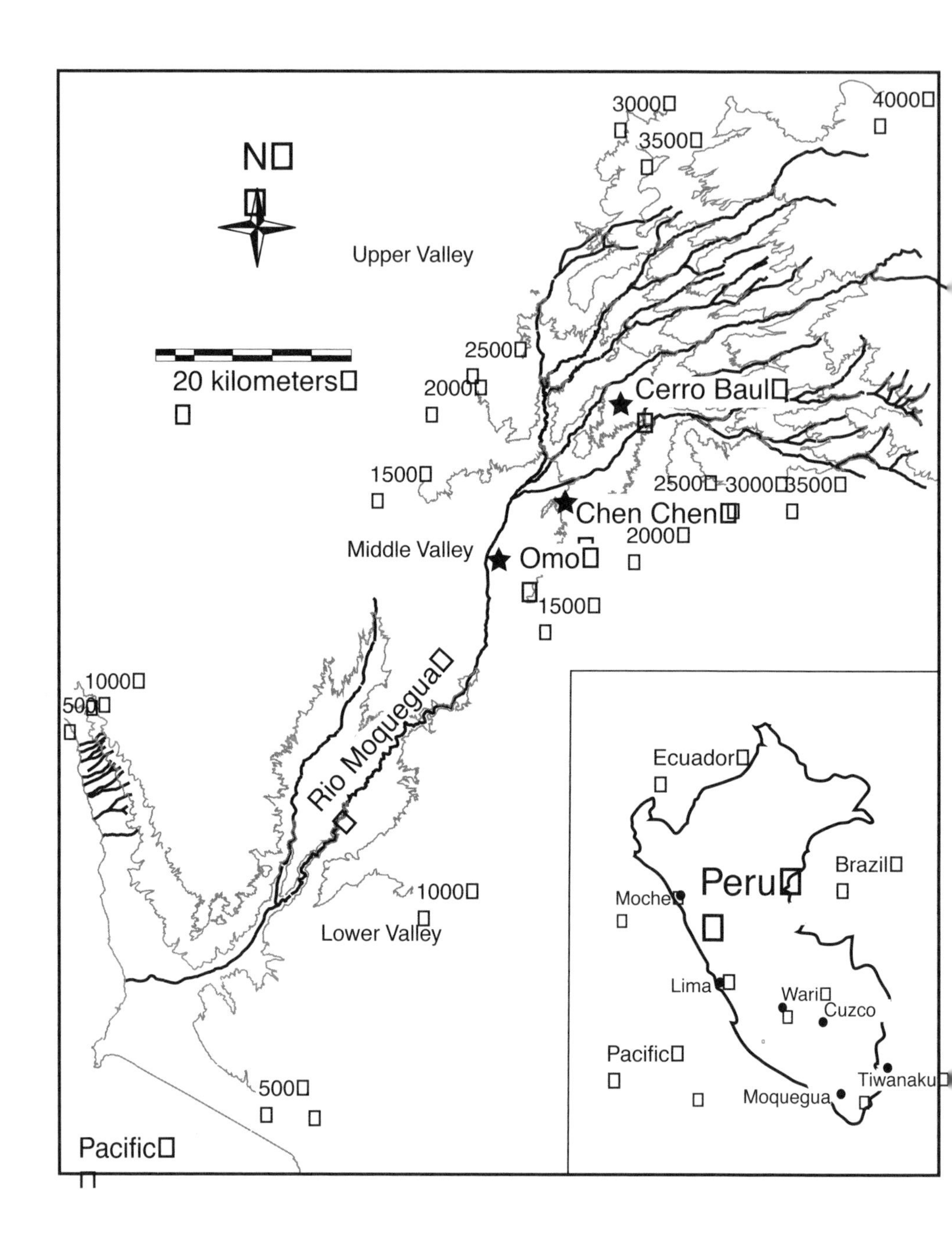

Figure 12.1. Location of the Moquegua Valley in southern Peru, as well as other sites mentioned in the text.

materialized either directly or indirectly through the development of large-scale irrigation systems with new mechanisms of social integration. These social mechanisms provide relief from the scalar stresses that often plague large systems as a result of competition among social factions. It is important to stress that highly centralized control is not a necessity for large-scale agricultural systems. Patricia Netherly (1984), for example, documents the complex relationship between social organization and irrigation for the Chimu and Inca field systems of north coastal Peru, with findings that do not support a direct association between political centralization and agrarian expansion.

Netherly's study was based in the large Chicama Valley, which was not a center of empire and is potentially characteristic of many coastal drainages. Alternatively, the immediately adjacent Moche Valley was an imperial center and housed the Chimu capital of Chan Chan. Here Michael Moseley and Eric Deeds (1982) argue for a correlation between irrigation system size and political unity. They stress consideration of local geography and environmental variables, including water availability, in determining if multiple canal systems can flourish in a single valley or if one system is likely to dominate. In the latter case, as they argue for the Moche Valley, the tendency for a valleywide political unit to emerge is the norm, whereas valleys without the same restrictions can support several independent polities with multiple irrigation networks. These particulars are important distinctions, and as Moseley and Deeds (1982:36) point out, labor availability and engineering skill are the critical social factors. From this perspective, labor capacity, in particular, is a key attribute of agrarian expansion, and Moseley (2001) has argued that the capitals of all major Andean polities mobilized substantial corvée labor to expand agrarian production in their adjacent heartlands.

Moving from capitals to commoners, as Charles Stanish (1994) says, farming households organized by kinship tend to underproduce in respect to their labor capacity. This risk-optimizing strategy maintains a reservoir of labor capacity for hard times and thus represents an ingenious adaptation to resource stress. It also implies that the production of a vast agricultural surplus is not the norm in these peasant economies. The existence of a labor surplus in these economic systems, adaptively accessed in times of resource stress, provides opportunities for exploitation under other conditions. Thus, the conversion of a labor surplus reserved for risk mitigation can become the fodder for the extension of elite control over economic systems (Stanish 1994).

Risk aversion by maintaining a labor surplus in the household is also complemented by the reluctance to become dependent on groups outside the household's social sphere. One of the limiting factors to the development of large-scale agricultural systems in peasant economies is the distrust of social groups outside a community or kin network for the basic necessities of production. Large, intensive irrigation systems may require community solidarity for maintenance. They also might compel upstream users to consider the rights and needs of downstream users, especially in times of water scarcity. Thus, given certain environmental constraints, large intensive agricultural systems are more likely to emerge when social networks are extensive or when households can form relationships to other communities. When the social group size is small, and integration with its neighbors is not being promoted, there is an effective impediment to irrigation system growth. When elite ideologies emerge that begin to encourage social integration in the interests of elite wealth consolidation, the relationship between the household and resource production is drastically altered.

THE HYDRAULIC HYPOTHESIS AND INTENSIFICATION

Karl Wittfogel's (1957) hydraulic hypothesis postulated that the management of irrigation systems was the prime causal force in the development of complex political systems. Although scholars largely reject the postulate, technological innovation in agricultural production, intensification of agrarian systems, and development of more complex sociopolitical forms have been documented as co-occurrences in many world regions. The mountainous spine of western South America, especially the dry deserts and high plains of Peru, Bolivia, and Chile, is an ideal world region in which to investigate these interrelationships. The development of several large political entities, the requirements of water management technologies for large-scale agricultural production, high-resolution proxies for past climate, and the excellent landscape preservation of features constructed over the past two thousand years are all advantages of the study area.

A key issue in this study is the role of intensification of agrarian resources. Ester Boserup (1965) argued that population pressure was the principal cause of intensification. She contends that environmental constraints on the type of agriculture are elastic. That is, people choose among a range of production methods and degrees of intensification. Their choice is conditioned by minimizing labor input to satisfy their needs. Since highly intensive systems usually have very high labor costs and yields have diminishing marginal returns, people do not choose highly intensive systems unless they are required to do so to meet their needs. Combining Wittfogel's and Boserup's arguments, increasing population pressure in an agrarian landscape with land scarcity leads to the intensification of agricultural systems, including irrigation. These new intensive irrigation systems required managers to operate effectively, and the growth of irrigation bureaucracies led to the development of state institutions.

Boserup and many of those who followed her hold climatic constraints constant in their assessments of agrarian intensification. Yet mounting evidence suggests climatic conditions that directly affect agricultural production were extremely variable over the course of the past two millennia in the Andes (Thompson 1985). Furthermore, a Boserupian model assumes land is in short supply, and that that is the primary stimulus for intensification. Given the nature of land scarcity inherent in the model, the principal form of intensification involves decreasing the fallow period by increasing labor inputs into each parcel of land. Thus, Boserupian models rely heavily on evaluating the fallowing cycle to explain intensification. Shorter fallow periods are often seen as a result of the intensification process (Boserup 1965). In some cases, however, other social factors may have played a role (Stone and Downum 1999), such as social production (Brookfield 1972), risk mitigation (Wilk 1997), as well as agency in social institutions (Stone 1996).

The opportunities for elite expansion of agrarian production, however, are predicated on several factors. One of the most important of these factors is the availability of resources to intensify production. In the dry Andean deserts these resources include not only labor but also water. Thus, it is critical to assess water availability for agrarian pursuits over the course of time in examining social relations of agrarian production. New high-resolution climate proxies for southern Peru have been produced and published in the last two decades, making this assessment possible.

CLIMATE CHANGE OVER THE PAST TWO MILLENNIA

The 1983 coring of the Quelccaya Ice Core (Thompson et al. 1985), near Cuzco, has presented a high-resolution data set of precipitation and climate that has been used extensively by archaeologists in examining past climate change (e.g., Binford et al. 1997; Shimada et al. 1991; Williams 2002). Using ice accumulation data from the cores, Thompson et al. have been able to reconstruct annual precipitation trends from the data, especially from core 1, which was not compromised by an unconformity in the stratigraphic layers pertaining to AD 1200 and earlier as was the summit core (Thompson et al. 1985:971). Although the data set has its limitations, it is an incredibly detailed reconstruction unparalleled in the Andean region for long-term precipitation trends. In summarizing the data set, Thompson et al. (1985) note dry events between AD 570 and 610, AD 650 and 730, and AD 1250 and 1310, all during the course of prehispanic state developments in the Andean highlands.

Recent lake cores (Abbott et al. 1997; Chepstow-Lusty et al. 2003) from the Titicaca Basin and the Cuzco region have confirmed the major dry events recorded in the Quelccaya record over the past 1,500 years. The data on dry events from the three climate proxies are summarized in Table 12.1. Long-term droughts around 900 BC, 500 BC, and AD 100 are reflected in both lake

Table 12.1. Climate proxy records for the Cuzco and Titicaca Basins.

Chepstow et al.	Abbott et al.	Thompson et al.	
900 BC	900–800 BC	drier	wetter
500 BC	400–200 BC		
AD 100	AD 1–300		
AD 550		AD 570–610	AD 610–650
		AD 650–730	AD 760–1040
AD 900–1800	AD 1000–1500	AD 1250–1310	AD 1500–1720
		AD 1720–1860	AD 1870–1984

Sources: Chepstow-Lusty et al. 1993; Abbott et al. 1997; and Thompson et al. 1985. Gray highlights show correlations between lake core and ice core drought proxies. Ice cores, counted from annual varve sequences, are much more precisely dated than the lake cores, whose dry stands are bracketed by two to three radiocarbon dates 1,000 years apart, and whose start and end dates are based on assumptions of uniform sedimentation rates.

cores. The early and late ice core data showing extreme dry spells (20 percent lower than mean) are also reflected in the lake core data, the former dated to around AD 550 in one core (Chepstow-Lusty et al. 2003) and the latter dated to AD 1000–1500 and AD 900–1800 in each of the lake cores (Abbott et al. 1997; Chepstow-Lusty et al. 2003). The temporal precision of the ice cores in dating this latter event is significantly better than the carbon dating bracketed lake cores, but both document the extended drought in the early first millennium AD.

In contrast to the drought periods, above average decadal precipitation in the ice cores are documented at AD 610–650, AD 760–1040, and AD 1500–1720 (Thompson et al. 1985). As water is a limiting resource in both rainfall-fed and irrigation-fed agricultural production, its increased availability provides additional resources to agrarian expansion and intensification of land use, especially in the arid climate regime of southern Peru. Thus, with increased rainfall come increased opportunities for manipulating labor resources to restructure agrarian landscapes. Doing so in drought times would risk the productivity of existent irrigation systems that rely on limited water resources to survive. This model does not imply that drought causes deintensification or disintensification (a la Brookfield) and wetter times cause intensification. Rather it acknowledges that resource availability plays an important role, as do sociopolitical forces, in intensifying agricultural productivity. Sociopolitical forces are often tied to increased pressure by elites to extract additional agricultural products from the landscape to finance craft specialists and other nonproducers who support a state ideological system.

The intensification of agricultural systems associated with increasing elite control creates new possibilities for large-scale changes to resource production.

In a system of multiple feedbacks between economic production and social control an enhanced opportunity for entrenched or newly emergent elites to extend their power base is presented. When these enhanced opportunities combine with an effective process of power consolidation, radical expansion and centralization of sociopolitical authority can occur. I assess this model with data from the southern Peruvian highlands, where an example of agrarian evolution is compared to sociopolitical developments from early small agrarian chiefdoms through the first expansive highland states and into the imperial expansion of the Inca.

AGRARIAN REGIMES IN SOUTHERN PERU

Two principal forms of terraced agriculture exist in the highland zones (1,500–4,000 masl) of the Pacific watersheds of southern Peru. These include terraced rainfall farming, restricted to areas above 3,500 meters above sea level, and terraced irrigation agriculture, which is present below 3,500 meters above sea level and predominates in the more rugged terrain between 2,000 and 3,500 meters above sea level. The former includes sectorial fallow regimes, long-term fallow regimes without sectors, and continuous rainfall agriculture (Guillet 1981:141). The area between 10 degrees and 17 degrees south latitude is dominated by the sectorial fallow regime (Guillet 1981:143), and the study area falls on the southern boundary of this belt.

Rainfall terracing, supplemented by extensive pastoral adaptations, is typically subject to the fallowing regimes that replace nutrient content through crop cycling and resting periods, although there are some limited examples of continuous rainfall agriculture (Guillet 1981:141). In highland Espinar, just on the eastern side of the Continental Divide in southern Peru, Benjamin Orlove (1977:94) notes that the sectorial cycle begins with tubers in the first year, followed by grains in the second year, followed by a three- to eight-year fallow. The elevation limits on crops restrict production to hardy high-altitude products like tubers and quinoa. Cultivated land is divided into several sectors, often divided by walls (Guillet 1981:141). Once a sector is planted in the two-year cropping cycle, it is left to fallow as another sector begins its cycle.

Maize, fruits, peppers, legumes, cucurbits, and coca only grow at lower altitudes in the zones restricted to irrigated agriculture in this arid climate. The only exceptions to this rule are the warm valley crops like maize that grow in the microenvironments on the shores of the great Lake Titicaca (Erickson 2000:324). The irrigated fields of the lower elevations have their nutrient content replenished by the materials washed in with canal waters and are not generally subject to fallowing or crop-cycling regimes (Guillet 1981:143). It is particularly this type of agriculture, based on irrigation of crops not generally available in the higher rainfall elevation zones, that characterizes the quichwa and yungas elevation zones between 1,000 and 3,500 meters above sea level.

Although flat land is a scarce commodity in these highland valleys, land itself is not. With appropriate modifications and enough water, the steep desert slopes can be turned into productive irrigated systems. In fact, these unimproved arid slopes are exploited for their sparse vegetative cover by peoples today as they were in the past. The cacti that thrive in these areas are used for fuel, and their fruits are consumed today as they were in the distant past. Seeds from several different cacti species have been recovered from domestic contexts at the Wari site of Cerro Baúl; they indicate that the cactus fruits were brought to the site, since these species do not grow on the mountain summit (Valencia and Goldstein 2004).

Intensification in the use of these landscapes involves dedicating additional water and labor resources to prepare them for intensified agrarian production. Since irrigated fields are multicropped and almost never fallowed, intensification of production cannot be measured by a decrease in the fallow regime, as traditional Boserupian models would indicate (e.g., Stone 1996). Rather, intensification involves the construction of new terracing or new irrigation canals to bring water and silts to land that had not been previously irrigated. It might also involve importing additional fertilizers to apply to existing terraces that could increase their yields.

AN ASSESSMENT OF AGRARIAN DYNAMICS IN THE QUICHWA ZONE OF MOQUEGUA, PERU

The Río Osmore, or Río Moquegua as it is often known, lies at approximately 17 degrees south latitude and 71 degrees west longitude in the sierra on the northern fringe of the Atacama Desert in southern Peru (Figure 12.1). The principal channel bifurcates into two main tributaries just upstream of Montalvo, where the Panamerican Highway crosses the river. The southern branch is the Tumilaca drainage and its tributaries, including the Quebrada Cocotea, Río Asana, and Río Capillune (Figure 12.2). The northern branch bifurcates into two principal tributaries near Estuquiña, the Río Huaracane, and the Río Torata. The northern Huaracane drainage is fed by the Sajena, Otora, Cueva Quemada, Porobaya, and Chujulay rivers. The Río Torata, located between the Tumilaca drainages and the Huaracane drainages, is a single course that flows directly from the highlands with very little dendritic tendency.

The region is extremely dry below 3,500 meters, and irrigation is the only way in which the desert can be cultivated. The sierra of the south-central Andes receives almost no annual precipitation. The winter months of June through September are hot, sunny, and dry with high diurnal temperature changes. The

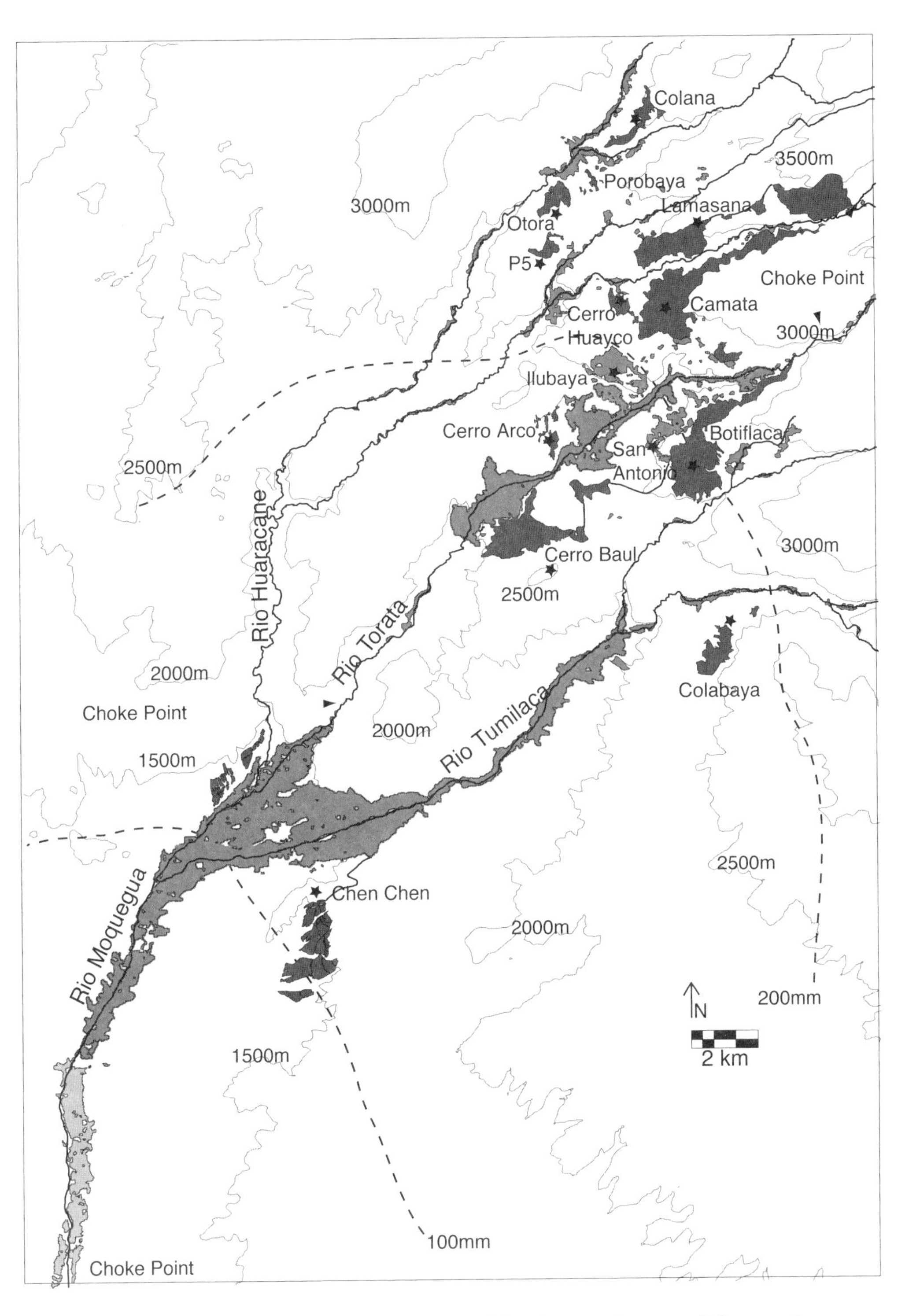

Figure 12.2. Ancient and modern irrigation systems in the middle and upper valley zones of Moquegua, Peru.

summer months of January through March are cloudier and slightly cooler and wetter. During summer the meager rains that fall at higher elevations begin to flow into the rivers that drain into the Pacific Ocean (ONERN 1976).

Water discharge rates vary greatly both seasonally and annually, however. Annual discharges ranged from 29,349,000 in 1959–1960 to 87,105,000 in 1954–1955, while within a single year discharge rates ranged from 0.48 to 80.0 cubic meters per second (ONERN 1976). This great disparity in discharge causes extreme shortages of water in drought periods and overabundance of water in wet spells. In fact, it is estimated that the annual discharge of the Moquegua is capable of supplying double the amount of water required for modern cultivation, but it is so variable that extreme shortages are typical at several times throughout the year (INP 1966).

Moquegua's drainage basin incorporates 3,480 square kilometers, of which 3,360 hectares were being cultivated as of 1972 (ONERN 1976). Most of this cultivated area is centered in the middle valley around the city of Moquegua (Figure 12.1). However, a substantial amount of land is being irrigated in the upper tributaries, especially in the Torata Valley. All this cultivation depends on consistent water discharge from the tributaries of the Río Moquegua.

The Moquegua drainage basin can be divided into three major zones: the upper, middle, and lower valley (ONERN 1976). These portions of the valley are bounded by bedrock choke points, which are deeply incised, erosion-resistant segments of the river bed at approximately 200, 1,500, and 3,000 meters above sea level. Choke points are important natural features in irrigation since they define and constrain the highest elevation at which a canal intake in a particular valley zone can exist. Choke points represent the upper limit of cultivation in any valley zone. The tripartite division of the basin is reflective of the degree of technology needed for farming. In the middle valley, land is relatively flat, with an average gradient of 2.5 percent, but in the upper drainage the gradient is much steeper. In the Torata Valley, for example, the river's gradient approaches 10 percent based on calculations from topographic maps.

Cultivation of steeper land requires greater labor input than flatter land. Based on cultivation areas as seen on aerial photographs and field survey, the Torata River is the most agriculturally productive of Moquegua's upper valley tributaries today. Field discharge measurements indicate that the Torata does not discharge more water than the other major tributaries of Moquegua. However, the Torata is the only tributary whose waters primarily irrigate the upper, rather than the middle, drainage centered near the city of Moquegua. This distinction is important, because the disparity between agricultural yield in the upper and middle drainage was probably a source of great conflict in the past. The late introduction of the technology used to irrigate the upper drainage also influenced the first major settlement of the Torata Valley.

EARLY AGRARIAN PRACTICE

Prior to AD 600 there was very little settlement in any of the upper valley tributaries of the Río Moquegua. The middle valley was heavily settled by Huaracane peoples (Goldstein 2000) by the Late Formative period (400 BC–AD 400). These societies were not hierarchically stratified. The settlement pattern consisted of small hamlets equidistantly spaced on the bluffs along the river floodplain with no settlement size hierarchy (Goldstein 2000). While leaders had access to some exotic trade goods from other south-central Andean groups, they did not live very differently from the rest of society (Goldstein 2000). No overarching political structure or discernible internal local polities characterized the sociopolitical configuration of this period.

The Formative agricultural tradition in Moquegua consisted of irrigating the flat floodplain of the middle valley zone with short canals. Agriculture in the upper valley was decidedly more difficult and probably inhibited settlement in this area. Indeed, very little evidence for any Huaracane settlement exists in the upper valley (Owen 1994). Since the floodplain farming of the Formative was so productive, it continued to be a principal source of agrarian production throughout Moquegua history. Today, however, any trace of Formative fields in the valley floodplain has been erased by subsequent cultivation. Even so, the 40 or so modern irrigation systems in this part of the valley do not exceed 60 hectares each and are fed by canals less than 3 kilometers in length (Moquegua Ministry of Agriculture IX Agrarian Region 1983). The entire middle valley floodplain can be and is irrigated with many independent systems at this scale or smaller.

AGRICULTURAL SYSTEMS AND THE FIRST STATES

Around AD 600 the expansive Wari state entered the Moquegua Valley and established a stronghold in the upper valley zone between the Torata and Tumilaca tributaries. With its center on the omnipotent mountain peak of Cerro Baúl (2,590 masl), the Wari colony in Moquegua radically transformed the settlement pattern of the valley (Williams 1997, 2002). Wari settlement patterns in Moquegua include a colony at Cerro Baúl, with subordinate villages around the great mesa. Third-level hamlets complete a local settlement hierarchy. Within settlements there were distinct differences between households (Nash 2002). Large-scale public architecture and associated elite households are spatially segregated; both only occur on the mountain summit of Cerro Baúl. Evidence for differential access to resources, full-time craft specialization, and pronounced social stratification are in evidence for the first time in the region (Moseley et al. 1991; Williams 2001; Williams et al. 2000). Furthermore, extensive exchange networks with areas outside the general region, and especially with the capital of Wari in Ayacucho, are well documented (Burger et al. 2000; Williams 2001). The upper valley was incorporated as a colonial enclave into a much larger political system. Elites on the summit of Cerro Baúl directed interaction with the outside and extracted a number of resources from the local area (Nash 2002; Williams 2001).

Concomitantly, the Wari established the first extensive agricultural systems in the upper valley (Williams 1997). The largest of these was the intertributary El Paso canal system, which drew water from the Torata River to irrigate the slopes of Cerro Baúl and Mejía. With a principal canal more than 14 kilometers in length, the canal brought water across the divide from the Torata watershed to the Tumilaca Basin (Williams 1997). In conjunction with its second arm, which flowed to the Quebrada Cocotea (Dayton et al. 2004), the El Paso canal was the longest canal system in Moquegua history. Today, only the partially functioning intervalley Pasto Grande canal under construction by the Peruvian government exceeds the El Paso system in length.

The Wari irrigation system also required construction of extensive terrace systems for the first time in Moquegua's history. The slopes on which the Wari irrigation system was built were gentler than later highland field systems; thus, the density of terrace architecture was also less. However, the total irrigated area is much larger, and the total labor investment in constructing the system is estimated at nearly 100,000 days (Table 12.2). This is a significant increase over nonterraced irrigated fields of the previous period. It also is significantly greater than most of the terraced irrigation systems from the succeeding period. Only under the Inca were more labor-intensive systems constructed.

During Wari times, for the first time, a single irrigation system with construction, maintenance, and production needs beyond the capacity of the traditional farming community (of 10 to 100 members) was built. The system integrated several Wari sites along its route, each several hectares in size and each housing hundreds of residents. The scale of the irrigation system itself could have provided food for more than 1,000 people (Table 12.2). The single canal that fed this system would have monopolized a majority of the flow of the Torata River. As the first canal to tap into the river just a few hundred meters below the upper valley choke point, it was well positioned to control the flow of water over the entire tributary. Because it integrated several towns along its length, intercommunity cooperation in water control is implied.

Wari began its colonization and irrigation intensification scheme during one of the strong drought periods between AD 570 and 610. Wetter periods prevailed between AD 610 and 650 and between 760 and 1040 (Table 12.1; Thompson et al. 1985). It is precisely during these wetter times that Wari field systems are developed to their greatest extent. Wari seems to reach peak populations in the valley between AD 700 and 800, given the dating of subsidiary sites like Cerro

Table 12.2. Agricultural and settlement system dynamics from AD 500 to 1532 in Moquegua, Peru

system	principal canal	area (ha.)	discharge capacity	site size (ha.)	dom. units	site pop	pop supported	source	terrace density (m/m^2)		construction costs (days)		man-years/ dom. unit
									density	m of wall	Guillet	Williams	Guillet
Middle Horizon (AD 600-1000)													
Cerro Baúl	14.2 km	324.7	393-436 l/s midpt		400	2000	1080	Torata	0.07	227290	97734.7	90916	0.7
Chen Chen	9.9 km	93	83 l/s		350	1800	310	Tumilaca	not terraced		na	na	
LIP abandoned (AD 1000-1450)													
P5	6.1 km	24	N/A	0.1	4	20	80	Huaracane (Porobaya)		16800	7224	6720	4.9
Otora (P7)	4.2 km	45.9	N/A	0.7	30	150	150	Huaracane (Porobaya)		96390	41448	38556	3.8
Colana (P3)	7.9 km	54.5	N/A	0.4	30	150	180	Huaracane (Porobaya)		114450	49214	45780	4.5
Co.Huayco	6.8 km	27.7	110 l/s	0.25	20	100	100	Huaracane (Chujulay)		58170	25013	23268	3.4
LIP still active (AD 1000 -)													
Porobaya (P1)		69.4			70	350	230	Huaracane (Porobaya)		145740	62668	58296	2.5
Ilubaya	7.2 km	93.4	388-485 l/s		60		310	Torata		196140	84340	78456	3.9
Cerro Arco	4.6 km	113	150 l/s		70		380	Torata		237300	102039	94920	4.0
San Antonio	5.5 km	71	75 l/s	0.84	60	300	300	Torata	0.05	149100	64113	59640	2.9
Colabaya	10.5 km	75	N/A	0.55	50	200	250	Tumilaca		157500	67725	63000	3.7
Late Horizon (AD 1475-1532)													
Lamasana	11.9 km	346.2	N/A		60		1150	Huaracane (Chujulay)		727020	312619	290808	14.3
Camata	6.7 km	340.6	837-1134 l/s	1.1	70	350	1140	Huaracane (Chujulay)		715260	307562	286104	12.0
Botiflaca	4.7 km	301.7	469-738 l/s	1	70	350	1000	Torata	0.21	633570	272435	253428	10.7

Note: Construction costs are calculated from terrace construction rates provided by Guillet (1987) and Williams (1995) per linear meter of terracing.

Mejía and sectors on the slopes of Cerro Baúl (Nash 2002). Thus, both increased labor and increased water were available to expand and intensify production. The elite social hierarchy was imported with the settlement, and differentiation between households, neighborhoods, and sites is distinguished in the earliest Wari contexts in Moquegua. These elite contexts were conspicuously using a much wider range of subsistence resources and were consuming them in greater quantities than their counterparts (DeFrance 2004; Valencia and Goldstein 2004). Elite agents were using increased resources to alter the means of production in their favor. The relationship between Wari's highest elites and their labor resources changed, however, by around AD 800.

TIWANAKU, WARI'S RIVAL

Perhaps as early as AD 600 colonists from the altiplano Tiwanaku cultural tradition had begun to inhabit Moquegua (Goldstein 1989, 2005). Not until well after AD 700, however, were they becoming well established in the middle valley and begun settling alongside the Wari in the upper valley. Their middle valley agricultural systems probably resembled those built by the Huaracane at first. However, in the eighth or ninth century AD a new wave of colonization changed the nature of Tiwanaku settlement in the middle valley. Larger towns were erected, and at the site of Chen Chen an extended canal system that irrigated the flat pampa above the valley floodplain was built near the modern city of Moquegua. By AD 800 Chen Chen was a small city in the low thousands of inhabitants that eventually became a vast necropolis of thousands of tombs (Owen 1997). The agricultural system on the pampa encompassed almost 100 hectares and was fed by a series of canals, the longest of which was 10 kilometers from its source in the Tumilaca River (Williams 2002). The canals that fed the river terraces above the floodplain around the confluence of the Moquegua tributaries near Chen Chen were also probably constructed in this period. These latter fields are still cultivated today, although the canals that feed them may not be the original channels constructed by the Tiwanaku.

The Tiwanaku fields were built on the flat alluvial terraces in the Middle Valley. They did not require terracing to farm, and their construction and maintenance requirements were thus significantly less. The Chen Chen system itself was also only a third the size of the Wari system. While it may have been dedicated to the production of specialized crops for export to the Tiwanaku heartland, if it was supporting a local population, it would only have provided enough food for some 300 people, far less than the 1,800 that were estimated to have inhabited the site (Table 12.1). The flat valley floodplain that had been in use for past millennia as prime farmland continued to provide the principal food resources to the Tiwanaku population, without a significant change in agrarian technology.

Under the Wari and Tiwanaku, for the first time since the inception of community irrigation systems in the valley more than 1,000 years before, new ecozones were being opened to agriculture, and they were being tamed in dramatically new ways. Long sinuous canals rising out of the valley floodplain that could carry most of a river's flow in the dry season were the new means of water distribution. Bonds of kinship and coresidence formed the goodwill between farmers in the small irrigation systems of the Formative. These larger irrigation networks of the Wari and Tiwanaku, which incorporated multiple communities, social classes, and hundreds or thousands of residents, relied on new mechanisms of social integration that left clear evidence of state colonies and elite control of production. They also reached their maximum extent under enhanced precipitation and greater availability of water resources.

AGRICULTURAL SYSTEMS IN LATE PREHISTORY

Sometime around AD 1000, state control over these colonies collapsed and both Chen Chen and Cerro Baúl were abandoned (Goldstein and Owen 2001; Williams 2002). The concomitant abandonment of the irrigation systems that brought water to the outskirts of these centers accompanied irrigation and settlement shifts back to smaller towns and single community irrigation systems. The exact causes of this shift are still being investigated, although the creation of social factions and scalar stresses on the system may have played a role (Williams 2002). Pioneer communities moved away from the Middle Horizon population centers to create small-scale irrigation systems in the other tributaries of the Moquegua drainage, specifically the Otora (Stanish 1992), Chujulay, and upper Tumilaca tributaries. These pioneers are likely descendants of local peoples that carried on the Tiwanaku ceramic traditions and were certainly influenced by their interactions with Wari, as demonstrated through household and irrigation structures derived from Wari.

Pioneer communities were soon joined and replaced by migrants from other regions, such as the altiplano groups that inhabit the Otora period sites that Stanish (1992) documents. Coastal polities that emerged at the end of the Middle Horizon, like the Chiribaya, also sent small colonies to the upper reaches of the Moquegua tributaries. The local Middle Horizon holdovers and small numbers of these coastal and highland settlers dominate the early Late Intermediate period (LIP) landscapes (AD 1000–1250). While many of these communities exist in the upper valley landscape, those associated with abandoned agricultural systems are few. Most of the lands they cultivated are still under cultivation today, and it is difficult to assess the extent of the ancient cultivation systems without seeing the ancient irrigation infrastructure. The communities of P5 and P7 in the Otora drainage are exceptions originally documented by Stanish (1985). Our research used Global Positioning Systems (GPS) to document these systems with the same precision used to document Wari and Tiwanaku systems in order to make direct comparisons. The results mirror Stanish's earlier findings that these systems were small-scale pioneering settlements. Table 12.1 indicates that the number of people capable of being supported by these systems is small and directly correlated to the estimated populations of the associated sites. They display the self-sufficient, labor surplus conserving, peasant economies predicted by Chayanov's rule (Stanish 1994).

By AD 1250 new agricultural systems much higher on the hill slopes and now heavily defended communities begin to emerge, replacing the post–Middle Horizon systems seen earlier. These peoples, referred to as Estuquiña (after the type site near the city of Moquegua), inhabit the many independent communities throughout the Moquegua drainage by AD 1350. The principal difference between this Estuquiña-dominated period (AD 1250–1450) and the preceding period is the construction of new terrace systems that are higher upslope with more investment in terrace infrastructure and pronounced defensive settlement. In neither case did the level of political integration extend beyond the village of 100 to 200 inhabitants. Village leaders existed, and trade with both the altiplano and the coast continued to be strong. However, very little evidence of public architecture, except for the defensive walls and moats that surrounded the villages, is present in Estuquiña times. Some differentiation in household size is present in these communities, but the easy access to exotic resources, the control of specialized producers, and the drastically different living standards documented for the Middle Horizon settlement at Cerro Baúl are not present in either the early or late LIP.

The community systems of Colana (P2), originally described by Stanish (1985), and Cerro Huayco (Owen 1994) are the two principal Estuquiña systems that were not occupied during the succeeding Inca period but were associated with completely abandoned canal systems.

An analysis of these systems provides a model for immediate pre-Inca land-use strategies in the valley. Populations of the single settlements associated with these irrigation systems ranged from 20 to 30 domestic units housing 100 to 200 inhabitants. While site size is generally under 0.5 hectare, other Estuquiña sites whose full extent is not unambiguously pre-Inca ranged up to 1 hectare in size, with estimated populations of up to 350 people. Site populations also tend to be directly correlated with irrigation system size, such that surplus production of agricultural foodstuffs is not as pronounced as it was in the Middle Horizon.

The onset of a centuries-long drought began in the early second millennium AD. The drought is most pronounced in the ice core records between AD 1250 and 1310, dates that are broadly coincident with the emergence and expansion of Estuquiña settlement patterns and agricultural systems. The balkanization of settlements into small, self-sufficient communities prohibited the accrual of large labor resources. Likewise, the severe drought reduced water resources substantially in the drainage. These patterns persist until c. AD 1475, when Inca rule had become established in the valley.

Many Estuquiña irrigation systems continued to be cultivated under the Inca, however, and many are still in use today. A sample of those Estuquiña systems are still used today; most of them were also probably cultivated by the Inca (see Table 12.2). In the Otora region the irrigation systems associated with Estuquiña populations under Inca rule more closely hug the valley bottoms, as is the case for modern agriculture there today; high elevation systems, like Colana, drop out of use (Stanish 1992). In the Tumilaca tributary on the southern watershed the high elevation Colabaya system is maintained and perhaps even expanded slightly under Inca rule, although the defensive settlement of Porobaya is replaced with a smaller Inca enclave (Ribeiro 1995).

AGRICULTURAL PRODUCTION UNDER THE INCA EMPIRE

Inca hierarchy is well documented. In Moquegua, for the first time since the Middle Horizon, large-scale public architecture associated with truly distinctive elite residence reemerges. The relatively large storage facility at Camata, perched on a hilltop in the midst of agricultural fields and visible from all around, indicates a state institutional presence (Stanish and Pritzker 1990). The site of Sabaya contains evidence for a *kallanka* (hall) and an *usnu* (platform), hallmarks of Inca public architecture (Bürgi 1993).

The key expansion in Inca period agricultural systems is seen in the central tributaries of the Moquegua in the areas surrounding the Inca administrative center of Sabaya, the tambo of Camata (Lamasana, Camata canals), and the town of Torata Alta (Botiflaca canal). These are all centered on the Torata and Chujulay drainages, where the main Inca road to the altiplano passes. In each of these cases Inca funerary structures, architecture, or other diagnostic infrastructure reflect ultimate Inca use of the relevant irrigation system. Furthermore, canal excavations in each of these systems indicate that ultimate abandonment of the systems just preceded the eruption of the Huaynaputina volcano in AD 1600 (Williams 1997).

These three field groups, two of which have previously been identified as Inca imperial systems (Mathews 1989), provide an opportunity to examine systems whose final use was under Inca control. Like the Colana and Cerro

Huayco systems that were clearly Estuquiña in date, these three systems were unambiguously last used under the Inca, with no further modification by later peoples.

These terrace systems tend to be long and continuous, following the contour of the land and not an individualized land-tenure system as they do in the modern period. Irrigation infrastructure utilizes a drop canal technology that pulls water off the main canal in straight lines, giving the systems a grid-like structure. These drop canals accelerate the movement of water off the main channel and represent high erosion regimes. When coupled with holding tanks, however, most of which are located at the distal ends of irrigation canals, they permit the irrigation of greater land areas owing to the ability to expand irrigation at the lower ends of the system. While holding tanks have been documented as early as AD 1000 to 1200 (Stanish 1985), it is only in these late systems that multiple large holding tanks located below extensive terraced fields are present in a single system. The Inca systems bear the mark of a corporate governance policy in the distribution of water and the maintenance of infrastructure.

The construction and maintenance of the Inca related systems required supracommunity levels of organization. Long continuous terraces require consistent maintenance over a large area—something beyond the ability of a household or small community. The high terrace density of a system would have required an enormous amount of labor to construct. Using modern estimates of terrace construction, the three centralized Inca systems would have required 300,000 person-days each to build, and perhaps 5 to 10 percent of that time annually to maintain (Table 12.2). Planting, irrigating, weeding, and harvesting are not included in this sum. These three systems each required three to five times the labor of any other documented system in the data set. Yet the settlements associated directly with these systems do not have three to five times the number of people (Table 12.2).

Inca elites were able to marshal substantially more labor than their predecessors. The construction of these systems to their full expansion probably took place in the late fifteenth century AD, immediately after the Inca conquered Moquegua. They were completed just as the south-central Andes were emerging from a prolonged period of drought and entering a period of above-average precipitation that began around AD 1500. Increasing water availability also likely played a role in the resource acquisition strategy of the Inca elite. It is also clear, however, that the life span of the Inca irrigation was short. Two of the irrigation systems were definitely abandoned by AD 1600, as dated by the canal infilling with Huaynaputina ash. Their abandonment was likely associated with depopulation created by European disease pandemics that strained labor resources. Changes in the elite administrative structure that accompanied the Spanish conquest also contributed to the demise of these indigenous organizations. Although water continued to be available, both management and labor resources declined, and abandonment of the irrigation systems was inevitable.

DISCUSSION

In this chapter I have argued that increases in irrigation system size and intensification of landscape resources is most notable during periods of increased sociopolitical organization and increased availability of water. Both cases of supralocal political organization in the history of the study area took place during a recovery from a pronounced period of drought. In both cases, the period of maximum expansion and intensification of agrarian resources co-occurred with above-average rainfall regimes. Under both the Wari and Inca, the organization of labor from more than one community was critical to constructing and maintaining the state agricultural systems. Likewise, increased water resources supported agrarian expansion without affecting existing production.

The study area was in neither case the heartland of empire. Sociopolitical organization was imported from outside, not an in situ development. In this respect the study area does not comment directly on Wittfogel's hydraulic hypothesis for the origins of sociopolitical complexity. It does, however, suggest that intensification of agrarian resources in the highland Andean zones depends on supracommunity levels of political organization centered in the high sierra. The only times such organization is evident in the Moquegua region is during the Wari and Inca regimes. The Moquegua data also suggest that intensification correlates with increased availability of a critical resource: water. Would it have been possible for the Wari agricultural adaptation to have been so successful if water resources were not increasing during its early development? Could it have maintained its active presence in the region for 400 years if drought stress prevented further expansion? The archaeological record suggests that continuing expansion was one of the keys to the system. As I have argued elsewhere (Williams 2002), it was probably not drought that caused the eventual collapse of these highland valley systems at the end of the Middle Horizon but rather factionalism among the constituent social groups.

In the case of the Inca, the increasing volume of water was a factor. Their development of the upland valleys may have been contingent on water availability, as increasing upland irrigation would have negative effects on the fields around the town of Moquegua and midvalley Inca settlements in times of drought. The Inca systems were short-lived because of damage to the other critical resource labor (and the social organization of that workforce). Much of the Inca system was constructed between AD 1475 and 1525. The fallout of the smallpox pandemics that arrived with the first Europeans on the Caribbean shores preceded Pizarro's arrival in Peru in 1532 by several years. Impact on both laborers and government officials was devastating, with mortality rates of 50 percent and higher in some places. Spanish replacement of the Inca bureaucracy by 1534 and continuing pandemics led to the demise of both the indigenous labor force and the social organization of intercommunity collaboration necessary to the intensification endeavor.

Opportunities for the extensive reworking of landscapes in large-scale, intensive irrigation systems was a product of elite social agency when both labor and water resources were abundant. It should be noted, however, that small-scale intensification was also pronounced in the intervening periods of village autonomy when water was scarce and labor was limited to community-level resources. In the case of the Estuquiña, for example, it seems that competition for access to water between groups spurred the construction of defended settlements and terraced slopes high above the valley floor. Yet these systems were never intended for surplus production to support a nonproducing elite and were thus never of the scale seen in the imperial systems of the Middle or Late Horizons.

What is also clear from this data set is that drought periods prior to the onset of the first state systems in the study area were not catalysts in the development of intensification. The lake core records concur that drought regimes prevailed at the beginning of the first millennium BC, the middle of the first millennium BC, and the beginning of the first millennium AD (Table 12.1). Yet the concomitant changes in agrarian intensification are not present. Prior to the development of social mechanisms of labor organization beyond the village, intensified highland agricultural systems were not practical nor developed. In the case of the study area, agricultural labor integration beyond the village level did not occur without a political hierarchy.

CONCLUSION

The intensification of agrarian production in the Andean highlands relied on both mechanisms of labor organization and favorable climatic conditions as prerequisites. It was elite demand for greater production that put these resources into play in expanding agrarian systems. The principal variables at play in the highland valleys parallel the arguments made by Moseley and Deeds (1982) for the construction of supracommunity canal systems on the north coast. It should also be noted that unlike a traditional Boserupian model (1965), burgeoning populations did not seem to play a central role in the agrarian intensification enterprise. Rather, agrarian intensification accelerates when imported social hierarchies take advantage of local conditions to increase production to support an elite.

Boserup (1965) also argued that decrease in fallow length was often a proxy measure of intensification. In the case of highland irrigation agriculture, where fallow is rarely practiced in any case, the intensification model is best understood via the investment in hydraulic infrastructure, such as terraces and canals. The shift from nonterraced agriculture in the valley floodplains and on the gentle alluvial fans of the middle and lower valleys to the high slope terraces of the upper valley zones marks a significant increase in labor investment per hectare. This trade-off is rewarded with more efficient use of water, often reflecting a 50 percent savings in overall water use (see Williams 1997). In turn, more area can be cultivated, though at a proportionally higher labor cost.

Thus, the highland irrigation systems of the Andes differ substantially from the river valley systems of Mesopotamia or even the closer raised field systems of the altiplano or the irrigation systems of the valleys on the north coast of Peru. Yet labor capacity and organization were integral to all three enterprises. In the Andean case, water availability and climate models have also played an important role in assessing the causes of agrarian intensification. Those variables were important to agricultural intensification to the highland valleys of the Andes as well. Ultimately, however, the opportunities presented by favorable climatic conditions or labor integration strategies depend on an enterprising elite to take advantage.

REFERENCES

Abbott, M. B., M. W. Binford, M. Brenner, and K. R. Kelts

1997 A 3500 ^{14}C yr high-resolution record of water-level changes in Lake Titicaca, Bolivia/Peru. *Quaternary Research* 47:169–180.

Binford, M., A. Kolata, M. Brenner, J. Janusek, M. Seddon, M. Abbott, and J. Curtis

1997 Climate variation and the rise and fall of an Andean civilization. *Quaternary Research* 47:235–248.

Boserup, E.

1965 *The Conditions of Agricultural Growth*. Aldine, Chicago.

Brookfield, H. C.

1972 Intensification and disintensification in Pacific agriculture. *Pacific Viewpoint* 13:30–41.

Burger, R., K. M. Chavez, and S. Chavez

2000 Through the glass darkly: Prehispanic obsidian procurement and exchange in southern Peru and northern Bolivia. *Journal of World Prehistory* 14(3): 267–362.

Bürgi, P.
1993 The Inca Empire's Expansion into the Coastal Sierra Region West of Lake Titicaca. PhD dissertation, Department of Anthropology, University of Chicago.
Chepstow-Lusty, A., M. Frogley, B. Bauer, M. Bush, and A. Tupayachi
2003 A late Holocene record of arid events from the Cuzco region, Peru. *Journal of Quaternary Science* 18(6):491–502.
Dayton, C., M. Moseley, and P. R. Williams
2004 Wari agriculture in Moquegua. Paper presented at the 69th Annual Meeting of the Society for American Archaeology, Montreal.
DeFrance, S.
2004 Wari diet in Moquegua: The ordinary and the exotic. Paper presented at the 69th Annual Meeting of the Society for American Archaeology, Montreal.
Erickson, C.
1987 The dating of raised field agriculture in the Lake Titicaca Basin of Peru. In *Pre-Hispanic Agricultural Fields in the Andean Region*, edited by W. M. Denevan, K. Mathewson, and G. Knapp, 373–383. British Archaeological Reports, Oxford.
1993 The social organization of prehispanic raised field agriculture in the Lake Titicaca Basin. In *Economic Aspects of Water Management in the Prehispanic New World*, edited by V. Scarborough and B. Isaac, 367–424. JAI Press, Greenwich, Connecticut.
2000 The Lake Titicaca Basin: A precolumbian built landscape. In *Imperfect Balance: Landscape Transformations in the Precolumbian Americas*, edited by D. Lentz, 311–356. Columbia University Press, New York.
Farrington, I. S., and C. Park
1978 Hydraulic engineering and irrigation culture in the Moche Valley, Peru. *Journal of Archaeological Science* 5:255–268.
Goldstein, P.
1989 Omo, a Tiwanaku Provincial Center in Moquegua, Peru. PhD dissertation, Department of Anthropology, University of Chicago.
2000 Exotic goods and everyday chiefs: Long-distance exchange and indigenous sociopolitical development in the south central Andes. *Latin American Antiquity* 11(4): 335–361.
2005 *Andean Diaspora*. University Press of Florida, Gainesville.
Goldstein, P., and B. Owen
2001 Tiwanaku en Moquegua: Las colonias altiplánicas. In *Boletín de arqueología PUCP No. 5, Huari y Tiwanaku: modelos vs. evidencias, segunda parte*, edited by P. Kaulicke and W. H. Isbell, 139–168. Pontificia Universidad Católica del Perú, Lima.
Graffam, G.
1992 Beyond state collapse: Rural history, raised fields, and pastoralism in the south Andes. *American Anthropologist* 94:882–904.

Guillet, D.

1981 Land tenure, ecological zone, and agricultural regime in the central Andes. *American Ethnologist* 8(1):139–156.

1987 Terracing and Irrigation in the Peruvian Highlands. *Current Anthropology* 28(4):409–430.

INP

1966 *Proyecto de irrigación de Moquegua*. Feasibility report prepared by Instituto Nacional de Planificación and McCreary Koretsky Engineers, Lima.

Kolata, A.

1986 The agricultural foundations of the Tiwanaku state: A view from the heartland. *American Antiquity* 51:748–762.

Mathews, J.

1989 Dual systems of Inca agricultural production: Evidence from the Osmore drainage, southern Peru. In *Ecology, Settlement, and History in the Osmore Drainage*, edited by C. Stanish, D. Rice, and P. R. Scarr, 415–434. British Archaeological Reports, Oxford.

Moquegua Ministry of Agriculture IX Agrarian Region

1983 *Diagnóstico del distrito de riego Moquegua y sector de riego Carumas*. Ministerio de Agricultura, Convenio Dirección General de Aguas, Suelo e Irrigaciones, Programa Sectorial Agropecuario, Lima.

Moseley, M. E.

2001 *The Incas and Their Ancestors: The Archaeology of Peru*. Thames and Hudson, London.

Moseley, M. E., and E. Deeds

1982 The land in front of Chan Chan: Agrarian expansion, reform, and collapse in the Moche Valley. In *Chan Chan: Andean Desert City*, edited by M. Moseley and K. Day, 25–54. University of New Mexico Press, Albuquerque.

Moseley, M. E., R. A. Feldman, P. Goldstein, and L. Watanabe

1991 Colonies and conquest: Tiahuanaco and Huari in Moquegua. In *Huari Administrative Structure: Prehistoric Monumental Architecture and State Government*, edited by W. H. Isbell and G. F. McEwan, 21–40. Dumbarton Oaks, Washington, DC.

Nash, D.

2002 The Archaeology of Space: Places of Power in the Wari Empire. PhD dissertation, Department of Anthropology, University of Florida.

Netherly, P. J.

1984 The management of the late Andean irrigation system on the north coast of Peru. *American Antiquity* 49(2):227–254.

ONERN

1976 *Inventario, evaluación, y uso racional de los recursos naturales de la costa: Cuencas de los ríos Moquegua-Locumba-Sama y Caplina*. Oficina Nacional de Evaluación de Recursos Naturales, Lima.

Orlove, B.

1977 Integration through production: The use of zonation in Espinar. *American Ethnologist* 4(1):84–101.

Ortloff, C. R., and A. Kolata

1993 Change: Climate and collapse: Agroecological perspectives on the decline of the Tiwanaku state. *Journal of Archaeological Science* 20:195–221.

Ortloff, C. R., M. E. Moseley, and R .A. Feldman

1982 Hydraulic engineering aspects of the Chimu Chicama-Moche intervalley canal. *American Antiquity* 47(3):572–595.

Owen, B.

1994 Were Wari and Tiwanaku in conflict, competition, or complementary coexistence? Survey evidence from the upper Osmore drainage, Peru. Paper presented at the 59th Annual Meeting of the Society for American Archaeology, Anaheim, California.

1997 Informe de excavaciones en los sectores mortuorios de Chen Chen: Parte del proyecto rescate de Chen Chen, temporada de 1995. [Cemetery salvage excavations at Chen Chen], Report submitted to the Instituto Nacional de Cultura, Lima.

Ribeiro, M.

1995 The Tumilaca terrace complex: Evidence for the Inca transformation of the Late Intermediate Period political economy of the Osmore Basin, Peru. Paper presented at the 61st Annual Meeting of the Society for American Archaeology, New Orleans.

Shimada, I., C. Schaaf, L. Thompson, and E. Mosley-Thompson

1991 Cultural impacts of severe droughts in the prehistoric Andes: Application of a 1,500-year ice core precipitation record. *World Archaeology* 22(3):247–271.

Stanish, C.

1985 Post-Tiwanaku Regional Economies in the Otora Valley, Southern Peru. PhD dissertation, Department of Anthropology, University of Chicago.

1992 *Ancient Andean Political Economy*. University of Texas Press, Austin.

1994 The hydraulic hypothesis revisited: Lake Titicaca Basin raised fields in theoretical perspective. *Latin American Antiquity* 5(4):312–332.

2001 The origin of state societies in South America. *Annual Review of Anthropology* 30:41–64.

Stanish, C., and I. Pritzker

1990 Reconocimiento arqueológico en el sur del Perú. In *Trabajos arqueológicos en Moquegua, Perú*, Tomo 3, edited by M. Moseley and F. Cabieses, 167–176. Programa Contisuyo del Museo Peruano de la Salud and Southern Peru

Copper Corp., Lima.

Stone, G. D.

1996 *Settlement Ecology: The Social and Spatial Organization of Kofyar Agriculture*. University of Arizona Press, Tucson.

Stone, G. D., and C. E. Downum

1999 Non-Boserupian ecology and agricultural risk: Ethnic politics and land control in the arid Southwest. *American Anthropologist* 101:113–128.

Thompson, L., E. Mosley-Thompson, J. Bolzan, and B. Koci

1985 A 1500 year record of tropical precipitation in ice cores from the Quelccaya ice cap, Peru. *Science* 229(4717):971–973.

Valencia, R., and D. Goldstein

2004 Putting the food on the mesa: Paleoethnobotanical investigations at Cerro Baul, Part 1. Paper presented at the 69th Annual Meeting of the Society for American Archaeology, Montreal.

Wilk, R. R.

1997 *Household Ecology: Economic Change and Domestic Life among the Kekchi Maya in Belize*. Northern Illinois University Press, DeKalb.

Williams, P. R.

1995 Agricultural Hydraulics and the State in the Torata Valley, Perú. Unpublished master's thesis, Department of Anthropology, University of Florida.

1997 The Role of Disaster in the Development of Agriculture and the Evolution of Social Complexity in the South-Central Andean Sierra. PhD dissertation, Department of Anthropology, University of Florida.

2001 Cerro Baúl: A Wari center on the Tiwanaku frontier. *Latin American Antiquity* 12(1):67–83.

2002 A re-examination of disaster induced collapse in the case of the Andean highland states: Wari and Tiwanaku. *World Archaeology* 33(3):361–374.

Williams, P. R., M. E. Moseley, and D. J. Nash

2000 Empires of the Andes: A majestic frontier outpost chose cooperation over war. *Scientific American Discovering Archaeology* March/April:68–73.

Wittfogel, K.

1957 *Oriental Despotism: A Comparative Study of Total Power*. Yale University Press, New Haven, Connecticut.

Intensification, Political Economy, and the Farming Community

In Defense of a Bottom-Up Perspective of the Past

CLARK L. ERICKSON

Archaeologists scramble up tall terraces, wade through cold water in irrigation canals, hop over stone walls, and diligently search farmers' fields for significant concentrations of potsherds that can be registered as a site. In their search for sites, however, most tend to ignore the landscape and what it can show us about agricultural intensification. Contemporary archaeology is firmly rooted in the site concept. Rural sites are said to date agriculture through proximity, and the density, duration, and distribution of settlements are considered indirect evidence of the degree of agricultural intensification. However, we pay only lip service to agricultural fields and boundaries, pathways, roads, and shrines—all seem to be secondary to the goal of finding sites in the form of settlements and monuments. Agricultural features might be described and sketched on the back of site survey forms, but they are rarely discussed in final publications. Despite the term's current popularity, *landscape* still equates with environment, as it is considered merely the context of a site for most archaeologists.

If we are genuinely interested in issues of intensification and intensive agriculture, why do we ignore the most important landscape for directly addressing issues of prehistoric agriculture such as social organization, land tenure, labor organization, and rural lifeways? I would suggest that our perspective has been directed, and limited, by our own cultural background. Few of us grew up on farms or have colleagues that did. Although we are often surrounded by living farming traditions where we excavate and do settlement survey, we rarely pay attention to the farm life going on around us and ignore the relevant local historical and ethnographic literature. In this chapter I will discuss and address a number of explicit and not-so-explicit archaeological assumptions about farming, social organization, settlement patterns, and intensification, as well as their relationship to political economy. I will then show that these assumptions remain largely unsupported by ethnographic and historical evidence. I suggest that a landscape approach that generates models through general and specific analogy and then tests them against landscape signatures of intensive agriculture, agricultural intensification, social organization of labor, land tenure, and energetics can provide a more complete understanding of the rural agrarian past than by relying on sites and settlement patterns alone. I argue that bottom-up approaches are valuable alternatives to the traditional top-down approaches currently used by archaeologists.

INTENSIVE AGRICULTURE, AGRICULTURAL INTENSIFICATION, AND POLITICAL ECONOMY

In the 1960s Ester Boserup (1965) developed her theory of agricultural change. It was a powerful general processual model that attempted to explain agricultural change throughout space and time; and archaeologists, geographers, and historians quickly adopted the Boserupian perspective (e.g., Farrington 1985; Sanders et al. 1979; Spooner 1972; and many others).[1] Over time, some scholars became increasingly disillusioned with many of the basic assumptions of the theory (e.g., population pressure as the primary cause of change, the Law of Least Effort, and the Law of Diminishing Returns). Unconvinced that population pressure could adequately explain agricultural change, scholars began to examine the roles of risk management, innovation, diffusion of technological improvements, competition, agency, market demands, historical contingency, and culture. Historical, ethnographic, and archaeological research based on detailed case studies demonstrated that the predictions of the Boserup theory were rarely confirmed by empirical examples (e.g., Bronson 1972, 1975; Brookfield 1972, 1984, 2001; Erickson 1993, 1996; Kirch 1994; Morrison 1994, 1996; Netting 1993; and others), yet the original theory receives continued support by archaeologists (e.g., Redman 1999; Stone 1996; Stone and Downum 1999). Despite the serious criticism, core elements of the original Boserup hypothesis continue to permeate contemporary interpretations of intensive agriculture, agricultural intensification, and their relationship to political economy in the archaeological record.

More recently, archaeologists have framed the evolution of agricultural systems, prehistoric agricultural intensification, and intensive agriculture within political economy.[2] As it applies to intensification of agriculture, the political economy approach has been labeled the neo-Wittfogelian Perspective (Erickson 1993; Stanish 1994). Although they may deny that the organizational demands of intensive agriculture (large-scale irrigation in Wittfogel's original model) caused state formation and centralized despotic government, some archaeologists assume that the intensification process and resultant large-scale intensive agriculture would require elite involvement and management (e.g., Earle 1997; Johnson and Earle 1987; Kolata 1993, 1996, 2002; Scarborough 2003; Stanish 1994, 2003, 2004, this volume; and others). According to this perspective, leaders (the elite) have a vested interest in the smooth functioning and growth of agricultural production as the source of staple and wealth finance (i.e., surplus extracted as a form of payments or taxation). A related assumption is that farmers practicing the domestic mode of production will resist producing more than is needed for their subsistence needs, an assumption often justified by the Law of Least Effort or farmers' decisions about risk management (Chayanov 1966; Sahlins 1972). Thus, some have concluded that farmers are unlikely to generate surplus unless forced to by local leaders, chiefs, and/or kings (e.g., Stanish 1994). Elite demands for gifts, bribes, payments, or corvée labor are often institutionalized as tax, tribute, or rent and are enforced by legal sanctions and threats of violence. Variations on this theme stress elite encouragement and facilitation of farmers' surplus production by farmers through ideology, ritual sponsorship, and selective distribution of prestige goods and exotics rather than outright control and force (Johnson and Earle 1987; Kolata 1993; Stanish 1994, 2003, 2004, this volume). In most archaic state scenarios political leaders and their bureaucrats are assumed to provide the design, engineering, labor organization, management, and ideology for intensive agriculture. In this perspective agency is often attributed to the elite, while the common masses remain passive and faceless. This particular view about the relationship between intensive agriculture and centralized authority has become orthodoxy in contemporary archaeological applications of cultural evolution and political economy. The Boserupian idea of a continuum from extensive to intensive agriculture that has been assumed to map the cultural evolutionary stages as progress through bands to tribes and chiefdoms to state, is often recast as a continuum of political organization from simple to complex (e.g., Earle 1997, 2001; Johnson and Earle 1987; Redman 1999; Stanish 2003; and others).

The assumptions about the relationship between intensification and political economy that need critical reevaluation include the following:

Assumption 1. All large-scale, highly patterned farming systems are evidence of intensive agriculture and agricultural intensification.

Archaeologists assume that any large-scale, highly-patterned farmed landscape represents intensive agriculture and agricultural intensification (Figure 13.1). The term *intensive agriculture* is often applied superficially, nonspecifically, and uncritically to describe any agricultural system that is (1) on a large scale (what is "large" is poorly defined); (2) assumed to be labor intensive (or

Figure 13.1. Eroded landscape of precolumbian terracing and field boundaries that is still being farmed and modern settlement near Chisi, Lake Titicaca, Bolivia.

assumed to require considerable labor, and again, "considerable" is poorly defined); and (3) more formally structured than what is assumed to be extensive agriculture, such as dry-field or swidden agriculture (the obvious corollary to points 1 and 2). *Agricultural intensification* is an even more elusive term, but the concept usually is applied to any agricultural system that is expanding and/or becoming more complex and permanent in terms of landscape infrastructure.

Scholars of agricultural change have developed specific definitions for intensive agriculture and intensification and have identified physical signatures that are applicable in both archaeological and modern contexts. Most archaeologists confuse labor-intensive agriculture (high production yield with diminishing returns for labor invested per unit area of farmland) with intensive agriculture (continuously farming units of land with short or no fallow periods).[3] Many intensive agricultural systems (as defined by Boserup) are not necessarily labor intensive over the short or long term. In addition, the concept of agricultural intensification is often confused with agricultural expansion or extensification (Brookfield 2001:200).

Archaeologists often seem unaware of these definitions and the natural and social science literature that debates the merits of Boserup's assumptions about population pressure, Law of Least Effort, Law of Diminishing Returns, and the sequence of agricultural evolution (summaries include Brookfield 1972, 1984,

1986, 2001; Denevan 2001; Erickson 1993, 1996; Hunt 2000; Kirch 1994; Morrison 1994, 1996; Netting 1993). Comparative agricultural analyses that employ energetics, crop-production data, labor efficiency, cropping frequency, fallow cycles, sustainability, cost-benefit analysis, farmer decision making, and cultural context are available for most farmed areas of the world but are rarely consulted by archaeologists (examples for the Andes include Denevan 2001; Erickson 1996; Goland 1993; Hastorf 1993; Mayer 2002; Treacy 1994; and others). Merely identifying the presence of intensive agriculture tells us nothing about the process of agricultural intensification. To understand the process, documentation of previous states of the system is needed, and the temporal and spatial scales of analysis must be well defined.

Assumption 2. Large-scale intensive agriculture requires centralized sociopolitical organization in order to function. Corollary: Farmers and rural communities are incapable of creating and managing complex, regional-scale intensive agricultural works.

Wittfogel (1957) argued that the water-management requirements for large-scale irrigation agriculture drove specific state formation processes that resulted in despotic states and the Asian Mode of Production.[4] Although most contemporary archaeologists deny that water management caused the state, elements of Wittfogel's theory relating to the need for top-down management of intensive agriculture have subtly reappeared in political economy explanations of agricultural change. Large-scale, highly visible agricultural infrastructure (canals, dams, reservoirs, raised fields, terraces, silos, roads, walls, field markers, and other elements of the built environment) (Figure 13.2) are uncritically assumed to have been created through state initiatives and ideology (extorted gifts, redistribution, taxation, corvée labor) designed by elite designers, engineers, and administrators.

Historians, ethnographers, geographers, and some archaeologists have soundly criticized the assumption that centralized state political organization is a necessary condition for large-scale intensive agriculture (e.g., Butzer 1996; Denevan 2001; Doolittle 2000; Gelles 1995, 2000; Glick 1970, 1995:64–91, this volume; Hunt 1988, 1989, 1994, 2000; Isaac 1993; Lansing 1991; Mabry 1996, 2000; Mabry ed. 1996; Mitchell 1973, 1976; Mitchell and Guillet 1994; Netting 1993; Stone 1996; Trawick 2001, 2003; Treacy 1994; Wilkinson 2003, this volume). These scholars do not deny that hierarchy and alternative organizational structures, including heterarchy, exist and are often necessary for the functioning of intensive agriculture. These hierarchical and heterarchical structures can be found operating at the local and regional level through families, lineages, communities, moieties, and intercommunity cooperation, often outside of state control and interference. Many archaeologists have largely ignored these contributions from historical and ethnographic case studies.

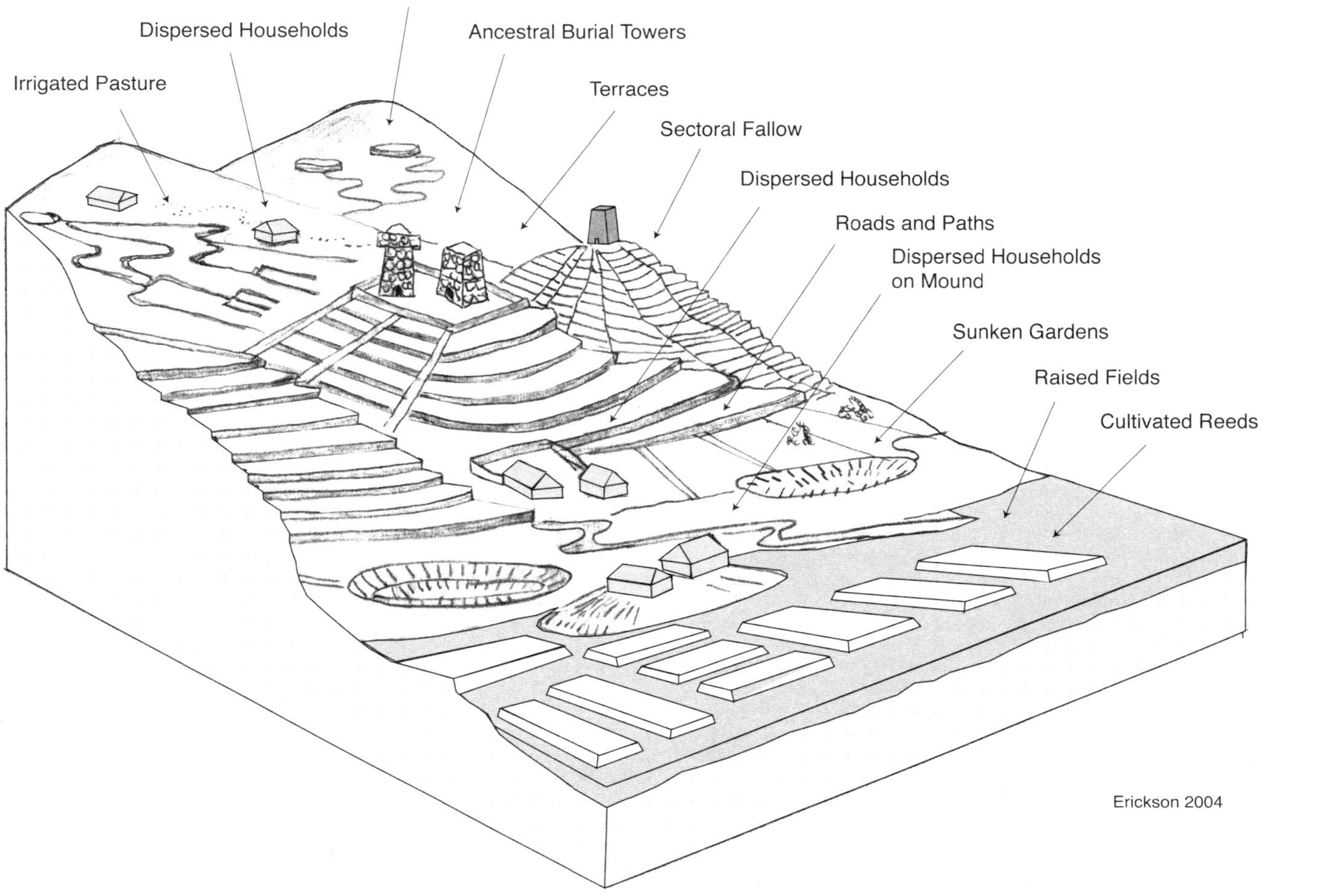

Figure 13.2. An idealized precolumbian cultural landscape in the Andean highlands of raised fields (*waru waru, suka kollu*), sunken gardens (*qocha*), terraces (*andenes*), irrigated pasture (*bofedales*), community burial towers (*chullpas*), paths, roads, and dispersed settlement (*llacta, marka, ayllu*). Drawing by C. L. Erickson.

Historical and ethnographic case studies of irrigation, the best-studied form of intensive agriculture, demonstrate that the social organization of irrigation is highly variable through time and space. Using cross-cultural comparison of irrigation societies at various levels of organization, Mabry and Cleveland (1996) show that locally organized irrigation systems usually outperform more centrally managed ones in terms of productivity (output per unit of land area over time) and efficiency (ratio of output per unit of input over time), in addition to being more stable over time. Numerous studies of state organization and management of intensive agriculture document poor performance, inefficiency, and environmental degradation (Brookfield 2001; Brookfield et al. 2002; Hunt 1988; Lansing 1991; Mabry 1996; Netting 1993; Scott 2001; and others). We may not conclude that locally organized intensive agriculture is always necessarily more successful, efficient, and environmentally friendly, but it is clear that centralized management is not necessary for intensive agriculture.

The assumption that local organizations cannot create and manage large-scale agricultural systems permeates most archaeological treatments of intensification and intensive agriculture (e.g., Kolata 1993, 1996, 2002; Stanish 1994, 2003, 2004 for Andean raised field agriculture). Many archaeologists are now willing to attribute a local origin and control to small-scale systems of intensive agriculture but still balk at attributing large-scale regional and interregional systems of intensive agriculture to farmer agency, skill, and knowledge.

Numerous case studies show that the assumption is false (Brookfield 2001; Brookfield et al. 2002; Denevan 2001; Doolittle 1984, 2000; Erickson 1993, 1996; Lansing 1991; Mayer 2002; Netherly 1984; Scarborough 2003; Trawick 2001, 2003; Treacy 1994; and others). Experimental archaeology and energetics studies clearly demonstrate that small groups organized at the family and community level are capable of constructing and maintaining intensive agricultural systems (Figures 13.3 and 13.4) (for raised fields: Erickson 1993, 1996; Erickson and Candler 1989; for terracing: Kolb 1997; Kolb and Snead 1997; Treacy 1994). Steven Lansing's *Priests and Programmers: The Engineered Landscapes of Bali* (1991) is a brilliant analysis of a huge integrated regional intensive agricultural system that operates without state control or direct involvement. Other clear examples of nonstate intensive agriculture include the irrigation systems of Mojave and Paiute hunting-gathering societies (Lawton et al. 1976; Steward 1930), the agroforestry management of Amazonian hunter-gatherers and small-scale farmers (Denevan and Padoch 1988; Posey and Balée 1989), and the raised fields of New Guinea "tribal societies" (Heider 1970; Serpenti 1965).

Figure 13.3. Potato harvest on a small family plot of raised fields in the community of Faon, Huatta, Peru. Remains of eroded precolumbian fields, canals, and settlement mounds extend to the horizon.

Figure 13.4. Potato harvest on communal raised fields by the community of Segunda Collana, Huatta, Peru. These reconstructed raised fields were part of an experiment in applied archaeology between 1981 and 1986.

Assumption 3. The coexistence of centralized political organization and intensive agriculture implies a causal or necessary relationship.

My critique of assumption 2 also applies to assumption 3. Textbooks on cultural evolution highlight intensive agriculture as an important trait of chiefdom and state-level societies (Earle 1997; Johnson and Earle 1987; Redman 1999; and others). Coexistence of intensive agriculture and complex societies is often confused with causation. Many, possibly most, intensive agricultural systems predate the appearance of political complexity characterized by the presence of chiefdoms or states.

Are the elite really interested in managing day-to-day farming matters? In some societies the answer is yes, in others no. There are clear archaeological cases where the state was directly involved in intensive agriculture and possibly the intensification process as well. Inca regional administrative and ceremonial centers and private estates are highly visible examples of the state's hand in creating intricate hydraulic infrastructure, transforming slopes into terrace walls and platforms, and establishing transportation networks and storage facilities (D'Altroy 2002; Hastorf 1993; Niles 1987; and others). Although highly visible and easily identified as Inca, these landscapes make up an insignificant portion of the total anthropogenic built landscapes of the Andes (Erickson 2000). In addition, the Inca often appropriated the preexisting landscape capital of hundreds of generations of farmers for their own use. Still, archaeologists uncritically attribute the transformation of Andean regional landscapes to the state, whether Inca or pre-Inca. Most scholars agree that Inca colonial policy was to expand agricultural production to new lands previously uncultivated (agricultural expansion) rather than to intensify agricultural production (agricultural intensification) per se (D'Altroy 2002; Murra 1980). The practical explanation for this practice is that the Inca did not want to disrupt what already existed as highly efficient production of local communities. I do not deny that in some cases agricultural intensification may have been an important factor in state origin, maintenance, and expansion. All states rely on production of surplus by peasant farmers for their existence. However, the details of how (and whether) state organization and intensive agriculture are causally related clearly vary from case to case.

Assumption 4. Agriculture evolves in a unilinear, stepwise pathway from extensive to intensive agriculture.

Boserup argued for a unilinear evolutionary continuum from extensive to intensive agriculture as measured by cropping frequency. Archaeologists have correlated this evolutionary scheme with that of sociopolitical organization. Simple societies are assumed to practice hunting-gathering-fishing subsistence strategies and/or extensive agriculture (primitive gardening, slash-and-burn agriculture, simple agroforestry), whereas complex societies practice intensive agriculture (irrigation, raised fields, and terracing) (e.g., Earle 1997; Johnson and Earle 1987; Redman 1999). Some scholars present the degree of agricultural intensification as a part of the definition of each cultural evolutionary stage and at the same time invoke it as a partial cause for the stage. This inherently circular argument assumes that the centralized state is a necessary condition for intensive agriculture and that centralized states are characterized by intensive agriculture.

Historical ecologists have questioned these unilinear cultural evolutionary assumptions (Balée 1989, 1994; Denevan 1992a, 2001; Erickson 1996; Lathrap et al. 1985). Some intensive forms of agriculture (agroforestry, house gardens, drained fields) appear early in the archaeological record (e.g., Denham et al. 2003; Neumann 2003, for New Guinea; Leach 1999, for the Pacific) while extensive agriculture such as swidden (slash and burn) may be a relatively late phenomenon (at least in the Americas owing to historical introduction of metal axes [Denevan 1992a, 2001; Lathrap 1977; Lathrap et al. 1985]). Ethnographic, historical, and prehistoric case studies show that most farming societies practice a wide range of extensive to intensive agricultural practices concurrently (Figure 13.2).

Extensive agriculture is assumed to have preceded intensive agriculture, but its archaeological existence remains much more elusive than intensive agriculture; indirect evidence such as pollen records, burning, and stone axes are sometimes alluded to (e.g., Piperno and Pearsall 1998). More commonly, the signature of nonintensive agriculture is identified as the "negative" of intensive agriculture: an absence of visible built environment infrastructure such as field boundaries, canals, or terraces.

Assumption 5. Farmers refuse to produce a surplus unless forced to do so by higher authority.

Boserup (1965) proposed that the Law of Least Effort underlay the evolution of agriculture. In other words, farmers will commit the bare minimum of labor effort toward agricultural production unless forced to do more (in her model, by population pressure). This basic assumption overlaps with one of the core assumptions of intensification in political economy. Marshall Sahlins (1972) and A. V. Chayanov (1966) proposed that farmers, left on their own, would not produce beyond their subsistence needs; thus, local leaders, chiefs, and kings must motivate farmers to produce a surplus for sustaining their activities, craft production, trade, rituals, and urban centers (Dietler and Hayden 2001; Earle 1997; Johnson and Earle 1987; Kolata 1993; Stanish 1994).

Robert Netting (1993) has presented a sound critique of the Sahlins and Chayanov assumptions. Most ethnographic or historical farming societies produce beyond the subsistence level and domestic sphere, and those that do not are probably under extreme stress and pressure beyond their control. The reasons farmers decide to produce more than needed for immediate subsistence are complex (Brookfield 2001; Brookfield et al. 2002; Denevan 2001; Kirch 1994; Netting 1993; Zimmerer 1996; and others). Scholars have documented past and present "noncomplex societies" that produced surpluses to meet a variety of social demands without elite motivation (Bender 1985; Brookfield 1984, 2001; Hastorf 1998; Lathrap et al. 1985; Netting 1993). The earliest monuments, long-distance trade, and intensive agriculture, oft cited as evidence of surplus production, appeared in nonhierarchical, stateless societies (e.g., Burger 1992, for South America; Shady Solís et al. 2001).

Assumption 6. Intensive agricultural production is more efficient if centralized and bureaucratized.

Archaeologists generally assume that the state provides cost-effective strategies for managing intensive agriculture. One goal of the state is to regularize the flow of agricultural surplus into state coffers. This often involves the development of bureaucracy for efficient tribute and tax collection but rarely direct involvement in agricultural production.

In fact, numerous scholars working with historical and ethnographic case studies show that when states meddle with traditional peasant farming, agricultural efficiency is often lost (Brookfield 2001; Lansing 1991; Netting 1993; James Scott 1998). In his cogent analysis of state mentality, Scott (1998) demonstrates how modern state experiments in collective agriculture, the Green Revolution, and other top-down centralized schemes have utterly failed.

Assumption 7. The presence of a settlement hierarchy is evidence of centralized political control and administration of agriculture.

Archaeologists assume that settlement hierarchies of three to four tiers are evidence of centralized political organization (chiefdom and state societies). The presence of higher order settlements within areas of intensive agriculture is believed to be the signature of direct state management and administration of rural communities (e.g., Kolata 1993, 1996, 2002; Stanish 1994, 2003, for the Andes). However, the relationship between variation in settlement pattern and intensification of agriculture remains poorly known and rife with black-box assumptions. The ethnographic and historical record can provide historically contingent and cross-culturally testable models for the relationships among settlement, farmer cooperation, labor, and intensive agriculture (Figure 13.5) (Brookfield 2001; Erasmus 1956; Goland 1993; Netting 1993; Stone 1996; and others).

Increasing political centralization does correlate with the archaeological record of changes in rural settlement patterns, domestic production strategies, architecture, and distribution and access to certain items of material culture. I do not deny that these changes can often be positive in the lives of common farmers. However, the possibility that hierarchical settlement patterns

Figure 13.5. A gathering of representatives from two communities in 1986 (Aymara farmers from Ilave on left, Quechua farmers from Huatta on right) to exchange ideas and experiences about rehabilitating and putting pre-Columbian terraces and raised fields back into production in Huatta, Peru.

(indicating a centralized political economy) and intensification are independent should also be considered. In Andean prehistory regular cycles of centralization and decentralization are more common than long periods of political stability (Marcus 1998; Moseley 2001; Willey 1991; and others). States are ephemeral, most lasting only several hundred years, with longer periods of smaller-scale local and regional sociopolitical organization. States come and go while rural farm life often continues relatively unaffected (as manifested in the lower rungs of settlement patterns, farming systems, material culture, and household architecture). For most farming peoples those periods without strong centralized states may have been ideal.

A LANDSCAPE APPROACH TO AGRICULTURAL INTENSIFICATION

Site- and settlement-based archaeology continue to dominate our view of the past (Figure 13.6). Landscape remains invisible to most archaeologists in their diligent search for what they see as the most relevant data: pottery scatters, settlements, standing architecture, and monuments. I believe that landscapes provide a radically different and productive approach to understanding the past and, particularly, to explaining intensification and intensive agriculture (Figure 13.7).

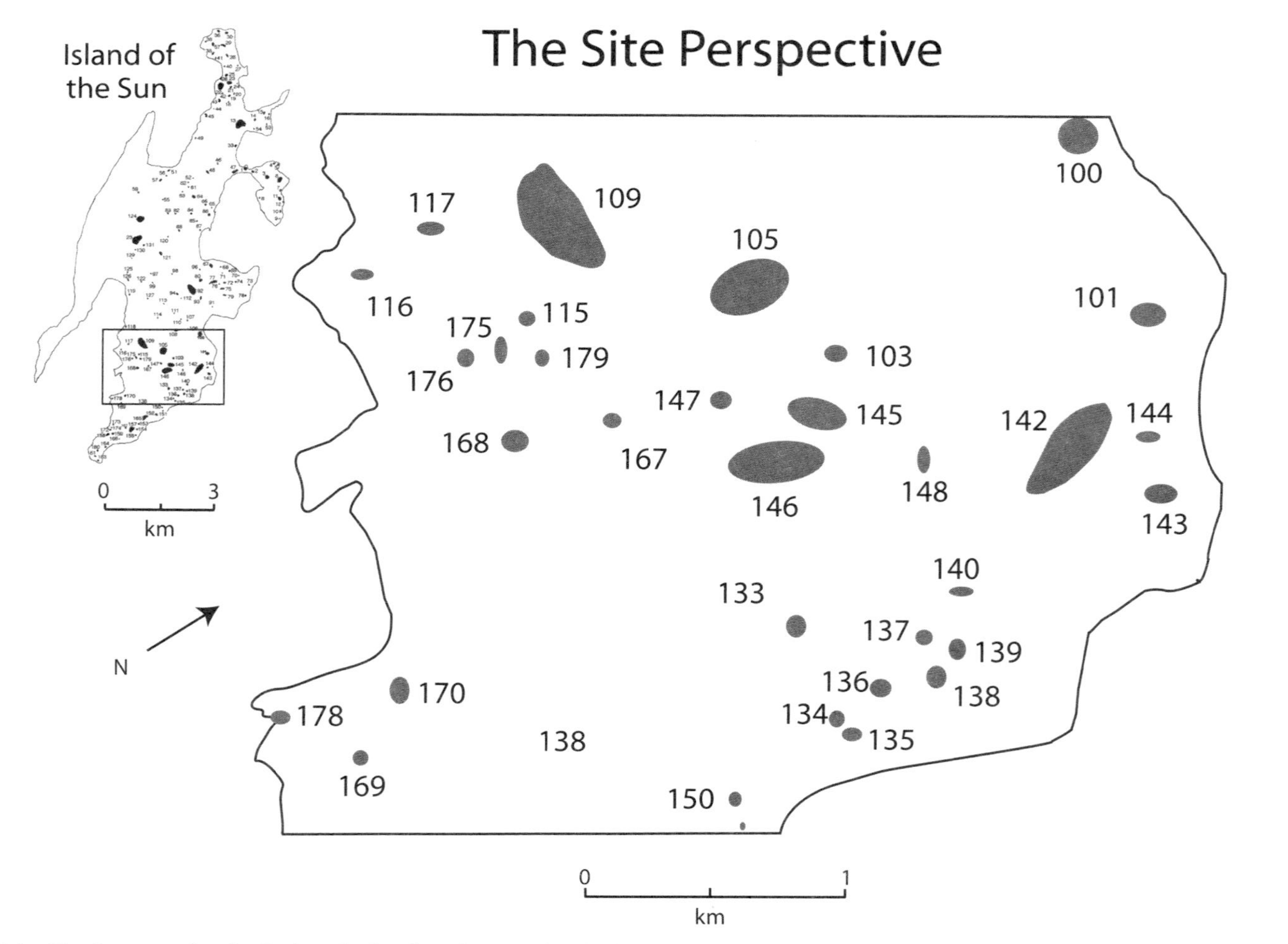

Figure 13.6. The site perspective: distribution of archaeological sites on the Island of the Sun in Lake Titicaca, Bolivia. Sites defined by artifactual surface debris and/or standing architecture during the settlement survey are mapped and plotted to be categorized by period, size, and function into a site hierarchy ranking that reflects the political economy and cultural evolutionary stage (adapted from Bauer and Stanish 2001:Map 4.4).

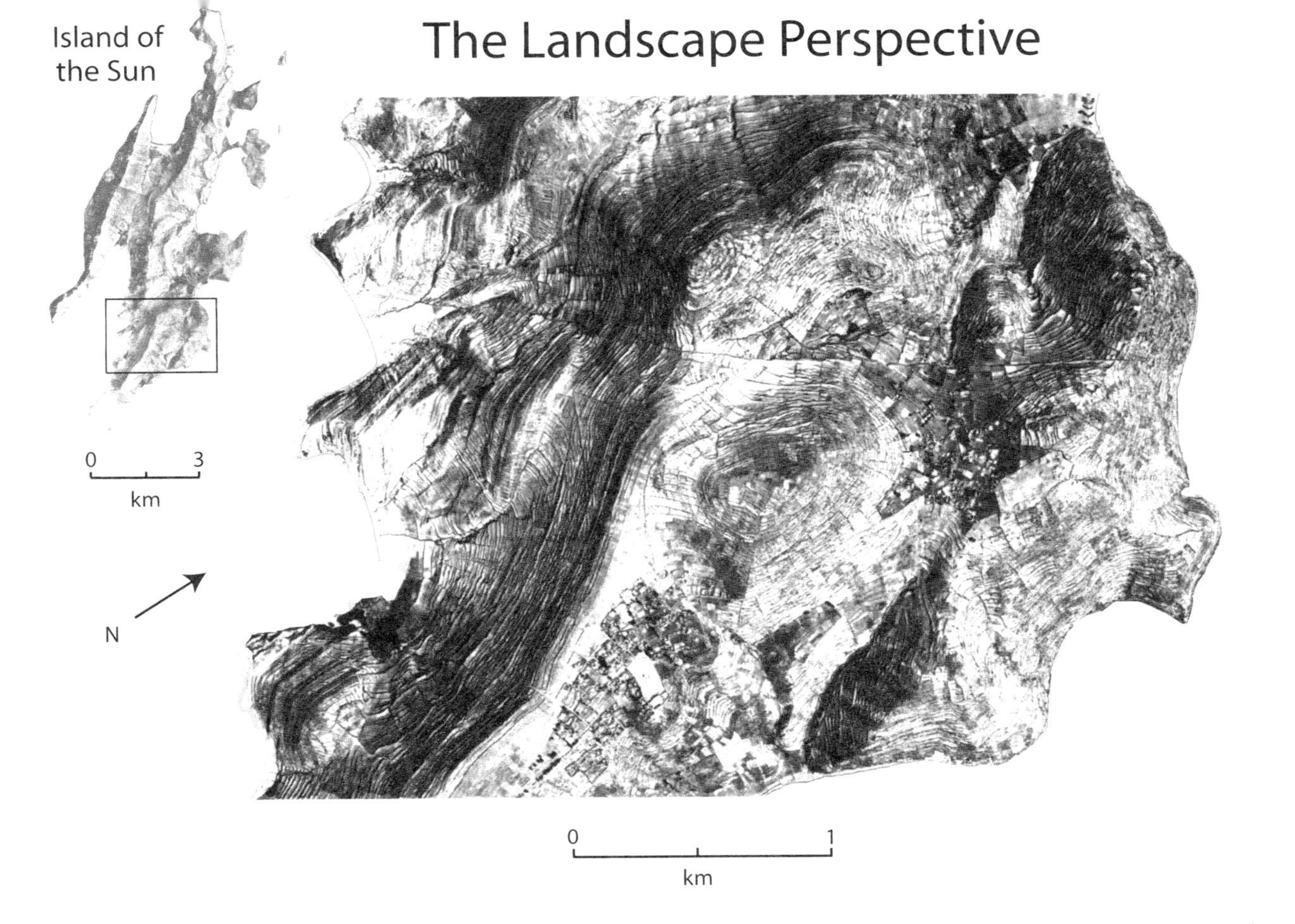

Figure 13.7. The landscape perspective: an anthropogenic, engineered landscape on the Island of the Sun in Lake Titicaca, Bolivia. The complex patterning and palimpsest of landscape features that cover nearly the entire island include thousands of stone-lined terrace fields on the steep slopes, walls running upslope and down marking field and community boundaries and sectorial fallow systems, webs of ritual roads and everyday paths, irrigation canals, corrals, and house gardens. By any measurement the engineering and labor invested in this built environment is monumental in scale.

I argue that the political economy approach denies agency to farmers and underestimates their knowledge and cumulative efforts in creating vast areas of anthropogenic landscape. I have contrasted the perspective of a settlement pattern analysis based on the site concept and informed by political economy, with a landscape perspective informed by indigenous knowledge systems, characterizing them as "top-down" vs. "bottom-up" approaches (Erickson 1993; Scarborough 1993). The archaeology of landscapes and historical ecology can provide a bottom-up farmer-centric perspective of the past. The bottom-up perspective (in contrast to the elite perspective of political economy) draws heavily on the works of Robert Netting, William Denevan, William Doolittle, Harold Brookfield, Stephen Lansing, Robert Hunt, Karl Zimmerer, Enrique Mayer, William Balée, Barbara Bender, Christopher Tilley, James Scott, Carole Crumley, Patrick Kirch, Kathleen Morrison, and others.[5] This approach is informed by theories of practice, structuration, agency, structuralism, and poststructuralism, and it relies on the concepts of space and place, historical ecology, indigenous knowledge systems, heterarchy, smallholders, landscape capital, inhabitation, resistance, and historical contingency. Its methods include energetics studies, experimental archaeology, ethnoarchaeology, multiscalar analysis, phenomenology, and pattern recognition. The archaeology of landscapes builds arguments from patterned physical evidence from the scale of activity area, to region, and beyond.

Intensive agriculture and the process of its intensification are natural subjects for landscape-based investigation. Most activities of farm life that are pertinent to intensive agriculture are not settlement based; rather, they occurred in that elusive gray zone imperceptible to archaeologists focusing primarily on sites (Figure 13.8). Because intensive agriculture and intensification are often associated with dense populations, heavy investments of landscape capital, formal bounding of fields and territories, and maintenance of ancestor cult architecture, their signature is permanently embedded in the physical landscape as highly patterned structures. These structures, in turn, channel the movements and actions of human actors on the landscape. As a form of built environment, the design, structure, and scale of landscapes of intensive agriculture are equal to or beyond the prehistoric architecture that we traditionally study (Figure 13.9). Landscapes are multicomponent, most having been occupied for thousands, if not hundreds of thousands, of years. Because the signature of intentional and unintentional human activities is so strong, pervasive, and sustained over long periods of time, landscapes are incredible palimpsests of both continuous traditions and abrupt disjunctures of habitation, land use, and sociopolitical systems.

Figure 13.8. Quechua farmers in Huatta, Peru, constructing raised fields using the Andean foot plow (*chakitaqlla*) to cut sod blocks elevating planting platforms and digging intervening canals. A stone or wooden bladed version of the tool was used prehistorically to create the Andean landscapes illustrated in the figures of this chapter.

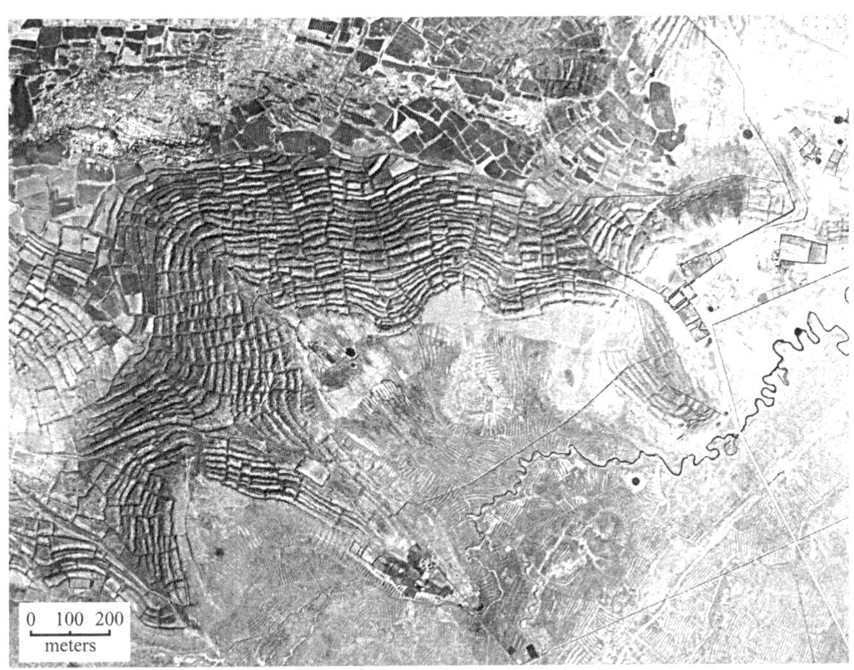

Figure 13.9. An integrated, engineered landscape of terraces (upper left) and raised fields (lower right). The landscape capital includes construction of many linear kilometers of terrace and field-boundary walls, movement of massive volumes of earth, and reworking of soil horizons to a depth of 1 meter or more.

Landscape and settlement archaeology continue to explore new ways of using the archaeological record to understand intensification. While the archaeological signatures of fallow periods, cropping intensity, labor, sustainability, demography (population size, density, pressure, and carrying capacity), farmer decision making, risk management, and social organization can be elusive, a focus on the archaeology of landscape, community, property, and everyday life and on historical ecology can provide new conceptual frameworks and methods (Balée and Erickson in press; Brookfield 2001; Canuto and Yaeger 2001; Denevan 2001; Doolittle 2000; Erickson 1999, 2003a, 2003b; Kolb 1997; Kolb and Snead 1997; Stone 1996; Whittmore and Turner 2002; Zimmerer and Young 1998; and others).

A landscape of intensive agriculture is best described as completely "anthropogenic." I consider the farmed landscapes of the Andes and Amazon where I work to be clear examples of designed, engineered, constructed, and humanized landscapes, a class of built environment and archaeological artifact that involved labor input over long periods, an accumulation of landscape capital, and multigenerational knowledge (Figure 13.10). Landscapes were built over temporally long scales, starting with systematic burning, forest and grassland management, and dispersal of plants and animals by hunting and gathering peoples 12,000 years ago in the Americas and hundreds of thousands of years ago in other parts of the world. Sedentism and farming brought new and massive transformations of the landscape. Complex agrarian landscapes are built through a process of accretion (Doolittle 1984). Farmers continuously make land improvements that are passed down to succeeding generations. In other cases farmers may have degraded environments through their transformation of the landscape (e.g., Denevan 1992b; Redman 1999; Stahl 1996; Whittmore and Turner 2002). Following Brookfield (Blaikie and Brookfield 1987; Brookfield 1984, 2001), I refer to these improvements as accumulated landscapes or landscape capital (also referred to as landesque capital).

A farmer-centric perspective on intensification highlights the long-term processes of design, construction, and imposing territoriality in the form of permanent field lines, paths, roads, field walls, boundary markers, plow patterns, field dimensions, and orientations (Figure 13.11). The concept of the palimpsest, analogous to stratigraphy, is useful for sorting out multiple meaningful cultural patterns and disjunctures embedded in anthropogenic landscapes. Similar to potsherds, chipped stone, and architecture, farmed landscape features are artifacts that can be described, dated, analyzed, and interpreted according to style, variation, patterning, context, distribution, and meaning.

While intensive agriculture has been the focus of study, the archaeological identification of extensive agriculture remains underdeveloped. The study of both extensive and intensive agriculture requires historically contingent

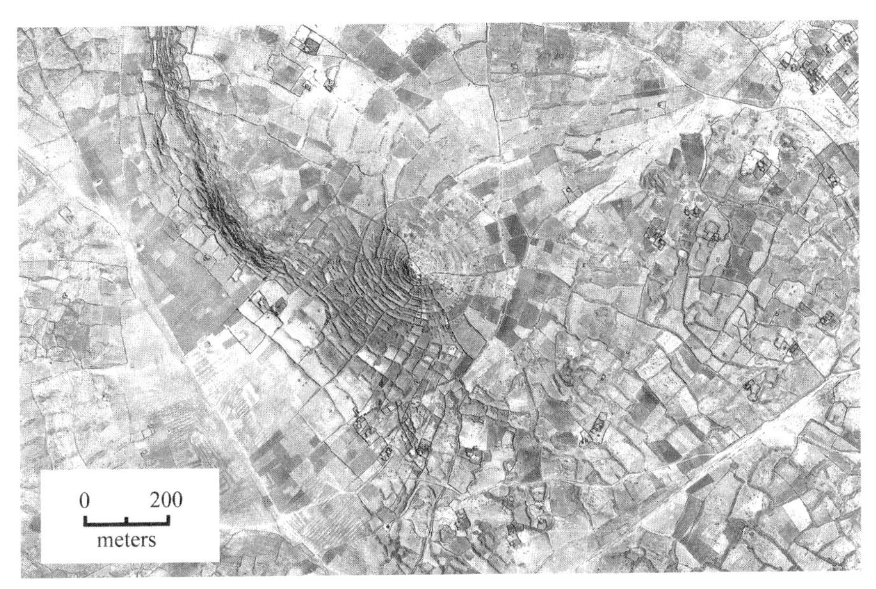

Figure 13.10. A highly patterned, anthropogenic landscape covered with a palimpsest of terraced and raised fields, boundary walls, paths, roads, and dispersed settlement that has been continuously inhabited for more than 10,000 years near Lake Umayo, Peru. Note the radial lines of walls from the low hill in the center of the image (aerial photograph of same landscape as Figure 13.9).

Figure 13.11. Terraces (upper half) and raised fields (lower half) at the edge of Lake Umayo, Peru. The linear walls radiating from the hill are divisions used by the community to rotate crops and fallow over seven-year cycles (ground view of same landscape as Figure 13.9).

and cross-cultural ethnoarchaeological, ethnographic, and historical studies on physical signatures and spatial dimensions of farm labor, land tenure, and property. To resolve these issues, we need better models of the physical archaeological signature of territory, community, property, and land-based lineage in settlement patterns and landscapes (Adler 1996; Bintliff 1999; Canuto and Yaeger 2001; Earle 2000; Hunt 1998; Hunt and Gillman 1998; Kolb and Snead 1997; Marcus 2004; Pyburn 1998; Stone 1996; Walker 2004).

I am not against cross-cultural comparison and application of general analogy to issues of agricultural intensification (excellent cross-cultural treatments include Adler 1996; Brookfield 2001; Denevan 2001; Earle 2000; Kirch 1994; Morrison 1994, 1996; Netting 1993; Stone 1996; and others). I do question the uncritical imposition of generic hierarchical models of political economy that are often derived from the largely discredited assumptions of Boserup, Wittfogel, and the "Mesopotamian Model" of political economy, settlement pattern, and cultural evolution that is applied to the rest of the world. In most areas local detailed ethnographic and historical data are more appropriate starting points for developing testable analogies, models, and hypotheses. I also resist interpretations that deny farmers credit for their works and engineering knowledge and skills. Vast anthropogenic landscapes are all too often attributed to the agency of elites or corporate groups without critical examination. Archaeologists have a unique opportunity to provide insights about the "people without history," but the opportunity is rarely realized through approaches that stress cultural evolution and political economy.

A PEOPLE-CENTRIC APPROACH TO AGRICULTURAL INTENSIFICATION

Who makes the decisions regarding intensive agriculture? In critiquing the political economy approach used by most archaeologists, I may have argued too strongly that all core decisions and strategy planning are made at the level of farm family and community. Ethnography and history show that both the local population and the state directly and indirectly affect the practice of farming. Farm families and communities tap into vast stores of multigenerational indigenous knowledge, social institutions of labor organization and management, ritual and symbolic systems, and capital embedded in the land from improvements contributed by their ancestors. Their decisions and behavior may or may not follow Chayanov and Sahlins-esque assumptions of political economists.

The bottom-up perspective is merely one perspective among many that are possible. It does not claim to provide all ultimate answers, but it does provide an important and useful counterbalance to some of the perspectives that have become so dominant in the field that they have blinded us to other possibilities.

I do not deny that there might be a relationship between political economy and intensive agriculture, but I do believe that the relationship should be demonstrated rather than simply assumed.

I do not argue for a romantic, idyllic view of happy, cooperative peasants. Real farming communities are rife with endemic tension, infighting, violence, and inequality. Contemporary and historical studies of rural farmers document how their difficult lives are characterized by exploitation, civil unrest, class conflict, land and civil rights struggles, and other problems. Intensive farming on ancestral lands, and the resultant investment in landscape capital, tends to tie farmers to their land. Intensive agriculture makes it easy for local leaders, governments, and expansive empires to control and extort taxes from farming communities (Carneiro 1970; Gilman 1981; Webster 1990; and others).

Peopling the past is a radical alternative to viewing farmers as faceless masses, the passive recipients of what the elite impose on them through direct coercion or state ideology. More important, this approach highlights the cultural links between the past and contemporary peoples of many of the regions of the world where we work (Figure 13.12). It also has important applied implications for empowerment, recovery of knowledge systems, sustainable technology and land use, indigenous land claims, ethnic identity, biodiversity, and the cultural revitalization of living communities (Erickson 2003b).

Figure 13.12. Members of a Quechua women's weaving cooperative resting after a day of constructing communal raised fields in Huatta, Peru.

Acknowledgments

I thank the participants of the Cotsen Advanced Seminar for their stimulating papers, discussion, and collegiality during and after the conference at the Cotsen Institute of Archaeology at UCLA. As editors, Joyce Marcus and Chip Stanish carefully read and provided critical editing of the manuscript. I also acknowledge my colleagues Jason Yaeger, Peter Stahl, Kay Candler, and William Denevan for discussion and editorial comments about the manuscript. A shorter, slightly different version of this chapter was presented in the symposium "Peopling Archaeology: Exploring Wendy Ashmore's Contributions to the Archaeology of Social Life and Cultural Landscapes," at the Annual Meeting of the American Anthropological Association, New Orleans, Louisiana, November 20–24, 2002. I thank the organizers, Cynthia Robin and Arthur Joyce, for the opportunity to participate. Critical comments that improved the essay were provided by the symposium discussant, Jerry Sabloff. In particular, I thank the people of Huatta, Coata, and Capachica in the Andean highland Peru and the Llanos de Mojos in the Amazonian lowlands of Bolivia.

NOTES

1. For excellent summaries of Boserup's hypothesis on agricultural intensification see Morrison (1994), Netting (1993), and Stone (1996).

2. Feinman and Nicholas (2004) provide a survey of archaeological approaches to political economy.

3. The distinction between intensive agriculture and labor-intensive agriculture is important. Both are germane to issues of sociopolitical evolution because the former relates to increased surplus potential with no necessary decrease in labor efficiency (a non-Boserupian assumption that characterizes many highly productive traditional farming systems) and the latter to increased surplus potential under decreasing labor efficiency with implications for the control of labor (a Boserupian assumption). These issues are discussed in detail in Netting (1993) and Erickson (1993, 1996).

4. Wittfogel (1957) clearly distinguished between hydraulic societies (those practicing large-scale irrigation that develop into centralized, despotic states) and hydroagricultural societies (those practicing small-scale irrigation that do not develop into centralized, despotic states) (see also Isaac 1993; Mitchell 1973, 1976; Price 1994; Scarborough 2003). Archaeologists applying the neo-Wittfogelian perspective to archaeological cases generally collapse the two categories.

5. See, e.g., William Balée (1994); Balée and Erickson (in press); Barbara Bender (1998); Harold Brookfield (2001); Brookfield et al. (2002); Carole Crumley (1994); William Denevan (2001); William Denevan and Christine Padoch (1988); William Doolittle (1984, 2000); Robert Hunt (1988, 1989, 2000); Patrick Kirch (1994); Stephen Lansing (1991); Enrique Mayer (2002); Kathleen Morrison (1994, 1996); Robert Netting (1993); James Scott (1998); Christopher Tilley (1994); Paul Trawick (2001, 2003); and Karl Zimmerer (1996).

REFERENCES

Adler, M. A.

1996 Land tenure, archaeology, and the ancestral Pueblo social landscape. *Journal of Anthropological Archaeology* 15:337–371.

Balée, W. L.

1989 The culture of Amazonian forests. In *Resource Management in Amazonia: Indigenous and Folk Strategies*, edited by D. A. Posey and W. L. Balée, 1–21. New York Botanical Garden, Bronx.

1994 *Footprints of the Forest: Ka'apor Ethnobotany: The Historical Ecology of Plant Utilization by an Amazonian People*. Columbia University Press, New York.

Balée, W. L., and C. L. Erickson (editors)

In press *Time and Complexity in Historical Ecology: Studies from the Neotropics*. Columbia University Press, New York.

Bauer, B. and C. Stanish

2001 *Ritual and Pilgrimage in the Ancient Andes*. University of Texas Press, Austin.

Bender, B.

1985 Emergent tribal formations in the American midcontinent. *American Antiquity* 50(1):52–62.

1998 *Stonehenge: Making Space*. Berg, New York.

Bintliff, J.

1999 Settlement and territory. In *Companion Encyclopedia of Archaeology*, edited by G. Barker, 505–545. Routledge, London.

Blaikie, P. M., and H. C. Brookfield

1987 *Land Degradation and Society*. Methuen, London.

Boserup, E.

1965 *The Conditions of Agricultural Growth*. University of Chicago Press, Chicago.

Bronson, B.

1972 Farm labor and the evolution of food production. In *Population Growth: Anthropological Implications*, edited by B. Spooner, 190–218. MIT Press, Cambridge, Massachusetts.

1975 The earliest farming: Demography as cause and consequence. In *Population Ecology and Social Evolution*, edited by S. Polgar, 52–78. Mouton, The Hague.

Brookfield, H. C.

1972 Intensification and disintensification in Pacific agriculture: A theoretical approach. *Pacific Viewpoint* 13:30–48.

1984 Intensification revisited. *Pacific Viewpoint* 25:15–44.

1986 Intensification intensified: Review of *Prehistoric Intensive Agriculture in the Tropics*, edited by I. Farrington. *Archaeology in Oceania* 21(3):177–180.

2001 *Exploring Agrodiversity*. Columbia University Press, New York.

Brookfield, H., C. Padoch, H. Parsons, and M. Stocking (editors)
2002 *Cultivating Biodiversity: Understanding, Analysing, and Using Agricultural Diversity*. ITDG Publishing, London.

Burger, R.
1992 *Chavin and the Origins of Andean Civilization*. Thames and Hudson, New York.

Butzer, K.
1996 Irrigation, raised fields, and state management: Wittfogel redux? Reviews of *Economic aspects of water management in the prehispanic New World*, edited by V. Scarborough and B. Isaac; and *The wet and the dry: Irrigation and agricultural intensification in Polynesia*, by P. V. Kirch. *Antiquity* 70(267):200–204.

Canuto, M., and J. Yaeger (editors)
2001 *The Archaeology of Communities*. Routledge, New York.

Carneiro, R.
1970 A theory of the origin of the state. *Science* 169:733–738.

Chayanov, A. V.
1966 *The Theory of Peasant Economy*. Edited by D. Thorner, B. Kerblay, and R. E. F. Smith. R. D. Irwin. Homewood, Illinois.

Crumley, C. L. (editor)
1994 *Historical Ecology: Cultural Knowledge and Changing Landscape*. School of American Research, Santa Fe.

D'Altroy, T.
2002 *The Incas*. Blackwell, New York.

Denevan, W. M.
1992a Stone vs. steel axes: The ambiguity of shifting cultivation in prehistoric Amazonia. *Journal of the Steward Anthropological Society* 20:153–165.
1992b The pristine myth: The landscapes of the Americas in 1492. *Association of American Geographers* 82(3):369–385.
2001 *Cultivated Landscapes of Native Amazonia and the Andes*. Oxford University Press, Oxford.

Denevan, W. M., and C. Padoch (editors)
1988 *Swidden-Fallow Agroforestry in the Peruvian Amazon*. New York Botanical Garden, Bronx.

Denham, T. P., S. G. Haberle, C. Lentfer, R. Fullagar, J. Field, M. Therin, N. Porch, and B. Winsborough
2003 Origins of agriculture at Kuk Swamp in the highlands of New Guinea. *Science* 301:189–193.

Dietler, M., and B. Hayden (editors)
2001 *Feasts: Archaeological and Ethnographic Perspectives on Food, Politics, and Power*. Smithsonian Institution Press, Washington, DC.

Doolittle, W.

1984 Agricultural change as incremental process. *Annals of the Association of American Geographers* 74:124–137.

2000 *Cultivated Landscapes of Native North America*. Oxford University Press, Oxford.

Earle, T.

1997 *How Chiefs Come to Power: The Political Economy of Prehistory*. Stanford University Press, Palo Alto, California.

2000 Archaeology, property, and prehistory. *Annual Reviews of Anthropology* 29:39–60.

2001 Institutionalization of chiefdoms: Why landscapes are built. In *From Leaders to Rulers*, edited by J. Haas, 105–124. Kluwer Academic/Plenum, New York

Erasmus, C. J.

1956 Culture structure and process: The occurrence and disappearance of reciprocal farm labor. *Southwestern Journal of Anthropology* 12:444–469.

Erickson, C. L.

1993 The social organization of prehispanic raised field agriculture in the Lake Titicaca Basin. In *Economic Aspects of Water Management in the Prehispanic New World*, edited by V. Scarborough and B. Isaac, 369–426. JAI Press, Greenwich, Connecticut.

1996 *Investigación arqueológica del sistema agrícola de los camellones en la cuenca del lago Titicaca del Perú*. Programa Interinstitucional de Waru Waru and Centro para Información para el Desarrollo, La Paz.

1999 Neo-environmental determinism and agrarian "collapse" in Andean prehistory. *Antiquity* 73(281):634–642.

2000 The Lake Titicaca Basin: A pre-columbian built landscape. In *Imperfect Balance: Landscape Transformations in the Precolumbian Americas*, edited by D. Lentz, 311–356. Columbia University Press, New York.

2003a Historical ecology and future explorations. In *Amazonian Dark Earths: Origin, Properties, Management*, edited by J. Lehmann, D. C. Kern, B. Glaser, and W. I. Woods, 455–500. Kluwer, Dordrecht.

2003b Agricultural landscapes as world heritage: Raised field agriculture in Bolivia and Peru. In *Managing Change: Sustainable Approaches to the Conservation of the Built Environment*, edited by J.-M. Teutonico and F. Matero, 181–204. Getty Conservation Institute in collaboration with US/ICOMOS, Oxford University Press, Oxford.

Erickson, C., and K. Candler

1989 Raised fields and sustainable agriculture in the Lake Titicaca Basin. In *Fragile Lands of Latin America: Strategies for Sustainable Development*, edited by J. Browder, 230–248. Westview Press, Boulder, Colorado.

Farrington, I. (editor)

1985 *Prehistoric Intensive Agriculture in the Tropics*. British Archaeological Reports, Oxford.

Feinman, G., and L. Nicholas (editors)

2004 *Archaeological Perspectives on Political Economies*. University of Utah Press, Salt Lake City.

Gelles, P.

1995 Equilibrium and extraction: Dual organization in the Andes. *American Ethnologist* 22(4):710–742.

2000 *Water and Power in Highland Peru: The Cultural Politics of Irrigation and Development*. Rutgers University Press, New Brunswick, New Jersey.

Gilman, A.

1981 The development of stratification in Bronze Age Europe. *Current Anthropology* 22:1–23.

Glick, Thomas F.

1970 *Irrigation and Society in Medieval Valencia*. Harvard University Press, Cambridge, Massachusetts.

1995 *From Muslim Fortress to Christian Castle: Social and Cultural Change in Medieval Spain*. St. Martin's, New York.

Goland, C.

1993 Field scattering as agricultural risk management: A case study from Cuyo Cuyo, Department of Puno, Peru. *Mountain Research and Development* 13(4):317–338.

Hastorf, C.

1993 *Agriculture and the Onset of Inequality before the Inka*. Cambridge University Press, Cambridge, UK.

1998 The cultural life of early domestic plant use. *Antiquity* 72:773–782.

Heider, K.

1970 *The Dugum Dani: A Papuan Culture in the Highlands of West New Guinea*. Aldine, Chicago.

Hunt, R. C.

1988 Size and structure of authority in canal irrigation systems. *Journal of Anthropological Research* 44(4):335–355.

1989 Appropriate social organization? Water user associations in bureaucratic canal irrigation systems. *Human Organization* 48(1):79–90.

1994 Response to Price. *Journal of Anthropological Research* 50:205.

1998 Properties of property: Conceptual issues. In *Property in Economic Context*, edited by R. C. Hunt and A. Gilman, 7–27. University Press of America, Lanham, Maryland.

2000 Labor productivity and agricultural development: Boserup revisited. *Human Ecology* 28(3):251–277.

Hunt, R. C., and A. Gilman (editors)
1998 *Property in Economic Context*. University Press of America, Lanham, Maryland.

Isaac, B.
1993 Asiatic mode of production, hydraulic hypothesis, and Oriental despotism: Some Comments. In *Economic Aspects of Water Management in the Prehispanic New World*, edited by V. Scarborough and B. Isaac, 429–471. JAI Press, Greenwich, Connecticut.

Johnson, A. W., and T. K. Earle
1987 *The Evolution of Human Societies*. Stanford University Press, Palo Alto,.

Kirch, P. V.
1994 *The Wet and the Dry: Irrigation and Agricultural Intensification in Polynesia*. University of Chicago Press, Chicago.

Kolata, A. L.
1993 *The Tiwanaku: Portrait of an Andean Civilization*. Blackwell, Cambridge, UK.

Kolata, A. L. (editor)
1996 *Tiwanaku and Its Hinterland: Archaeology and Paleoecology of an Andean Civilization, Vol. 1: Agroecology*. Smithsonian Institution Press, Washington, DC.
2002 *Tiwanaku and Its Hinterland: Archaeology and Paleoecology of an Andean Civilization, Vol. 2: Urban and Rural Archaeology*. Smithsonian Institution Press, Washington, DC.

Kolb, M. J.
1997 Labor mobilization, ethnohistory, and the archaeology of community in Hawai'i. *Journal of Archaeological Method and Theory* 4:265–285.

Kolb, M. J., and J. Snead
1997 It's a small world after all: Comparative analysis of community organization in archaeology. *American Antiquity* 64(4):609–628.

Lansing, S.
1991 *Priests and Programmers: Technologies of Power in the Engineered Landscape of Bali*. Princeton University Press, Princeton, New Jersey.

Lathrap, D. W.
1977 Our father the Cayman, our mother the gourd: Spinden revisited, or a unitary model for the emergence of agriculture in the New World. In *The Origins of Agriculture*, edited by C. Reed, 714–751. Mouton, The Hague.

Lathrap, D. W., A. Gebhart-Sayer, and A. M. Mester
1985 The roots of the Shipibo art style: Three waves on Imiríacocha, or there were "Incas" before the Incas. *Journal of Latin American Lore* 11(1):31–119.

Lawton, H. W., P. J. Wilke, M. DeDecker, and W. M. Mason
1976 Agriculture among the Paiute of Owens Valley. *Journal of California Archaeology* 3(1):13–50.

Leach, H. M.

1999 Intensification in the Pacific: A critique of the archaeological criteria and their application. *Current Anthropology* 40(3):311–339.

Mabry, J.

1996 The ethnology of local irrigation. In *Canals and Communities: Small Scale Irrigation Systems*, edited by J. Mabry, 3–32. University of Arizona Press, Tucson.

2000 Wittfogel was half right: The ethnology of consensual and nonconsensual hierarchies in irrigation management. In *Hierarchies in Action: Cui Bono?*, edited by M. W. Diehl, 284–294. Center for Archaeological Investigations, Occasional Papers No. 27, Southern Illinois University, Carbondale.

Mabry, J. (editor)

1996 *Canals and Communities: Small Scale Irrigation Systems*. University of Arizona Press, Tucson.

Mabry, J., and D. Cleveland

1996 The relevance of indigenous irrigation: A comparative analysis of sustainability. In *Canals and Communities: Small Scale Irrigation Systems*, edited by J. Mabry, 237–260. University of Arizona Press, Tucson.

Marcus, J.

1998 The peaks and valleys of ancient states: An extension of the dynamic model. In *Archaic States*, edited by G. M. Feinman and J. Marcus, 59–94. School of American Research Press, Santa Fe, New Mexico.

2004 Maya commoners: The stereotype and the reality. In *Ancient Maya Commoners*, edited by J. Lohse and F. Valdez, 255–283. University of Texas Press, Austin.

Mayer, E.

2002 *The Articulated Peasant*. Westview Press, Boulder, Colorado.

Mitchell, W.

1973 The hydraulic hypothesis: A reappraisal. *Current Anthropology* 14:532–534.

1976 Irrigation and community in the central Peruvian highlands. *American Anthropologist* 78(1):25–44.

Mitchell, W., and D. Guillet (editors)

1994 *Irrigation at High Altitude: The Social Organization of Water-Control Systems in the Andes*. American Anthropological Association, Washington, DC.

Morrison, K. D.

1994 The intensification of production: Archaeological approaches. *Journal of Archaeological Method and Theory* 1(2):111–159.

1996 Typological schemes and agricultural change: Beyond Boserup in precolonial South India. *Current Anthropology* 37(4):583–608.

Moseley, M. E.

2001 *Incas and Their Ancestors: The Archaeology of Peru*. Rev. ed. Thames and Hudson, New York.

Murra, J. V.
1980 *The Economic Organization of the Inka State*. JAI Press, Greenwich, Connecticut.

Netherly, P.
1984 The management of late Andean irrigation systems on the north coast of Peru. *American Antiquity* 49(2):227–254.

Netting, R.
1993 *Smallholders, Householders: Farm Families and the Ecology of Intensive Sustainable Agriculture*. Stanford University Press, Palo Alto, California.

Neumann, K.
2003 New Guinea: A cradle of agriculture. *Science* 301:180–181.

Niles, S. A.
1987 *Callachaca: Style and Status in an Inca Community*. University of Iowa Press, Iowa City, Iowa.

Piperno, D., and D. Pearsall
1998 *The Origins of Agriculture in the Lowland Neotropics*. Academic Press, San Diego.

Posey, D. A., and W. L. Balée (editors)
1989 *Resource Management in Amazonia: Indigenous and Folk Strategies*. New York Botanical Garden, Bronx.

Price, D. H.
1994 Wittfogel's neglected hydraulic hydroagricultural distinction. *Journal of Anthropological Research* 50:187–204.

Pyburn, K. A.
1998 Smallholders in the Maya lowlands: Homage to a garden variety ethnographer. *Human Ecology* 26(2):267–297.

Redman, C.
1999 *Human Impact on Ancient Environments*. University of Arizona Press, Tucson.

Sahlins, M.
1972 *Stone Age Economics*. Aldine, Chicago.

Sanders, W. T., J. R. Parsons, and R. S. Santley (editors)
1979 *The Basin of Mexico: The Ecological Processes in the Evolution of a Civilization*. Academic Press, New York.

Scarborough, V. L.
1993 Introduction. In *Economic Aspects of Water Management in the Prehispanic New World*, edited by V. L. Scarborough and B. Isaac, 1–14. JAI Press, Greenwich, Connecticut.
2003 *The Flow of Power: Ancient Water Systems and Landscapes*. School of American Research Press, Santa Fe, New Mexico.

Scott, J. C.
1998 *Seeing like the State: How Certain Schemes to Improve the Human Condition have Failed*. Yale University Press, New Haven, Connecticut.

Serpenti, L. M.
1965 *Cultivators in the Swamps: Social Structure and Horticulture in a New Guinea Society*. Van Gorcum, Assen.

Shady Solís, R., J. Haas, and W. Creamer
2001 Dating Caral: A preceramic site in the Supe Valley on the central coast of Peru. *Science* 292:723–726.

Spooner, B. (editor)
1972 *Population Growth: Anthropological Implications*. MIT Press, Cambridge, Massachusetts.

Stahl, P. W.
1996 Holocene biodiversity: An archaeological perspective from the Americas. *Annual Review of Anthropology* 25:105–126.

Stanish, C.
1994 The hydraulic hypothesis revisited: Lake Titicaca Basin raised fields in theoretical perspective. *Latin American Antiquity* 5(4):312–332.
2003 *Ancient Titicaca: The Evolution of Complex Society in Southern Peru and Northern Bolivia*. University of California Press, Berkeley.
2004 The evolution of chiefdoms: An economic anthropological model. In *Archaeological Perspectives on Political Economies*, edited by G. Feinman and L. Nicholas, 7–24. University of Utah Press, Salt Lake City.

Steward, J. H.
1930 Irrigation without agriculture. *Michigan Academy of Sciences, Arts, and Letters Papers* 12:149–156.

Stone, G. D.
1996 *Settlement Ecology: The Social and Spatial Organization of Kofyar Agriculture*. University of Arizona Press, Tucson.

Stone, G. D., and C. E. Downum
1999 Non-Boserupian ecologies and agricultural risk: Ethnic politics and land control in the arid Southwest. *American Anthropologist* 101:113–128.

Tilley, Christopher
1994 *A Phenomenology of Landscape: Places, Paths, and Monuments*. Berg, Oxford.

Trawick, Paul
2001 The moral economy of water: Equity and antiquity in the Andean commons. *American Anthropologist* 103(2):361–379.
2003 *The Struggle for Water in Peru: Comedy and Tragedy in the Andean Commons*. Stanford University Press, Palo Alto, California.

Treacy, J.
1994 *Las chacras de Coporaque: Andenería y riego en el Valle de Colca*. Instituto de Estudios Peruanos, Lima.

Walker, J. H.
2004 *Agricultural Change in the Bolivian Amazon*. University of Pittsburgh Latin

American Archaeology Reports, Pittsburgh.

Webster, G.
1990 Labor control and emergent stratification in prehistoric Europe. *Current Anthropology* 31:337–366.

Willey, G. R.
1991 Horizontal integration and regional diversity: An alternating process in the rise of civilizations. *American Antiquity* 56(2):197–215.

Whittmore, T. M., and B. L. Turner
2002 *Cultivated Landscapes of Mesoamerica*. Oxford University Press, Oxford.

Wilkinson, T. J.
2003 *Archaeological Landscapes of the Near East*. University of Arizona Press, Tucson.

Wittfogel, K.
1957 *Oriental Despotism: A Comparative Study of Total Power*. Yale University Press, New Haven, Connecticut.

Zimmerer, K. S.
1996 *Changing Fortunes: Biodiversity and Peasant Livelihood in the Peruvian Andes*. University of California Press, Berkeley.

Zimmerer, K. S., and K. Young (editors)
1998 *Nature's Geography: New Lessons for Conservation in Developing Countries*. University of Wisconsin Press, Madison.

Prehispanic Agricultural Strategies of Intensification in the Titicaca Basin of Peru and Bolivia

CHARLES STANISH

As a concept, agricultural intensification is an analytical tool for addressing specific theoretical questions in comparative anthropology. *Intensification* can be defined in a variety of ways, depending on the problem or problems to be addressed. The numerous contributions to this symposium illustrate the many useful ways in which this concept can be employed. Definitions can be specific, focusing on technology, types of crops, and/or types of energetic balances, depending on the questions asked. Alternatively, *intensification* can be broadly defined to address evolutionary patterns of cross-cultural change in a variety of historical contexts.

In this chapter I utilize a traditional definition of *agricultural intensification* that borrows from Ester Boserup (1965:43). She defines *intensification* broadly as a shift in land use that allows a "greater amount of food production than was previously possible in any given area of land." To this definition I would add a regional component and explicitly note that a "given area of land" includes the entire area where a political unit extracts agricultural resources. By *political unit* I mean any group of people bound by some rules of cooperation. In short, *agricultural intensification* is a change in the production of domesticated plants and/or the raising of animals in a politically bounded region that increases the yield of resources.

In premodern agriculture, that is, prior to industrialized economies, agricultural intensification is generally achieved through higher labor inputs.[1] These higher labor inputs may be direct and involve greater labor investment in the cultivation of the land itself. An example would be an increase in the number of people from any political group working the land or herds, or, conversely, an increase in the total average per capita labor by that group. Alternatively, these greater labor investments can be indirect, invested in technologies that can increase the efficiency of agricultural production. An example would be labor invested in manufacturing plows, axes, hoes, and so forth that increases the per capita labor efficiency of the farmer or herder.

A second means of intensification is the creation of economies of scale with specialized labor organizations. An example of this would be large-scale irrigation building or raised field construction. While individual households could certainly build such constructions, the efficiencies created by mobilizing large numbers of workers for specialized tasks vastly increases the total per capita output for the same number of hours worked. These hours can also result in more per capita output for the same amount of labor time (Stanish 2004). In either case the intensified production implies either more work for people or requires a new kind of specialized labor organization that increases economic efficiencies but involves a loss of autonomy over production. Agricultural intensification entails both economic and social benefits and costs to individual farmers. As such, these costs and benefits shift as the social landscape and physical environment change through time.

The modern debate on the relationship between intensive agricultural systems and political organization most likely begins with Julian Steward's publications in 1955. Steward referred to "irrigation civilizations" that depended on large-scale canal constructions to feed large populations.[2] Steward saw the development of irrigation technology as a factor that increased the carrying capacity of the land. Following Karl Wittfogel, some kind of theocratic political organization would have been necessary to manage the complex agricultural systems (Steward 1955a, 1955b). Following this Malthusian logic and classic cultural ecological theory, populations would continue to increase as an independent variable. This increase would put pressure on the resource base and therefore force new cultural changes, particularly economic and technological ones.

Karl Wittfogel took this "hydraulic hypothesis" much further, arguing that state bureaucrats were a requisite for the proper functioning of irrigation systems.[3] In his view household-level and village-level labor organization and technological skill were insufficient to manage irrigation systems. The control of the irrigation technology essentially provided the elite of these civilizations with the ability to control their populations.

Throughout the 1960s and early 1970s a host of ethnographic and historical studies demonstrated that the hydraulic hypothesis was wrong. Even Steward ultimately conceded that the control of irrigation agriculture was not the cause of complex societies in arid lands (and see Mitchell 1973). In fact, virtually no theorist today would suggest that the original hydraulic hypothesis, as formulated by Wittfogel, is correct.

I have maintained, however, that the rejection of the hydraulic hypothesis went too far (Stanish 1994). The hydraulic hypothesis has several components, and each has separate theoretical implications. Each part has been subjected to empirical testing with varying results. One component—that villagers in nonstate societies were not capable of maintaining complex agricultural systems—is clearly incorrect. The idea that repressive states (oriental despotism) must necessarily develop by controlling irrigation systems fails the empirical test of comparative anthropology as well. However, one component of the hydraulic hypothesis—that there is a link between political complexity and agricultural intensification—cannot be discounted. There is a fairly strong empirical correlation between these two variables that deserves greater investigation by archaeologists and ethnologists.

Instead of focusing on Wittfogel's book, and his poorly disguised political agenda, we should concentrate on the tradition created by Steward, which theorized causal links between polity and agricultural intensification. We can address this important question by (1) testing the underlying progressive evolutionary assumption in Steward and later cultural ecology that societies adopt or add increasingly more complex intensification techniques through time, and (2) reformulating the definition of agricultural technologies to include not just irrigation canals but a variety of other water-management technologies. A major empirical task that we face is to trace the evolution of agricultural intensification and political evolution in areas around the world. The theoretical issue that we can address with empirical data is this: do agricultural strategies shift in a generally progressive manner, with an ever-increasing reliance on more intensive techniques, or do people employ a suite of techniques in different ecological and cultural contexts?

Fine-tuned studies of long archaeological sequences permit us to define much more precisely how agricultural production was intensified. I now turn to a case study for which we have sufficient data to conduct this empirical task.

THE CIRCUM-TITICACA BASIN

When the first Europeans began their explorations and conquest of the vast South American continent in the early sixteenth century, they encountered one of the greatest preindustrial states in all of human history. The Inca Empire was by far the largest and longest expansionist state in Andean history. It ranks as one of the largest preindustrial empires in world history. Tawantinsuyu, or "Land of the Four Quarters," as the empire was then known, covered an area that stretched from modern Ecuador to central Chile. The "jewel in the crown" of Tawantinsuyu was Collasuyu, or the land of the Colla people who inhabited the Titicaca Basin. The immediate Titicaca lake region was the heartland of the Collas.

The Titicaca Basin is a huge region that constitutes a largely autonomous area of prehistoric state formation (Stanish 2001, 2003). The entire area that falls into the circum-Titicaca cultural area is more than 400,000 square kilometers, covering tropical forests, snowcapped peaks, vast alpine plains, rugged mountain slopes, coastal deserts, and rich marine littorals. From this perspective we can view the Titicaca region as one of the major centers of indigenous cultural evolution around the world, providing us with an important comparative data set to understand and model cultural evolution.

The Incas were the last political power in a long string of indigenous groups that lived in the Titicaca region. Human occupation goes back at least 10,000 years, with the appearance of low-density hunter-gatherers. The first occupation of the Titicaca Basin began in the Early Archaic period, circa the eighth millennium BC. This hunter-gatherer-forager-fisher lifeway lasted until around 2000 to 1500 BC in some places but generally until around 1300 BC in most. At that time people gradually started to settle in permanent villages and to rely on agriculture and lake resources.

From around 1300 to 800 BC some moderately ranked simple chiefly societies developed in the region. The period from 800 BC to AD 500 was a time of complex chiefly development in some parts of the basin. This period was characterized by some dynamic cycling (Marcus 1992), with some polities expanding and contracting over time. At peak periods of complexity some of these polities had some of the attributes of early states.

An early state society known as Tiwanaku evolved by AD 500 or so, and by AD 800 the Tiwanaku state had established colonies or trade relationships over the entire circum-Titicaca Basin, essentially defining the south-central Andean cultural region. Tiwanaku collapsed between AD 1000 and 1200. By about 1300 the Titicaca Basin was divided into a series of chiefdoms, referred to by Spanish chroniclers as *señoríos*. The last indigenous Andean state to control the Titicaca Basin was the Inca, which conquered the region in the late fifteenth century AD. After a generation or more of intense political negotiation and military incursions, the Inca conquered the region but did not completely subjugate the local populations. Rebellions continued up to the Spanish conquest.

In the last generation we have accumulated a fine body of data on the prehistory of the circum-Lake Titicaca region. We have extensive settlement and excavation data that allow us to define a number of cultural patterns through time. We can therefore use these data to trace the evolution of land use in the immediate Titicaca Basin. These data permit us to test the degree to which factors such as population pressure, environmental change, technological change, elite organization, and so forth are significant in anthropological models of agricultural intensification.

Figure 14.1. South America.

The Titicaca Basin proper, as defined by its hydrological boundaries, covers approximately 50,000 km^2. This vast region, about twice the size of Belize, was home to complex cultural developments over time, including one indigenous first-generation state and countless complex chiefdoms and smaller polities. The Titicaca region is a huge geological basin that sits between the two mountain ranges of the Cordillera Real to the west and the Cordillera Blanca to the east (Figures 14.1 and 14.2). As mentioned above, the cultural region that defines ancient Collasuyu is about eight times larger (see Stanish 2003).

The high altitude of the region, more than 3,800 m, alters what would otherwise be an intertropical climatic zone to a more alpine character (Dejoux and Iltis 1991:11). Mean annual precipitation in the Titicaca Basin is quite

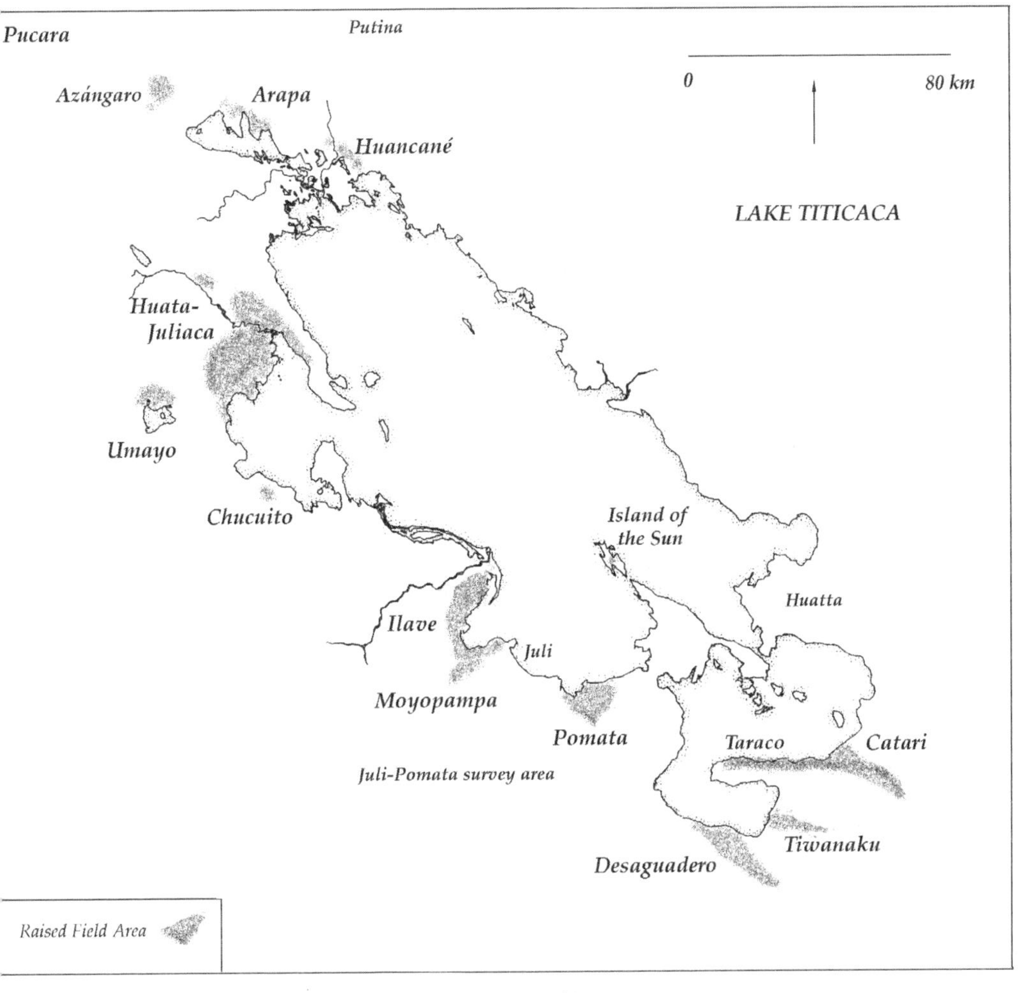

Figure 14.2. Titicaca Basin, showing names mentioned in text.

variable, ranging from approximately 500 mm to 1,500 mm per year (Roche et al. 1991:87). Rainfall is concentrated in less than six months, with December through March receiving precipitation of more than 100 mm per month. The driest months are June through September, with some months virtually rainless. According to Roche et al. (1991:86) median annual temperatures vary between 7° and 10°C. The lake itself has an ameliorating effect on the cold, and median temperatures are often higher than 8°C at the lake edge, a temperature that would be higher than expected if the mass of water was not present (Boulange and Aquize 1981). As a general rule temperatures are warmer near the lake and become progressively colder away from the water, a function of both water mass and altitude.

Figure 14.3. Aboriginal vegetation near Lampa.

The geographer Pulgar Vidal (n.d.) divides the Titicaca Basin into two broad agricultural and ecological regions called the *suni* and *puna*. The *suni* is located between 3,800 and 4,000 masl. It is characterized by rolling hills cut by steep gullies and low, flat plains called *pampas*. The region is largely denuded today, a result of millennia of human land use. It is likely, however, that for much of the early prehistory, stands of low trees were found in the gullies and other protected areas. Figure 14.3 shows a modern stand of indigenous trees in the Lampa region. The higher and drier puna is located between 4,000 and 4,800 masl. The *suni* represents the upper limit of plant agriculture, whereas the puna is a grazing zone for the extensive camelid herds owned by many Titicaca Basin peoples. The traditional altiplano crops, most notably chenopods and tubers, do not grow above the *suni* zone in any meaningful quantity. The *suni* is the richest area and is where most modern and prehispanic settlement is located.

The water of Lake Titicaca is saline, with salt concentrations being a bit higher in the south, where the lake drains through the narrow outlet of the Río Desaguadero. The water is therefore not usable for agriculture. This presents obvious problems for farmers. Rains and groundwater provide the bulk of the water for agriculture today and did so in the past. Terraces trap runoff, and canals from arroyos (or *quebradas* as they are known there) channel water to the terraced fields. To get access to additional water, the farmer must build canals from the freshwater springs or tap into the groundwater that collects in front of the lake. It is therefore necessary to move water from subterranean sources and springs to farming areas utilizing canals and raised canals or aqueducts.

There are other classifications of the Titicaca region. Carl Troll divides the higher vegetative region into the *puna brava* and the *puna* (Troll 1968:48). The Troll classification is an ecological one, useful for the study of plant communities. The puna brava is located between 4,500 and 5,300 masl. Vegetation is intermittent, composed of plants adapted to a short growing season (Graf 1981:353). The puna is located between 3,800 and 4,500 masl in the Troll classification. Likewise, the Tosi (1960) classification lists eight zones for the Titicaca region, a typology based on the Holdridge system.

The classifications of Pulgar Vidal, Troll, and others were not designed for anthropological research. As suggested, the basic *puna/suni* distinction of Pulgar Vidal is a good first approximation of the broad agricultural/pastoral zones and is the most useful of the existing classifications. From a cultural perspective, however, there are a variety of distinct geographical zones that are important to our understanding of the prehistory of the region. A very useful alternative to the ecological classifications are indigenous categories of land. Years of fieldwork have reinforced the observation that Aymara farmers and herders possess an extremely sophisticated and subtle understanding of their environment that differs in some important aspects from those of professional agronomists and geographers.

Perhaps the first published attempt at constructing a typology of soils and land types is that of Harry Tschopik. In an important section in his review of Aymara culture Tschopik (1947:513) distinguishes four types of arable land classified by the Aymara farmers themselves. Tschopik worked in the area immediately outside Chucuito, but numerous references in his text indicate that he accumulated data from throughout the region. The four land-use categories are as follows:

1. Valley-bottom fields. Tschopik's informants said that these soils are considered the best in the region. They are located at the base of the many *quebradas* that cut through the hills toward the lake.

2. Lake-edge fields. Lake-edge fields are considered to have the second-best soils for agriculture in the region. Today, in a number of lake-edge regions, canals are used to augment the fields.

3. Hillside fields. These extensive areas have soils that are thin and rocky, according to Tschopik's informants. These areas are extensively terraced today and are cultivated on a long-fallow system.

4. The flat pampas away from the lake shore are considered the worst soils, according to Tschopik. Although Tschopik says that irrigation is not practiced in this region, I have noted canals in such areas.[4] These fields are used today

largely for animal pasturing, and there are very few cultivated fields in the flat pampas, except for those near rivers. The canals provide water to pastureland and for some marginal agriculture.

A study by Luperio Onofre (in Stanish et al. 1997) provides a typology of soil and land types by Aymara farmers in the Juli area. Onofre defined nine types based on a variety of factors. His work outlines these types and demonstrates a very subtle and sophisticated understanding of agricultural land and soil by contemporary Aymara farmers. The perception and the reality of these land types most likely were important settlement determinants in the past, as they are today.

Overall, the most important animal product of the *puna* in the prehistoric past is the camelid, particularly the llama and the alpaca (see Flannery et al. 1989). Camelids provide wool and meat and serve as pack animals. The virtually unique capacity of the Titicaca Basin to support such large camelid herds has contributed to its position as a major center of complex society in the Americas prior to European conquest. Archaeological middens in the region are replete with camelid bone. Fish, both river and lake varieties, constitute the second most common resource in archaeological middens, followed by guinea pig and some deer.

LAKE TITICACA BASIN AGRICULTURAL TECHNIQUES

A number of agricultural practices used in the Titicaca Basin today were utilized in the past as well. The following list includes all of the major agricultural types that would have been utilized prior to the Spanish conquest. Techniques added since the sixteenth century include greenhouses, fish raising in weirs, use of fertilizers, pumps and wells, and deep plowing. These are not relevant and are therefore not discussed here. Likewise, use of lake resources is a very important part of the ecological landscape exploited by peoples from Archaic times onward. We can always assume that any lakeside settlement was also exploiting lake resources. Fortunately for this analysis, we can assume that the productivity of the lake resources—basically fish and reeds—was an invariant constant throughout history that does not affect the interpretation of the data here.[5] The agricultural techniques are presented here from the least intensive to the most intensive.

Herding. The most extensive form of production in the Titicaca region was camelid herding. Large herds can be kept by a fairly small number of households. However, grazing requires very large areas of land, and population densities are quite low in grazing areas. Grazing is a very low-risk strategy, but it precludes population nucleation on any significant scale. As in most areas of the world, there are very strong social and economic links between herders and farmers. Most farmers keep a few animals that feed off the agricultural stubble or fallowed land. Farmers also maintain family ties (real or fictive) to herders in the puna that provides for reciprocal exchange of goods from each zone.

Cocha. Cocha are small, round depressions dug into the ground of the flat pampas (Figures 14.4 and 14.5). They can reach groundwater and hold water during the dry season. During the rainy season they swell into small lakes. Farmers dig small furrows that radiate out of the cocha, and they are cultivated. Cochas today are confined to small areas in the region. There are, however, numerous examples of out-of-use cochas throughout the region. Unfortunately, it is virtually impossible to date cochas. However, cochas are restricted geographically to the low pampas near the lake where the groundwater table is high. They therefore can be considered part of the agricultural system utilized in a narrow zone near the lake or in the low floodplains of the major rivers.

Rain-fed terrace agriculture. This is referred to as dryland farming in other parts of the world. Rain-fed agriculture is practiced on hillsides with low terracing (Figure 14.6). These terraces are also built in gullies where springs are found or where water collects in the rainy season. Most of the terraces are cultivated with tubers and chenopods. In the low pampas, particularly near the lake, the growth of wild grasses is encouraged for pasture. Terrace agriculture is higher risk than herding. It is dependent on rainfall, of course, but the farmer also has to constantly balance the benefit of planting earlier or later than

Figure 14.4. Large cocha (foreground) with series of smaller cochas in the upper Huancané Valley, near Huatasani.

Figure 14.5. Large cocha in the Pucara Valley.

Figure 14.6. Terraced hill jutting into reed beds, Lake Arapa. Photo by A. Umire.

normal and getting a better crop versus the risk of losing that crop to frost or lack of rain. Today, and most certainly in the past, farmers spread their risk by altering the time of planting for the particular fields that they own.

Irrigated terraces and pampas. Irrigation canals are used today to water the low pampas (Figure 14.7) and to feed water from springs to some terraces (Figure 14.8). Most terraces are found above canals, but a few areas of the region have irrigated terraces at the base of hills (Figure 14.9). Canalized rivers (Figure 14.10) are associated with prehispanic settlements. Numerous canals are slightly elevated off the ground by a meter or two. We refer to these, somewhat liberally, as aqueducts. Today, many of these raised canals have been converted to causeways (Figure 14.11) needed for the periodic inundations that characterize the areas below 3,820 meters above sea level (Figure 14.12). These are found throughout the Titicaca Basin, and most are not in use. Irrigation reduces the risk of catastrophic drought by tapping into perennial water sources (most groundwater comes from aquifers that recharge over decades). However, irrigation agriculture is more labor intensive and requires intercommunity cooperation.

Raised fields. The lowlands near the lake and along the rivers supported raised field agriculture in the past. Raised fields are agricultural constructions built in swampy land to improve planting conditions. Essentially, they are large mounds of earth, raised above water level and designed to provide a moist planting surface (Figure 14.13). Raised fields were built in a variety of physical forms (see Erickson 1988; Lennon 1983). They were concentrated along the lake edge, along rivers, and in the low, swampy pampas near the lake and in the river

Figure 14.7. Canals and canalized rivers in the pampa between Lake Arapa and Lake Titicaca. Photo was taken from Cerro Luriata. Note extensive agricultural fields in the pampa.

Figure 14.8. Small canal in pampa near Asillo.

Figure 14.9. Small canal feeding lake-edge fields. Note reed beds at lake edge.

Figure 14.10. Canalized section of Río Ramis, near Taraco.

Figure 14.11. Causeway near Taraco. The causeway is also used as a levee. It is likely that this causeway was used in the past as a low aqueduct.

Figure 14.12. Inundation of habitation area near Saman during recent rise in lake level.

floodplains. Raised fields, more than any other agricultural technique, require some degree of cooperative labor (Figure 14.13). They are most efficiently built and maintained with labor groups larger than a few households. Raised fields are also the most intensive form of agriculture. While it is true that they require about the same high initial labor inputs to construct as irrigation, they have considerably higher regular maintenance costs.

Although there is no indigenous use of raised field agriculture today, archaeological evidence indicates extensive use of this technology in the past.[6] Factoring in the use of raised field agriculture alters the optimal land-use categories for prehispanic populations. According to Harry Tschopik (1947) the flat pampas away from the fields are the worst land category for agriculture. Ironically, the low pampas, converted to raised fields, become the most productive land. This fact, of course, is very significant in the agricultural history of the region.

PREHISTORIC AGRICULTURAL INTENSIFICATION IN THE TITICACA BASIN

In the last two decades we have accumulated survey and excavation data from a number of areas in the Titicaca Basin, including the far south (Albarracin-

Figure 14.13. A small set of abandoned raised fields near Asillo. Informants say that the fields have not been used in 20 years.

Jordan and Mathews 1990; Hastorf 1999), the southwest and west (Stanish et al. 1997), the Island of the Sun and Moon (Bauer and Stanish 2001; Stanish and Bauer 2004), the north (Plourde 1999), and the northeast (Neira 1962, 1967). These data permit a reconstruction, albeit subject to future testing and refinement, of the agricultural history of the region.

Systematic data from the Juli-Pomata region provide quantitative details on land use through time. These survey data can be used to define agricultural land use through time with criteria and assumptions as outlined in detail in Stanish (1994). The geographical location of settlements in the survey area indicates the agricultural strategies employed by the populations in each period. Raised fields represent massive alterations of the landscape by humans and leave very large archaeological markers (Erickson 1999). Canal location and site association with raised fields provide a means of assessing their use through survey data. Furthermore, habitation sites above 4,000 meters were almost certainly populated by pastoralists because agriculture is not possible in the region.

The optimal distribution of settlement for raised field agriculture is one in which habitation sites ring the edge of the fields. Settlement data therefore allow us to define periods in which raised fields were utilized. Likewise, certain stratigraphic and architectural relationships on the landscape allow us to determine whether an agricultural feature was or was not in use during a particular period. A looted tomb (with diagnostic sherds on the surface) that is intrusive into an agricultural canal indicates that the canal was not in use at the time of tomb construction. Architectural and geographical associations of sites with other agricultural features provide a means of dating water-management features as well. The total area of habitation serves as an index for relative population size per period. In short, the data allow us to define the percentage of the population in each geographical area and therefore the economic focus of the entire settlement system per period. These combined data allow us to reconstruct the agricultural history of the region with reasonable precision.

EARLY FORMATIVE

The development of the first permanent settlements defines the Early Formative period and dates to circa 2000–1500/1300 BC. These settlements were small, undifferentiated hamlets located near optimal resource zones. Many of the earliest hamlets had earlier Late Archaic occupations. In fact, the transition from the Late Archaic lifeways to the Early Formative was not characterized by any major changes in location or apparent economic focus (see Stanish et al. 2002). Pottery was added to their technology, probably associated with tuber preparation and storage, and horticulture or small-scale agriculture increased in importance as settlements became less mobile.

Settlement data from the Island of the Sun, located in the far southern Lake Titicaca, indicate that Early Formative sites were located near springs or naturally inundated areas near the lake edge. There is no evidence for raised field agriculture or canals or terraces during this period. Likewise, in the Juli-Pomata, Early Formative period sites cluster near the richest low, swampy areas near the lake. The settlement pattern again represents a resource catchment-optimization strategy of mixed farming, herding, and wild resource exploitation. We also see that the transition to settled societies in this region was very gradual in the Juli-Pomata region as well. In the north, preliminary survey data are similar to the patterns in the south and west. The Early Formative settlements favor the river floodplain and paleo-lakeside locations. These data likewise suggest that nonintensive, rain-fed agriculture near the rivers was one component of subsistence during this period, along with lake and river exploitation, herding, and wild plant collecting as part of a broad subsistence strategy mix. The settlement pattern in the north also is best explained as a catchment-optimization one with political or other social factors relatively unimportant.

From the settlement and excavation data in the region we can characterize the Early Formative period subsistence system as simply an extension of the

Late Archaic one with the addition of pottery within a context of very modest population growth. This was a period of extensive agricultural production, little sociopolitical differentiation, and little labor-organizational complexity.

MIDDLE FORMATIVE

Around 1300 BC in some areas of the Titicaca Basin, people created simple chiefly societies. These societies were characterized by a two-tiered settlement system of a number of small habitation settlements and a few large centers. The centers or major sites had small but complex architectural constructions. The smaller sites were modest hamlets and small villages, usually less than 1.0 hectare. The period from 1300 to 500 BC is referred to as the Middle Formative and defined as the time when simple chiefly societies were the dominant political organization in the region (Stanish 2003). Political organization varied over the landscape, of course. In some areas agriculture was not as significant as the reliance on wild foods. Economic surpluses were small, and there is a corresponding lack of complex political organization. In other areas, however, chiefdoms developed as the dominant type of political organization. Here there is evidence of agricultural intensification and precocious sociopolitical development.

Survey data from the Island of the Sun, in the southern basin, strongly suggest that agricultural terraces were first utilized during the Middle Formative, probably around or slightly after 800 BC. This is illustrated by the settlement changes seen between the Early Formative and Middle Formative periods. In the Early Formative, sites were all located near springs in low swampy areas or on cliffs immediately above the lake. The interpretation of these settlement data, along with excavation data (Stanish et al. 2002), is that people were exploiting the wild resources of the area—fish, reeds, wild deer—plus some nonintensive horticulture. In the Middle Formative period most of these sites were reoccupied. However, there was a major expansion into the island where only terrace agriculture was possible. This shift in settlement from a focus on springs above the lake to higher land that is terraced today indicates a shift to terrace agriculture (Figure 14.14).

There is also settlement evidence for the use of raised fields on the Island of the Sun in the Middle Formative period. A small area of fields is found in the Challa Bay, on the northern side of the island. A series of causeways and raised fields are most closely associated with major Middle Formative and Upper Formative sites (Stanish and Bauer 2004).

In the Juli-Pomata region the vast majority of the Middle Formative sites were in the lower *suni*, where farmers practiced rain-fed terrace agriculture, as seen today in Figures 14.15 and 14.16. The data also suggest the use of raised field

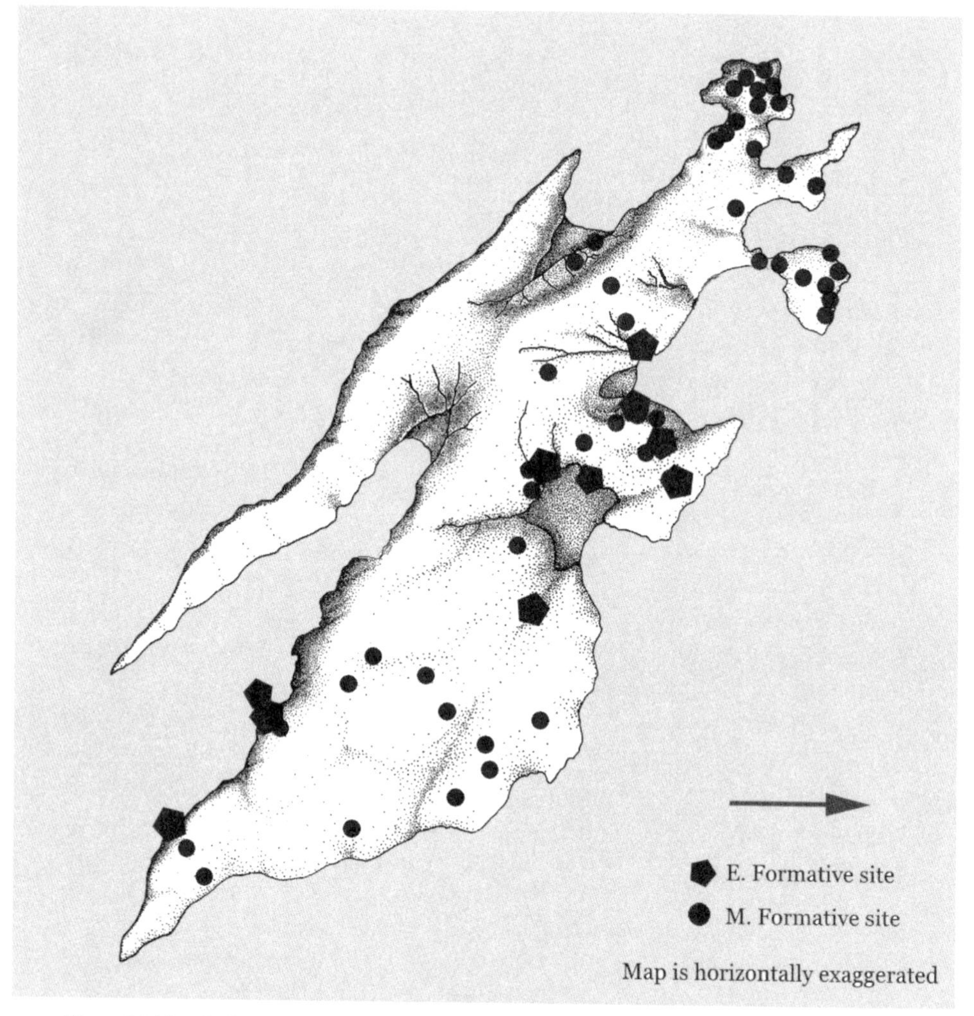

Figure 14.14. Settlement pattern map of Early Formative and Middle Formative sites on the Island of the Sun, Bolivia. Note the shift from lake edge and spring to the interior, where dry farming on terraces is necessary.

agriculture in this region at this time as well. Table 14.1 illustrates the percentage of the population located in the raised field areas relative to higher pasture sites and field terraces in the *suni*. While low, relative to later periods, there was most likely some raised field land use in the region at this time. Furthermore, two of the major Middle Formative period sites—Palermo and Sillumocco-Huaquina—are directly associated with large raised field sections and aqueducts. Clark Erickson (1988) has documented extensive use of raised fields in this period in the northern basin, as has Gray Graffam (1992) in the south Titicaca Basin.

Figure 14.15. Dry-farming area with quinoa in foreground.

Figure 14.16. Dry-farming area with potato plants. Note extensive hillside terracing in background.

Table 14.1. Summary data for agricultural land use in the Juli-Pomata area.

	Early Sillumocco	Late Sillumocco	Tiwanaku	Altiplano	Late Horizon	Early Colonial
Total population (ha)	23.04	32.72	62.86	74.16	178.49	153.75
Total number of sites	25	19	41	140	242	224
Mean size of all sites	.92	1.72	1.53	.53	.73	.69
Total sites in raised fields	11	12	17	44	48	43
Population index of raised field zone (ha)	9.49	22.71	35.74	21.04	25.15	15.18
Mean size of sites in raised fields	.86	1.89	2.10	.48	.52	.35
Population index of raised fields as % of total	41	69	57	28	14	10
Total sites in non-raised-field *suni*	11	6	21	75	143	124
Population index of non-raised-field *suni* (ha)	10.20	8.01	24.53	42.64	118.62	99.89
Mean size of sites in non-raised-field *suni*	.93	1.33	1.16	.57	.83	.80
Population index of non-raised-field *suni* as % of total	44	24	39	57	66	65
Total sites in puna	3	1	3	21	51	57
Population index of puna (ha)	3.40	2.00	2.59	10.48	34.72	38.68
Mean size of sites in puna	1.13	2.00	.86	.50	.68	.68
Population index in puna as % of total	15	6	4	14	19	25

In the Putina region Aimée Plourde (1999) mapped and excavated a site called Cachichupa. This site has a substantial Middle Formative occupation. There is massive nonagricultural terracing at the site that dates to the very beginning of the Middle Formative period (Plourde, personal communication 2002). Although these terraces were not designed as agricultural features, their existence clearly indicates that terrace construction was part of the Middle Formative period technological repertoire.

Excavations of Middle Formative period domestic areas indicate substantial use of camelid and fish (de la Vega 1997; Plourde, personal communication 2002). Settlement data from the Juli-Pomata region and the Huancané-Putina river valley suggest Middle Formative use of cochas as well (Stanish 1994; Stanish et al. 1997).

In short, in the Middle Formative period we have good evidence for the use of herding, cocha farming, raised fields, canals, aqueducts, and intensive lake exploitation. All of the agricultural techniques of prehistoric Titicaca agriculture were present by 800 BC, if not earlier.

UPPER FORMATIVE

By 500 BC some areas developed complex chiefly societies characterized by multiple site size hierarchies, regional polities, complex architecture, fancy pottery production, and the importation of high-value goods. The Upper Formative dates from 500 BC to AD 400. Around 200 BC, two major polities emerged in the region and represented the culmination of the late Upper Formative period complex chiefly development: Pucara in the north and Tiwanaku in the south. These centers pulled in populations from surrounding areas and created the largest settlements in the region. At its height Pucara covered at least 100 hectares and probably reached 150 or more. Tiwanaku during the late Upper Formative, known locally as the Qeya period, was probably equivalent in size and complexity to Pucara. Since Pucara collapsed and was essentially abandoned at the end of this period, we have a good understanding from the surface of the architecture, size, and complexity of this site. Late Upper Formative Tiwanaku, in contrast, was subsequently covered by later massive urban constructions, and we cannot easily identify the size and complexity of the settlement during this period.

In the Juli-Pomata region there was a significant shift to the raised field areas during this period (Table 14.1). The most intensive use of raised fields occurred during this period. Three sites with complex corporate architectural constructions were built in the area. Two of these were architecturally associated with a formal complex of agricultural features, including the construction of aqueducts that fed the raised fields, a canalization of the river, and a formal raised field layout with satellite settlements. By architectural association I mean that the habitation and agricultural features were integrated. The canal at one site was placed precisely between the fields and the edge of the habitation area. Likewise, the river was canalized away from its original discharge area to make room for the field systems. Later occupations built tombs and settlements on top of the fields, indicating that they were not in use at that time.

In the northern Titicaca region, settlement data from the Huancané region indicate a correspondingly complex agricultural system with canalized rivers, raised fields, aqueducts, and other canals all architecturally associated with a Middle and Upper Formative period site that includes elaborate corporate constructions. Similarly complex agricultural systems are found at the site of Maravillas in the north and other settlements in the northern Titicaca Basin.

Around AD 300–400 the site of Pucara collapsed as a major regional center, while the site of Tiwanaku continued to grow. Systematic survey data in the Pucara Valley (Cohen n.d.) indicate a dispersal of the population into small hamlets. Raised fields continued to be used near the lake during this time by local polities that I have earlier referred to as Early Huaña (Stanish 2003), but there was a clear shift to extensive agricultural systems in the immediate Pucara area.

TIWANAKU PERIOD

Around AD 600 the Tiwanaku peoples expanded out of their capital in the southern Titicaca Basin. The nature of Tiwanaku political economy remains poorly known (Stanish 2002), but we know that they at least gained control of a number of colonies and territories outside of their heartland. Tiwanaku was the first indigenous state in the Andes south of Cusco. It was contemporary with the Wari state, and it flourished two or three centuries later than the Moche culture on the coast, the first state in South America. Tiwanaku collapsed around AD 1100. The collapse of this state correlates with a major drought in the region, although there had been earlier droughts during Tiwanaku expansion.

In the Juli-Pomata area the total *percentage* of the population in the raised field areas decreases from the previous period, although total population living near raised fields increased 50 percent from the Upper Formative. I interpret this to indicate that raised field agriculture had reached its maximum capacity, at least in this study area. A subjective review of the topography of the area supports the hypothesis that there was no more room for raised field construction. In this context Tiwanaku peoples expanded the next most intensive techniques available to them, irrigated and rain-fed terrace agriculture. A mere 4 percent of the population lived in the pasturelands.

In the southern basin in the Tiwanaku heartland, Tiwanaku raised field agriculture dramatically expanded. The work of Juan Albarracin-Jordan, Gray Graffam, John Janusek, Alan Kolata, James Mathews, Charles Ortloff, and Matthew Seddon (Graffam 1992; Janusek and Kolata 2000; Mathews 1992; Ortloff and Kolata 1993; Seddon 1994; Seddon and Janusek 1994; and others) indicates an expansion of fields into the Tiwanaku and adjacent valleys to the north and south. Kolata (1993:201) estimates that at its height the Tiwanaku state was cultivating 19,000 hectares of fields. While his population estimates of between 285,000 and 1,485,000 are based on questionable assumptions (most notably his short to nonexistent fallow time and assumed contemporaneity of almost all fields), his data still point out that a huge population could have been fed in the southern three valleys of the Titicaca Basin at the time of Tiwanaku ascendancy. Even if we allow for more reasonable fallow periods and land use and abandonment over time, we still have conservative population estimates of more than 100,000 people in the Tiwanaku core area alone. In the entire Titicaca Basin we can conservatively estimate populations of at least one-half million during Tiwanaku times.[7]

Data from the north Titicaca Basin near Huancané provide fascinating insights into Tiwanaku period political strategies for agricultural intensification. A few kilometers northeast of Huancané is a local, complex, and contemporary site known as Huancawichinka, a small valley with a series of agricultural constructions including canals, fields, and a reservoir. There is a major Tiwanaku settlement system in this valley. It is almost certain that the population at Huancawichinka continued to work the complex field system around their site while the Tiwanaku state established an enclave a few kilometers away. There is no evidence of Tiwanaku influence at Huancawichinka, but the surface evidence at least suggests a Tiwanaku-contemporary occupation that I have called Late Huaña (Stanish 2003). Certainly, the agricultural resources at Huancawichinka are substantially better than in the Tiwanaku enclave. At the present time, without excavation data, the best interpretation of these settlement patterns is that Tiwanaku people were permitted or forced their way into an adjacent valley near Huancawichinka and exploited an independent water resource. This small enclave maximized the resources in the area with a sophisticated agricultural complex while the local population along the Huancané River continued to intensify their resources independently.

Tiwanaku ceased to be a regional political force by the twelfth century AD, possibly earlier. According to Charles Ortloff and Alan Kolata (1993) the proximate cause of Tiwanaku collapse was a drought that began around the middle of the eleventh century and lasted for two centuries or so, to around AD 1300. This interpretation has been challenged by Clark Erickson (1999). However, it would appear that this drought did indeed have severe consequences for the continued utilization of raised field agriculture. Throughout the southern Titicaca Basin there are numerous abandoned canals and aqueducts that run from now-dry quebradas and springs to fossil raised fields. They are rare in the north basin, where the topography precludes their efficient use. These can only be interpreted as a response to the drought, bringing fresh water to the raised fields from increasingly more distant and less productive rivers and springs.

ALTIPLANO OR LATE INTERMEDIATE PERIOD

The collapse of Tiwanaku political organization and the prolonged drought of the twelfth century provided the context for the emergence of numerous smaller polities in the Late Intermediate or Altiplano period. These Aymara *señoríos* were characterized by internecine warfare, a dispersed settlement system, the collapse of regional exchange, and a general collapse of centralized political structures (see Frye 1997).

These small polities relied extensively on camelid herds, as well as rain-fed terrace agriculture. While earlier cultures most certainly were pastoral as well, people during the Altiplano period greatly intensified the use of the puna grazing lands and created a much more dispersed settlement system to maintain the herds. The quantitative data from the Juli-Pomata area illustrate the dramatic shift in land use. Population in the puna grazing areas increased threefold as a percentage and increased by a factor of seven in the total number of sites (Table 14.1). Even though there were modest population increases in this period, settlement in the raised field areas dropped almost in half, both in percentage and in absolute terms. Furthermore, numerous canals and raised fields were used as architectural platforms for cemeteries and habitation sites, suggesting that the drop in raised field use was even greater. That is, it is most likely that people moved into previous raised field areas as the water resources contracted, and they used the area like they do today for grazing. While raised field use in post-Tiwanaku times is documented in other areas of the basin as well, it was most likely on a household or at best village level of organization, not a state or regional level as in the previous period (but see Graffam 1992 for an alternative view). In short, raised fields increasingly became less viable in the drier environment of the post-Tiwanaku periods. Populations dispersed throughout the region and refocused subsistence activities toward the pasturing of animals and terrace agriculture.

INCA PERIOD

The Inca occupation of the Titicaca Basin represents the first nonindigenous conquest of the region in its history. The Collasuyu region was one of the most important territories in the Inca state. It was perhaps one of the richest provinces in the empire. More than any period, the Inca occupation represents a strong political control in the region, with a state capable of moving populations, imposing peace, facilitating exchange, and organizing production.

Analysis of the Late Horizon or Inca period settlement data demonstrates several patterns. First, raised field agriculture virtually disappeared during the Late Horizon. Settlement data indicate a shift away from the raised field zones in the survey area to locations in the rain-fed terrace areas. Second, there was a substantial shift to the puna pasturelands from earlier levels. The shift to the puna lands was significant, particularly when compared with earlier figures. By the Late Horizon this figure had increased to almost 20 percent.

In fact, about one-fifth of the population lived in the pasture grazing lands, a pattern that emphasizes the importance of camelid wool and meat in the Inca economy. Third, the Late Horizon settlement pattern is heavily weighted to terrace agricultural and lakeside urbanized areas, suggesting a maximization strategy in the region designed to produce and move commodities and to locate populations in optimal agricultural land.

These data support both Graffam's (1990:248–249) and Ortloff and Kolata's (1993) arguments that the fields were economically unfeasible by the time of the Inca conquest. The Juli-Pomata settlement data reflect this changed ecological situation. Less than 15 percent of the population lived in the raised field areas during this period, and most of that population can be accounted for by the presence of a major Inca road that runs through the pampas in areas of former raised fields.

In short, the Inca Empire moved into an environment in which raised field agriculture was impossible, except in a very few restricted spots. They intensified the Altiplano period agricultural strategies focusing on rain-fed agriculture, some irrigated terraces, and pastoralism. The Incas also dramatically increased exchange with regions outside of this region, particularly to the east and west, where agricultural production was optimal (Coben and Stanish in press; Wachtel 1982).

The Island of the Sun has a small area of raised fields that was one of the exceptions that proves the rule (Bauer and Stanish 2001; Stanish and Bauer 2004). In the southern Kona Bay a small complex of canals, fields, and reservoirs was most likely utilized by the Inca. The Island of the Sun is warmer than the mainland areas because of higher ambient temperatures provided by the lake. Where raised fields were indeed possible, the Inca utilized them as best they could.

The Inca settlement pattern was one of a high quantity of dispersed, small sites plus the establishment of large urban settlements on the road system. This strongly bimodal settlement distribution indicates an imperial system of rural agricultural development combined with urban centers where craft production was intensified. The earlier Tiwanaku state had housed craft specialists (Janusek 1999). But the Inca strategy was to disperse these urban centers around the region. These centers were much smaller than Tiwanaku, with the largest being around 80 hectares in size and some as small as five hectares. However, we have identified scores of Inca urban settlements in the Titicaca region during their occupation, indicating a conscious effort to disperse these productive centers (Stanish 1997, 2003).

DISCUSSION

The data from the Titicaca Basin permit us to trace the evolution of agricultural land use and intensification over more than two millennia. These data unequivocally indicate a correlation between political complexity and agricultural intensification. However, that pattern was not one of a slow, progressive increase in technology and labor investment through time. Rather, the pattern was a rapid one in which preexisting technologies were differentially utilized to optimize the political and ecological environment of any given period.

All of the agricultural techniques were developed as early as 800 BC, probably before. As political organizations became more complex, farmers adopted the more labor-intensive strategies at the expense of less-intensive ones within the limits of the physical environment. When less-complex societies followed more complex ones, agricultural strategies shifted to earlier patterns. It is not surprising that the Middle Formative and Altiplano period agricultural strategies are strikingly similar, even though they were employed in different cultural, ecological, and demographic contexts. In each period chiefly societies relied heavily on the pasturing of animals and a strong focus on terrace agriculture. Raised fields were used but most likely by household or village level farmers and not in any formal way.

During the Tiwanaku and Inca periods, the time of maximum political complexity in the region, agriculture was the most intense in the historical sequence. The drought of AD 1100–1200 most likely contributed to the collapse of Tiwanaku. However, and this is extremely important, the survey data strongly indicate that there was no demographic collapse that accompanied the Tiwanaku state collapse. In fact, there was a slight population increase in the Juli-Pomata region and probably a leveling or slight increase in the Tiwanaku Valley itself (Albarracin-Jordan and Mathews 1990; Stanish et al. 1997). What *did* collapse was the political structure that was able to provision the capital city. The post-Tiwanaku settlement landscape was one in which populations did not decline but merely dispersed, relying on a much more extensive economic system. Therefore, the drought did not lower the carrying capacity of the land. Rather, it slowly destroyed the viability of one kind of land use—raised field agriculture—as an intensification option. In the Altiplano period people once again adopted more extensive, risk-averse, systems of agricultural production and were able to maintain the Tiwanaku period population levels.

The Inca and Altiplano periods were characterized by very similar physical environments. However, the economic strategies of the latter represented a low-risk strategy of a wide mix of land use and economic behaviors. The Incas, in contrast, faced with a very similar ecological environment, intensified commodity production and agricultural terrace production to their maximum.

They also intensified camelid production to its highest level in the history of the basin, up to that point in time. While the physical environment was indeed a major factor in land use, the intensity of that land use depended on the political structure that was, or was not, capable of amassing labor. The Inca Empire took the same environment and overcame the obstacles through their ability to mobilize labor. They intensively produced food through a more complex system of camelid pastoralism (probably state herds that created efficiencies through economies of scale) and rain-fed terrace farming. They also created economies of scale in commodity production and interregional exchange.

In short, we see a shift in agricultural strategies over time, depending on the political organization of the region. I conclude that the patterns of land use in the Titicaca region demonstrate that intensification involved changes in the labor organization of existing technologies that were all in place quite early and not the introduction of new technologies. This observation even holds for the Inca Empire, where the major changes were in labor organization (commodity specialization, economies of scale, divisions of labor, and so forth) and not the introduction of new technologies.

Agricultural intensification in the Titicaca Basin is best understood as a type of elite political strategy to maximize production within the existing ecological and social constraints. Population pressure was not a direct factor; however, minimum population densities were most likely a necessary cause in the development of the various agricultural technologies by at least 800 BC, if not earlier. Agricultural deintensification (a la Brookfield), in contrast, was a result of altered physical environments and the breakdown of complex political organizations. In the absence of strong political organization, households and villages opted for strategies that were optimal in the particular physical and social environment in which they existed. A shift from raised fields to pastoralism, after Pucara collapse in the Early Huaña period in the north and in the Altiplano or Late Intermediate period in the south basin, illustrates these strategies.

Timothy Earle (1987, 1997:86–87) argues that this kind of agricultural intensification was part of the political economy, the dynamic component of wealth production in complex societies that provides the elite with the material means to maintain their positions. The data here support his observation that agricultural intensification is a strategy to increase surplus production and not a direct response to population pressure. Likewise, Spencer et al. (1994:139) specifically see agricultural intensification as an elite strategy to solidify chiefly power and not as a response to demographic pressure. The data from the Titicaca Basin support this general theoretical position that agricultural intensification is a strategy used by populations to increase production under certain circumstances, both ecological and political. Political elites that can mobilize labor will adopt higher risk strategies when possible. When political organization is strong, agriculture is intensified. When political organization is weak, less risky, more extensive strategies are utilized.

The causal link between political organization and agricultural intensification is complex. In the Titicaca region the Inca were able to intensify agricultural and other economic productive activities to their highest level. In contrast, the immediate pre-Inca peoples who lacked a complex political organization maintained a very low risk, extensive agricultural system even though the environment was essentially the same. There is an undeniable link between political organization and agricultural intensification. Defining the precise causal relationships between these two variables will require more fine-grained chronologies and creative methodologies that can help define this classic question in anthropological archaeology.

Acknowledgments
I thank several people, including Joyce Marcus, participants of the second Cotsen seminar, and anonymous reviewers for providing comments on this chapter. Timothy Earle provided comments on an earlier published paper on raised fields that have been useful in this essay. Short sections of this chapter have appeared in other publications and are included here with the permission of the publishers. Finally, I thank Lloyd Cotsen for making this seminar series and publication possible through his generous support of our endowments.

The fieldwork that led to the collection of the data described in this article was supported over the years by The National Science Foundation, the Wenner-Gren Foundation for Anthropological Research, The Cotsen Institute of Archaeology, the Committee on Research of the Faculty Senate of UCLA, Scott Waugh, Dean of Social Sciences at UCLA, Charles Steinmetz, Debby Arnold, and several anonymous donors. The support of all is greatly appreciated. Any errors of interpretation or fact remain the responsibility of the author.

NOTES

1. With industrialization, the use of machines that use fossil fuels substitutes for human energy.

2. As is fairly well known, Steward cited Karl Wittfogel's work on China, published some 20 years earlier, that explicitly linked irrigation and political centralization. Steward, however, formulated these ideas into anthropological and comparative concepts. Wittfogel's later publications (1957 in particular) owe much to Steward's theoretical synthesis.

3. Strictly speaking, the "hypothesis" is actually a "model," as used by most anthropological archaeologists. I retain the term for historical consistency.

4. Field observations by the author in the Juli area 1988–1995, Pomata area 1991–1995, and several areas in between these two towns from 1991 to 1995.

5. This is an acceptable assumption since fish utilization by netting was probably very old. We have data from the Island of the Sun in Bolivia indicating that people were exploiting the lake with reed boats by at least 1500 BC, before any form of intensive agriculture existed. Human populations were well below carrying capacities up to the Inca period. The lake is extremely deep in some spots where native fish bred (up to 225 meters), and it would not be possible for indigenous populations with prehispanic technologies to seriously alter the availability of native fishes. Therefore, lake productivity can be considered an invariant constant that does not affect interpretations of agricultural land use over time until the introduction of nonnative fish species and more efficient boating technologies in modern times.

6. To my knowledge virtually all raised field production today is sponsored by outside funding agencies. According to local informants raised field production is not economically worthwhile without subsidies.

7. These figures are consistent with nineteenth-century and early-twentieth-century population estimates. David Forbes (1870:200–202) lists a figure of between 750,000 and 870,000 Aymara for Peru and Bolivia, a figure that Tschopik (1947:504) considered too high. Later, José Marroquín (1944:1) noted that the Peruvian Department of Puno had 600,000 people in the 1940s. Likewise, Tschopik suggested figures of approximately 500,000 to 750,000 Aymara speakers between the mid-nineteenth century to 1935 (Tschopik 1947:504, 506), basing this estimate in part on a manuscript by La Barre, who reported a figure of around 600,000 people living in the area in 1935 (see La Barre 1948).

REFERENCES

Albarracin-Jordan, J., and J. E. Mathews
1990 *Asentamientos prehispánicos del valle de Tiwanaku*. Producciones CIMA, La Paz.

Bauer, B., and C. Stanish
2001 *Ritual and Pilgrimage in the Ancient Andes*. University of Texas Press, Austin.

Boserup, E.
1965 *The Conditions of Agricultural Growth*. Aldine, Chicago.

Boulange, B., and J. E. Aquize
1981 Morphologie, hydrographie et climatologie du Lac Titicaca et de son bassin versant. *Revue d'hydrobiologie tropicale* 14(4):269–287.

Coben, L., and C. Stanish
2005 The Inka occupation of Carabaya, Peru. In *Advances in the Archaeology of the Titicaca Basin-I*, edited by C. Stanish, A. Cohen, and M. Aldenderfer, 243–266. Cotsen Institute of Archaeology at UCLA, Los Angeles.

Cohen, A.
n.d. *Survey in the Pukara Valley*. Preliminary report, manuscript in possession of author.

Dejoux, C., and A. Iltis
1991 Introduction. In *El Lago Titicaca*, edited by C. Dejoux and A. Iltis, 11–16. ORSTOM/HISBOL, La Paz.

de la Vega, Abel Edmundo de
1997 *Característica de la re-ocupación Tiwanaku en el sitio de Sillumocco-Huaquina, Juli (Puno)*. Tesis Licenciatura, Universidad Católica "Santa María," Arequipa.

Earle, T.
1987 Specialization and the production of wealth: Hawaiian chiefdoms and Inka power. In *Specialization, Exchange, and Complex Societies*, edited by E. M. Brumfiel and T. K. Earle, 64–75. Cambridge University Press, Cambridge, UK.
1997 *How Chiefs Come to Power: The Political Economy in Prehistory*. Stanford University Press, Palo Alto, California.

Erickson, C.
1988 *An Archaeological Investigation of Raised Field Agriculture in the Lake Titicaca Basin of Peru*. PhD dissertation, Department of Anthropology, University of Illinois at Champaign-Urbana. University Microfilms # 8908674, Ann Arbor.
1999 Neo-environmental determinism and agrarian "collapse" in Andean prehistory. *Antiquity* 73:634–642.

Flannery, K. V., J. Marcus, and R. G. Reynolds
1989 *The Flocks of the Wamani: A Study of Llama Herders on the Punas of Ayacucho, Peru*. Academic Press, San Diego.

Forbes, D.
1870 On the Aymara Indians of Bolivia and Peru. *Journal of the Ethnological Society of London*, n.s., 2:193–305.

Frye, K. L.
1997 Political centralization in the Altiplano period in the southwestern Titicaca Basin. In *Archaeological Survey in the Juli-Desaguadero Region of Lake Titicaca Basin, Southern Peru*, edited by C. Stanish, E. de La Vega, L. Steadman, C. Chávez J., K. L. Frye, L. Onofre M., M. Seddon, and P. Calisaya Ch., 129–141. Fieldiana Anthropology, Field Museum of Natural History, Chicago.

Graf, K.
1981 Palynological investigations of two post-glacial peat bogs near the boundary of Bolivia and Peru. *Journal of Biogeography* 8:353–368.

Graffam, G. C.
1990 *Raised Fields Without Bureaucracy: An Archaeological Examination of Intensive Wetland Cultivation in the Pampa Koani Zone, Lake Titicaca, Bolivia*. PhD dissertation, University of Toronto, Toronto.

1992 Beyond state collapse: Rural history, raised fields, and pastoralism in the south Andes. *American Anthropologist* 94(4):882–904.

Hastorf, C. (editor)

1999 *Early Settlement at Chiripa, Bolivia.* Contributions of the University of California Archaeological Research Facility, Berkeley, California.

Janusek, J. W.

1999 Craft and local power: Embedded specialization in Tiwanaku cities. *Latin American Antiquity* 10(2):107–131.

Janusek, J. W., and A. L. Kolata

2000 Prehispanic settlement dynamics in the Rio Katari Basin. In *Tiwanaku and its Hinterland: Archaeology and Paleoecology of an Andean Civilization*, vol. 2, edited by A. L. Kolata. Smithsonian Institution Press, Washington, DC.

Kolata, A.

1993 *The Tiwanaku: Portrait of an Andean Civilization.* Blackwell, Cambridge, UK.

La Barre, W.

1948 *The Aymara Indians of the Lake Titicaca Plateau, Bolivia.* American Anthropological Association, Menasha, Wisconsin.

Lennon, T. J.

1983 Pattern analysis of prehispanic raised fields of Lake Titicaca, Peru. In *Drained Fields of the Americas*, edited by J. P. Darch, 183–200. British Archaeological Reports, Oxford.

Marcus, J.

1992 Dynamic cycles of Mesoamerican states. *National Geographic Research & Exploration* 8:392–411.

Marroquín, J.

1944 Medicina aborigen Puneña. *Revista del Museo Nacional* 13(1):1–14.

Mathews, J. E.

1992 Prehispanic Settlement and Agriculture in the Middle Tiwanaku Valley, Bolivia. PhD dissertation, Department of Anthropology, University of Chicago.

Mitchell, W.

1973 The hydraulic hypothesis: A reappraisal. *Current Anthropology* 14(5):532–534.

Neira Avendaño, M.

1962 Informe preliminar de la expedición arqueológica al altiplano. *Kontisuyo: Boletín del Museo de Arqueología e Historia de la UNSA.* Universidad Nacional San Andrés, Arequipa, Peru.

1967 Informe preliminar de las investigaciones arqueológicas en el Departamento de Puno. *Anales del Instituto de Estudios Socio Económicos* 1, no. 1. Universidad Técnica del Altiplano, Puno, Peru.

Ortloff, C., and A. Kolata
1993 Climate and collapse: Agro-ecological perspectives on the decline of the Tiwanaku state. *Journal of Archaeological Science* 20:195–221.
Plourde, A.
1999 The role of inter-regional exchange in Pucara and Tiwanaku state formation and expansion, northeastern Titicaca Basin, Perú. Paper presented at the Annual Meetings of the Society for American Archaeology, Chicago.
Pulgar Vidal, J.
n.d. *Historia y geografía del Perú: Las ochos regiones naturales del Perú*. Universidad Nacional Mayor de San Marcos, Lima.
Roche, M. A., J. Bourges, J. Cortés, and R. Matos
1991 Climatología e hidrología de la cuenca del lago Titicaca. In *El Lago Titicaca*, edited by C. Dejoux and A. Iltis, 83–104. HISBOL, La Paz.
Seddon, M. T.
1994 Excavations in the Raised Fields of the Río Catari Sub-Basin, Bolivia. Unpublished master's thesis, Department of Anthropology, University of Chicago.
Seddon, M. T., and J. W. Janusek
1994 Recent research on the organization of agricultural production in the Río Catari Basin, Bolivia. Paper presented at the 59th Annual Meeting of the Society for American Archaeology.
Spencer, C., E. Redmond, and M. Rinaldi
1994 Drained fields at La Tigra, Venezuelan Llanos: A regional perspective. *Latin American Antiquity* 5(2):119–143.
Stanish, C.
1994 The hydraulic hypothesis revisited: A theoretical perspective on Lake Titicaca Basin raised field agriculture. *Latin American Antiquity* 5(4):312–332.
1997 Nonmarket imperialism in a prehispanic context: The Inca occupation of the Titicaca Basin. *Latin American Antiquity* 8(3):1–18.
2001 The origins of the state in South America. *Annual Review of Anthropology* 30:41–64.
2002 Tiwanaku political economy. In *Andean Archaeology I, Variations in Socio-Political Organization*, edited by W. H. Isbell and H. Silverman, 169–198. Kluwer Academic, New York.
2003 *Ancient Titicaca: The Evolution of Complex Society in the Titicaca Basin of Peru and Bolivia*. University of California Press, Berkeley.
2004 The evolution of chiefdoms: An economic anthropological model. In *Archaeological Perspectives on Political Economies*, edited by G. Feinman and L. Nicholas, 7–24. University of Utah Press, Salt Lake City.
Stanish, C., and B. S. Bauer (editors)
2004 *Archaeological Research on the Islands of the Sun and Moon, Lake Titicaca Bolivia:*

Final Results from the Proyecto Tiksi Kjarka. Cotsen Institute of Archaeology Press, Los Angeles.

Stanish, C., R. Burger, L. Cipolla, M. Glascock, and E. Quelima

2002 Evidence for early long-distance obsidian exchange and watercraft use from the southern Lake Titicaca Basin of Bolivia and Peru. *Latin American Antiquity* 13(4):444–454.

Stanish, C., E. de La Vega, L. Steadman, Cecília Chávez J., K. L. Frye, L. Onofre, M. Seddon, and P. Calisaya Chuquimia

1997 *Archaeological Survey in the Juli-Desaguadero Area, Lake Titicaca Basin, Peru*. Fieldiana Anthropology, n.s., 29. Field Museum Press, Chicago.

Steward, J.

1955a *Irrigation Civilizations: A Comparative Study; a Symposium on Method and Result in Cross-Cultural Regularities*. Pan American Union, Washington, DC.

1955b *Theory of Culture Change*. University of Indiana Press, Bloomington.

Tosi, J.

1960 *Zonas de vida natural en el Perú*. Organization of American States Technical Bulletin No. 5. Washington, DC.

Troll, C.

1968 The cordilleras of the tropical Americas. Aspects of climate, phytogeographical and agrarian ecology. In *Geo-ecology of the Mountainous Regions of the Tropical Americas*, edited by C. Troll, 15–65. Geographisches Institut der Universität, Bonn.

Tschopik, H.

1947 The Aymara. In *The Andean Civilizations*, edited by J. Steward, 501–574. *Handbook of South American Indians*, Vol. 2, J. Steward, general editor, Smithsonian Institution Press, Washington, DC.

Wachtel, N.

1982 The mitimaes of the Cochabamba Valley: The colonization policy of Huayna Capac. In *The Inca and Aztec States: 1400–1800*, edited by G. A. Collier, R. I. Rosaldo, and J. D. Wirth, 199–229. Academic Press, New York.

Wittfogel, K.

1957 *Oriental Despotism: A Comparative Study of Total Power*. Yale University Press, New Haven, Connecticut.

Discussion

INTENSIFICATION AND THE POLITICAL ECONOMY

A CONTEXTUAL OVERVIEW

VERNON L. SCARBOROUGH

> Instead of searching for universal states of intensification, and universal causes, research now needs to be directed toward delineating the actual paths of intensification and examining such contextually specific factors as the organization of labor, the role of surplus production, mobility strategies, markets and trade.
>
> —Kathleen D. Morrison, The intensification of production

By the time the reader arrives to this point, he or she will have some sympathy for anyone attempting to summarize a book of such far-reaching import. Not only are the chapters well written, but they are also engaging, challenging, and representative of current anthropological theory grounded in carefully retrieved and interpreted archaeological data. Although each chapter is a stand-alone contribution, the editors have done a fine job of inviting authors capable of examining the broad issues and cross-disciplinary currents influencing "agriculture, polity, and society," while at the same time providing material focus from discrete regions of the globe. The organization of the book allows us to reflect on several overlapping themes that affect our views of political economy and definitions of intensification. At the risk of repeating some of the comments made in the editors' introduction, let me emphasize what I view as the book's major strengths.

More than a decade ago Morrison (1994) thoughtfully critiqued Ester Boserup's (1965) brilliant theory in *The Conditions of Agricultural Growth* as that theory was (and is) incorporated into archaeological interpretation. Because

Morrison's review is as fresh, applicable, and pithy today as it was then, I will not repeat it here, but I do suggest that the reader return to it for a broad assessment of what *intensification* really means. Nevertheless, because this volume devotes considerable time and energy to this condition of land, labor, and water management within and between ancient societies, this is where I wish to start.

Kinds and degrees of intensification are the underpinnings of social complexity. Intensification means more resources made available from the same-sized labor pool and/or from the same area of land—or perhaps body of water, if harvesting fish. But as Morrison emphasizes, this is not a simple equation. Boserup held that population growth, as the independent variable, precipitated agricultural intensification through shortened fallow lengths and technological innovations that permitted more food for an ever-increasing labor force. Regularly incorporated technological invention might initially buffer workers from the drudgery of lengthened work schedules and from the harshness of the undomesticated environment. However, if technology was not suitable for the extraction and processing of resources in the quantities or qualities necessary, then labor was required to work harder or more efficiently. This very simple set of equations continues to inform implicitly our views of the past, especially when examining complex society. Because archaeology has been wedded to "things," or material data sets—an understandable and acceptable analytical condition—we emphasize the role of technology as the agent for increased production when "pushed" by a set of outside vectors like population pressure but that also include climatic change and the complex social issues or decisions enmeshed in the political economy.

Perhaps as a consequence of more time and energy devoted to regional topographic and environmental surveys in concert with settlement pattern studies, archaeologists are displaying more interest in the entire engineered landscape (Ashmore and Knapp 1999; Bender 1993; Bradley 2000; Hirsch and O'Hanlon 1995; Scarre 2002). The observation that humans have occupied and altered to varying degrees nearly every corner of the world (Denevan 1992, 2001; Scarborough 2003a)—though seldom in an uninterrupted or sustainable manner—has opened archaeological data sets and interpretations to assessments of just how completely a society manages its resources across an altered and diverse set of built environments. Although we have invested considerable time and energy in attempting to identify the primary cause for intensification, much less effort has been directed to the nuanced conditions that generate and maintain intensification. By embracing key aspects of the myriad forces influencing agricultural change, archaeologists are now placing less emphasis on technological innovation.

Labor organization has frequently been noted as another primary component in the intensification process. Unfortunately, it is a much more difficult set of variables to quantify when compared to the material facets of technology. Nevertheless, in many ancient societies without major technological enhancements, labor organization likely led to greater social complexity that was as profound as changes stimulated by any other vector. Elsewhere (Scarborough 2003b) I develop the concept that societies evolve, in concert with their initially colonized environment and associated resources, a set of "economic logics." One socioeconomic pathway is an emphasis on labor relationships that are played out within family and extended family bonds that modify an occupied environment or geographic locality in an incremental manner over several generations. Careful attention is taken to sustain the productivity of the landscape in the context of labor assignments and community labor divisions. This accretional model for the early state is apparent in the fragile though resilient ecologies of the semitropics and is typified by the ancient Maya of lowland Central America, a society that sustained uninterrupted longevity for more than 1,500 years. I have suggested that resource-specialized communities distributed throughout the rainforest and associated environs existed and that they were self-organized in their use of microenvironments or ecological patches but linked to economic and political requirements of a larger region. Under these conditions centralized nodes or marketplace "cities" arose to coordinate intercommunity exchange in their sizable open-air plazas. This complex evolutionary trajectory produced a set of highly interdependent labor relationships grounded on extraction and production of resources at the resource-specialized community level in spite of a highly dispersed settlement design. Furthermore, the social cohesion generated by ideology and ritual created regional integration within and between communities and cities (Scarborough 1998). Although seldom measured in terms of ideological parameters, intensification is archaeologically identifiable by pyramids, ziggurats, and other ritually imbued architecture or artifacts (cf. Trigger 1990).

The political economy is generally understood as the way producers (people who are making their living) are affected by those in positions of political authority. It frequently acknowledges elite control of limited resources and how the elite gain and maintain access to primary resources at the expense of an agriculturally dependent support population. Through tax, tribute, or corvée labor the elite affect intensification processes by raising the level of productivity of the immediate field systems. Just how they extract and concentrate resources and how they separate those resources from others is the traditional realm of the political economy. Demographic, climatic, and surplus demands alter and affect aspects of this equation in concert with technological innovation, which is also frequently stimulated by these changes. Elsewhere (Scarborough 2003b) I refer to aspects of this hierarchical emphasis as "technotasking," an economic logic often affected by rapid and exploitive changes. Clearly, this socioeconomic and sociopolitical adaptation is predicated on hierarchical order and control. Nevertheless, the political economy must have a sustaining population that acts and reacts to manipulating elites. Although frequently less visible in the archaeological and historical record than palace intrigues or wars commanded by posturing kings, the everyday interdependency within and between villagers, townspeople, and the majority of urbanites—and their indirect linkages to the elite quarter—are of fundamental importance.

Because societies make a living on the backs of this labor pool, labor organization merits more attention for a complete understanding of the political economy. Although elites may direct aspects of socioeconomic and sociopolitical sustainability and order, it is the relatively slow, self-organizing community of laborers in concert with their evolving engineered landscape, human ecology, and the complex mix of social and environmental feedback that shape complex society and the archaic state. Heterarchical labor organization, when flexible and receptive to most internal and external social and environmental contingencies, will either absorb or reject aspects of technological innovation, depending on whether the social and environmental conditions warrant it. Overtly, labor organization is the independent variable, although a complicated set of interdependencies underlies its form and substance. Unless catastrophic, the effects of demographic change, too, will little alter labor organization in a society adapted to the regular interaction of people and the rapid dissemination and accommodation of information, services, and goods. Elsewhere (Scarborough 2003b) I refer to this economic logic as "labortasking."

The chapters in this volume treat aspects of the above trajectories influencing early statecraft. From my vantage elements of all these economic logics affect each of the case studies provided herein, though the authors clearly differ in their orientations and emphases. Adams, Kirch, Lucero, Williams, and Stanish tend to support a top-down view of how landscapes are altered. Labor organization is little implicated except that elite coercion of a subordinate sector of society rapidly got the job done. Although Robert McC. Adams (1955, 1966) was one of the first and most erudite to debunk Karl Wittfogel's (1957) claim that irrigation was the prime mover for social complexity, here he indicates the role of the tertiary state as a hegemonic landscape colonizer via grand hydraulic works in redirecting the seasonal fury of the Tigris River to otherwise wasted desert settings. Coupled with engineering feats along the southeastern Euphrates drainage, the sixth-century Sasanians constructed some of the most ambitious irrigation systems known for the time. The Nahrwan Canal created agricultural lands over an area the size of Connecticut, lands that were otherwise sparsely watered. The incentive for this massive, top-down investment was to raise revenues from colonizing farmers while circumventing the establishment of private estates that frequently compromised the tax and tribute base otherwise made available to the state. The costs to the Sasanian state were enormous—a set of expenses likely perceived as a necessary income stream in its contestations with the Roman-Byzantium Empire. The price was the collapse of the Sasanian Empire in the seventh century. Adams's chapter in no way challenges his earlier assessment of multiple causes for the origins of complex society, but it does make room for subsequent hegemonic empire building and statehood based on a highly centralized and exploitive land and water use adaptation.

Patrick Kirch is perhaps the most prolific and articulate of the new human ecology school practitioners. Like most of the contributors in this volume who support a technotasking economic logic, his work is grounded in Julian Steward's early work. Kirch is wedded to Brookfield's (1984) intensification vs. innovation dichotomy for explaining agricultural change, an interesting separation but one that masks the fundamental regulatory principles that identify the social relationships of production. Whether short-term fallow ("cropping cycle intensification"), significant and lasting alterations to the built environment ("landesque capital intensification"), technological innovations, or the social behaviors stimulating aspects of domestication ("genetic innovation"), the processes of intensification are directly linked to a demographic trigger. Like several other human ecologists (Sanders et al. 1979; cf. Stanish, this volume; Williams, this volume), Kirch distances himself from Wittfogel's hydraulic society and its determinism while implicitly accepting Boserup's.

Kirch's theoretical orientation draws on his well-received book, *The Wet and the Dry* (1994), in which several Polynesian chiefdoms are shown to augment their resource holdings through predatory aggression by expanding from a "dry" setting associated with an initial long-term cropping cycle of slash-and-burn into a "wet" or developed landscape of pondfields and irrigation. He examines Hawaii and its emergence from the "dry," first as a predatory chiefdom and then as an archaic state by forcefully taxing neighboring islanders who invested in "wet" pondfield agriculture. Although Kirch makes reference to heterarchical relationships, they are identified as competitive contests for prestige and land stimulated by something akin to "competitive opposition" (Sahlins 1961; cf. Sahlins 1955) as I interpret it. He makes reference to a preconquest exchange sphere in which bark cloth and perhaps yams from the dry zones were traded for pigs or taro from the wet zones. What is not explained is the co-occurrence of complexity and pondfield irrigation in Polynesia.

The argument that intensification begins in the short-fallow swiddens does not incorporate the significance of the exchange of goods and ideas among these highly circumscribed, closely interfacing islanders—chiefdom groups frequently occupying the same small island. One must ask, why do swidden chiefdoms choose to intensify? Kirch implicitly argues that they have no choice; it is a matter of a spiraling population's demand for more food. But Kirch provides no real evidence for overpopulation in the dry swidden zones, as opposed to the wet irrigated zones. I would argue that population growth was surely a factor in complexity and social ordering, but just as important were the innovations created by the major landscape modifications established by the pondfield societies. New ways of coordinating time and delegating activities culminated in altered regulatory principles, principles that directly influenced the fragile drylanders' decision-making designs as well. Because of the limitation of their environs, the latter were unable to redirect their landscape adaptations. But in an attempt to maintain the same kinds of sociopolitical exchange systems among groups on a small island as they did before there was pondfield irrigation, the drylanders opted for production surpluses comparable to those of the highly productive pondfield dwellers. In accord with Kirch's definition of intensification, the short-fallow swidden community aligned itself with warlords and predatory expansion to maintain its surplus equity when compared to the productivity of the pondfield dwellers.

There is much to recommend Kirch's careful attention to environmental limitations and population pressures influencing the decisions people make. But it is the underlying regulatory principles (Marx's notion of social relations of production) that direct the course of decision making. The fact that short-fallow swidden societies evolved predatory tendencies is based less on population pressure—implicit here and in Kirch's book—and more on an attempt to maintain an earlier tradition of islandwide exchange and equity based on a thousand years or more of swidden harvesting.

The question for Polynesia should not be why intensification developed first in the short-fallow dry lands, but how pondfield irrigation was accepted by a set of traditional slash-and-burn societies. From a Western economically deterministic perspective the answer is obvious: it produces more food more efficiently. But what about time and resource allocation (including fishing resources), land tenure, social structures, and the deeply rooted and traditional regulatory principles that had been in place for a thousand years? From my perspective it was these fundamental social adjustments that precipitated the later "intensification" that Kirch associates with the short-fallow swidden expansionists.

Lisa Lucero presents the case for agricultural intensification and the political economy for the ancient Maya using a somewhat deterministic model of the environment. She takes a top-down view, arguing that Late Classic kings controlled those ancient engineered landscapes in which making a living was initially very difficult but that those landscapes were subsequently made productive through the efforts of people over generations. Because of the time and energy invested in these landscapes, Lucero suggests that coercive elites could tax or control surpluses given the agriculturalist's costs associated with relocating (cf. Gilman 1981). She further argues that most communities in the Maya lowlands were self-sustaining and only answerable economically to themselves, especially where resources were abundant along the limited number of permanent drainages and within their rich alluvium. However, I would suggest that the drainages Lucero notes were actually less desirable for sizable populations from the outset because of capricious flooding regimes and unpredictable soil productivity given the vagaries of the alluvial sediment source. Although semitropical settings are unusual for the kinds of ecological complexities manifested (i.e., when compared to other locations for the archaic state), these settings are not so unique as to prevent patches of regional abundance for human use within a mosaic of diverse microenvironments.

The flaw in Lucero's presentation is the lack of a developed diachronic dimension to her model. Beginning in the Preclassic, perhaps as early as 1000 BC, early colonists began the long and daunting trajectory of incrementally modifying what many of us interpret as an inhospitable tropical landscape. Dunning (Dunning et al. 1999, 2002) and I (Scarborough 1993, 2003a) have argued that this extended period of landscape alteration was stimulated by a complicated set of variables (weighted differently through time), with shallow permanent lakes in the interior of the Yucatán Peninsula attracting pioneer populations especially during the protracted drought season. Subsequent sediment infilling induced by slash-and-burn land clearance, coupled with possible climatic desiccation at the end of the Late Preclassic period (AD 100), forced a new set of social and biophysical adaptations. It is these variables that Lucero neglects to associate with subsequent Late Classic populations (AD 700), a highly sedentary labor pool living on a slowly modified landscape with relatively uninterrupted occupation for at least 600 years. The elevated reservoirs at the sizable urban nodes (convex microwatershed) represent a history of cumulatively altered landscapes made especially attractive through time. Although Lucero incorporates several environmental variables into her model, she does not emphasize the primary role of the internally draining karstic depressions, or *bajos*, that constitute more than a third of the peninsula nor their role in the environment and their impact on settlement choice. Although an economic and ecological legacy for Mesoamerica (Scarborough 2005), shallow lakes and swamplike settings require considerably more attention from archaeologists for a complete understanding of intensification (and for understanding the political economy nearly everywhere in the world). Years ago Andrew Sherratt (1980) mentioned the significance of these settings to early sedentary societies along the southern Tigris-Euphrates drainages, and before Saddam Hussein's brutal relocation policies and the draining of the perennial swamps, the Shiite marsh dwellers commanded vast reaches of these rich biophysical environments in a highly productive manner. The role of heterarchy, both in terms of social and ecological interdependencies, accommodated these changes. In other words, shallow lakes and swamplike settings often cultivate a cooperative set of interactions based on the time and energy costs invested in maintaining water-saturated fields, field systems frequently drawing from the same visible water source, thus making overuse or harmful use of the agricultural resource immediately apparent to one's neighbors. The proximity of individuals to one another and their well-watered fields accentuates a defined sense of community—a counter to aspects of the tragedy of the commons and formal irrigation (Scarborough, in preparation).

Williams's contribution shares several similarities with Stanish's penultimate chapter. Both examine aspects of Wari, Tiwanaku, and Inca statecraft and their expansion over agricultural landscapes; they view significant surplus as something only elite wealth consolidation accommodates. Both contributors have powerful data sets from their survey regions. Williams takes care to introduce the reader to the role of climatic reconstruction, indicating the effects of both abundance and drought on expanding state irrigation systems. Although Williams alludes to the role of exchange and social relations, there is little room in his chapter for the influence of community intra- and interdependencies—a pivotal concern of Erickson's chapter, which is also on South America. Drawing on Stanish's position (1994), Williams suggests that common laborers underproduced to maintain a labor surplus for risk aversion in accommodating the unpredictable sets of tasks that arise in a household and small community. This model argues for a "closed corporate community" organization and readily lends itself to hierarchical manipulation when elites at a regional level initiate colonization efforts. This community structure suggests that the political economy is preadapted to top-down manipulation and to control by way of converting surplus labor into surplus food or other tangible resources. Williams correlates wet climatic oscillations with state expansion, periods in which labor produced more from a more productive landscape. Surpluses from these colonized settings fueled the state's regional agendas. During times of drought the state withdrew from these holdings and allowed the smallholder to revert back to the ways of an underproductive peasant/farmer.

This position draws heavily from A. V. Chayanov (1966), but Williams suggests that ameliorating climatic conditions correlate with state expansion and with surplus labor and laborers who were put to the task of exploiting a wetter, more productive environment. This argument inches closer to Wittfogel's notion of a hydraulic state. As Adams (1966, 1978) stated much earlier in his career, statecraft is complex and subtle in its formation and expansion. Although Williams points out a set of interesting co-occurring variables, other variables are less well developed. Nevertheless, I do not challenge Williams's salient and hard-won observations.

Stanish's chapter complements Williams's but does not focus on climatic variables—nor does it deny their import. Stanish develops the view that elites are the arbiters of social change, and as the influence peddlers of society they drive the political economy and effect modes of intensification. Like Williams, Stanish distances himself from demographic triggers but emphasizes the productivity of "economies of scale" as stimulated by state expansion and colonization. States and their hierarchical administrators amass and direct legions of laborers (some of them specialized) that affect the intensification process. Stanish's mention of pastoralism as a significant economic component and its later incorporation into Inca state economy lends itself to Adams's frequently cited 1978 article addressing the resiliency (as opposed to stability) of any long-lived complex society. Although Stanish makes a compelling case for the ebb and flow of state control and levels of agricultural intensification, less time is directed to community organization and those bottom-up self-organizing economic principles that undergird all societies, even states. Nevertheless, his is a well-articulated case study clarifying conventional wisdom about the political economy and intensification.

The chapters by Wilkinson, Morrison, Miller, Thurston, Glick, Marcus, Feinman, and Erickson have another set of similarities from those discussed thus far. As Kathy Morrison suggests, the many dimensions of intensification make an assessment of the processes as important as the causes. Most of the chapters I have identified as top-down explanations for agricultural intensification and the political economy focus on causality. Causality, like the search for origins, can inform our understanding of process, but it tends to be an elusive exercise, especially when treating complex issues. This second group of chapters does not emphasize causal triggers. Morrison's chapter explicitly examines the pathways and consequences of intensification using data before, during, and especially after the collapse of Vijayanagara. Like Erickson (this volume) and Denevan (1992), Morrison indicates that the engineered landscape is nearly always an inherited environment significantly modified by earlier groups and cultures, especially in the case of the archaic state. These modifications can be inspired by practical environmental limitations experienced by a pioneering complex society but modified incrementally or radically by the new settlers or new group. Other landscape changes may be effected by cultural nuances as slight as food preferences that affect the choice of which grain is grown, which, in turn, results in permanent landscape alterations like rice paddies or elaborate canal systems where previously there were none. She makes the point that intensification is not an all-or-nothing proposition. The mosaic of landscapes—produced naturally and altered culturally—makes for a highly variegated environment, permitting intensification in some localities and not in others, though generally speaking the closer one is to a sizable urban aggregate, the more broadly will intensification be extant. The role of ideology in the decision-making process of land and water use is examined, too, a point that Marcus (this volume) subsequently elucidates. The cultural value afforded Hindu reservoir construction (as opposed to canalization efforts) clearly influenced landscaping—her "oceans of dharma" (Morrison 2005). Morrison's underlying comment about memory and the role of ancient religious sculpture—where once a reservoir or sluice gate functioned—captures both the lag in ideological systems and the effect past interpretations of landscape use have on the present.

Wilkinson articulates the position that most sustainable irrigation systems are those controlled and directed by the smallholder (after Netting 1993). Using ancient Yemen as his case example, he argues that when the state attempts to construct large, planned, and elaborate water-management systems, those systems are subject to many degrading variables that promote deterioration as a result of environmental neglect and/or sociopolitical caprice. Following state collapse or major political realignment, it is the smallholders and their intimate knowledge of the immediate surroundings that lasts. What Wilkinson refers to as the "guiding hand" or top-down elite planning and control is really the evolved effects of bottom-up self-organization (my term, not his). The single exception seems to be the loosely knit confederation of tribes associated with the Sabaean "state." During the first millennium they built the huge Marib Dam. And though the dam and the state went the way of all complex societies—the dam was breached and with it, its irrigation potential—the smallholders continued. Expedient dam and irrigation systems associated with extensive valleywide terracing as early as 4000 BC reveal the resilience of the villager when pitted against the periodic and forceful wadi floodwaters.

However, the Yemenite case must be contextualized, especially when compared to the other Middle Eastern example presented herein. The Marib Dam supplied water through a short canal segment to nearby field systems that consisted of less than one-four hundredth of the area provided by the waters of the Nahrwan Canal (Adams, this volume). The Yemenite terrace and floodwater systems highly constrained any attempt at truly grand levels of state complexity. In many ways its expedient organization was preadapted to a heterarchical socioeconomic and sociopolitical order. Self-organization within a wadi system required flexibility and was likely complemented by pastoral group organization. Although Wilkinson alludes to Robert Fernea's (1970) classic ethnography, the fundamental labor relationships between these groups are not assessed.

Miller's chapter is an overview of land- and water-use strategies of the ancient Indus "state" (contra Possehl 1998). She contextualizes her interpretations well and clearly admits that the limitations of the data sometimes prevent unflinching support for her positions. Nevertheless, she carefully formulates her arguments in presenting levels of agricultural potential along the Indus. Miller's chapter complements Morrison's work in that both recognize the role of multiple agricultural systems operating in varying proportions on a landscape. Though Morrison is more sensitive to ideological factors, both researchers give labor and land availability considerable attention. Given the "faceless" character of Harappan elite, Miller's point about a "consensual authority" (Thomas Glick's [1970] terms, not Miller's) that serves to organize aspects of land and water rights is appealing. Thomas Park's (1992, 1993) treatment of "common ownership of a portfolio of lands" might also be a helpful approach when examining vast seasonal floodplains in proximity to many of the largest Harappan cities. Park speculates that his ethnographic cases from Senegal may be relevant to the kinds of labor and land organization that precipitated the early Pharaonic state in Egypt. I have thought that this model might have greater applicability along the turbulent lower Tigris or portions of the Indus. In addition to an implicit set of heterarchical relationships embedded in these labor interdependencies, the political economy here (like that noted for the ancient Maya) offers a different view of social organization. Although I will have more to say later, these examples challenge and debunk the dominant economic logic or myth of a "tragedy of the commons" (Harding 1968).

Thurston makes good use of ethnohistorical accounts in revealing the nuanced character of top-down vs. bottom-up models to explain agrarian intensification. Although she frequently dichotomizes her arguments, she does so almost as a foil to emphasize the varied adaptations made by states, as well as those made by politically autonomous regions through time and space. Local history and what she calls "agency" affect statecraft as much as state or elite coercion influences the countryside. Her comments on the role of ethnicity as a function of local history, too, further indicate Thurston's awareness of the many variables that must be assessed to interpret the role of elite control over a landscape. In the final analysis, however, her assessment of statecraft is that of a hierarchical order in which a weak political organization (identified by an absence of a history to accommodate resistance) results in conquest and state control. Whereas Williams emphasizes the role that climatic conditions played in state expansion, Thurston suggests that cultural intangibles like history, ethnicity, and resistance movements help to identify whether or not a state will control a region. Her emphasis on these regional heterarchical relationships in the context of a top-down view of statecraft makes Thurston's contribution somewhat difficult to categorize—and this is a good thing.

Glick, like Adams, is an old hand in water circles. As a historian he approaches the question of intensification, and most specifically irrigation, from the vantage of text—an approach Thurston has also heavily incorporated. In this volume Glick revisits aspects of his classic irrigation study of Spanish medieval Valencia (Glick 1970), and he provides additional context and comparison. As mentioned above, he emphasizes the successes embedded in local irrigation system control, and its highly integrative social structures that preadapt a community to a way of interacting with the environment and one another. Small "irrigation communities" develop resilient socioeconomic and sociopolitical strategies from their everyday water management experiences that establish societal rules and traditions applicable to several realms of community activity and stability. Although Glick attributes his views of economy to his mentor Maass (Maass and Anderson 1978), his ideas have deeper roots in Karl Polanyi's (1957) presentation of the institutional economy (Halperin 1994). Glick rightly emphasizes that "the most significant changes historically . . . have been in land tenure" (p. 166). His references to ethnographic New Mexico and Yemen are especially timely and nicely complement Wilkinson's contribution.

Glick's emphasis on a bottom-up approach to irrigation systems does leave room for state manipulation of the built environment. He identifies the challenges that develop between administrative elites and their constituencies. Operating at the community level, Glick points out by way of example the widening of social distances between municipal elites and their water users when traditional irrigation schemes are phased out and replaced by more "efficient" subterranean pipe systems. Significantly, Glick does not neglect the role of ritual performance necessary for a complete discussion of water, land, and labor use. His transcripts of litigants (and the adjudication of complaints) in courtroom dramas are an aspect of tradition couched in public display that were designed to ameliorate conflict and foster community cooperation. Ritual and ideology are frequently viewed as more ethereal than in the examples presented, but such repeated, predictable, and often symbolic posturings are nothing less than reified ritualized performances. Glick's contribution aids in identifying the mundane behaviors, activities, and interdependencies associated with elements of labortasking economies.

Feinman's contribution, like others in this volume, does not use the opportunity to vilify Wittfogel's hydraulic society, although the latter's deterministic shortcomings are not ignored. Nevertheless, Wittfogel can be credited with being the first social scientist to focus a career exclusively on the significance of irrigation and the origins of state. Feinman examines the recently surveyed Tlacolula arm of the Valley of Oaxaca (Kowalewski 2003) and demonstrates that as the driest sector of the semiarid valley, it became one of the most densely occupied zones in Oaxaca during the Classic and especially the Postclassic periods. Using data from the site area of El Palmillo, Feinman suggests that elevated domestic terraces likely accommodated maguey and cactus varieties at some distance from the valley floor, where maize harvests were unpredictable and not abundant. The model that develops is a classic heterarchical one suggesting the exchange between "resource-specialized communities" (cf. Scarborough and Valdez 2003). During the Postclassic collapse of Monte Albán the Tlacolula arm remains vibrant, perhaps based on exchanging goods manufactured from cactus fiber and juice (cloth or cordage and alcohol) in exchange for maize or related foodstuffs. Feinman shows that changing fortunes in the cultural web of a circumscribed region with varied microenvironments can shift the focus from state intervention to networks of interaction between communities.

Marcus provides a dimension to the discussion of intensification, and most specifically the role of water management in developing complex society, that is not generally examined and is certainly not the focus of past ancient-water-system studies. By broadly addressing water-management studies throughout Mesoamerica, she reveals the significant effects of ritual and religiosity on how landscapes are perceived, used, and altered. Since water was a scarce commodity, if only seasonally in most Mesoamerican regions, it played a major role in both ideology and cosmology. The interplay between what can be technologically controlled (canals, reservoirs, terraces) and what can be ritually controlled (imagined and enacted through a corpus of ritualized beliefs and behaviors) recognizes the full dimensions of an engineered and "imagined" landscape. Marcus explicitly opens the door to a new and vibrant way of thinking about the built environment, one grounded in what archaeologists and human ecologists have labored to establish, then connects us to the world of the mind and its culturally embedded set of decision-making strategies for coping with an unpredictable and imperfect landscape (cf. Marcus and Flannery 1994). The role of ritual in water management challenges many of our views about primal causality for the origins of complex society and its "man-land" adaptations (after Steward). As Morrison implies, the search for processual understanding will be more rewarding and probably more meaningful than the search for causality. Both intensification and the political economy evolve in the context of traditions grounded in the spheres of technological breakthroughs, demographic pressures, climatic and ecological conditioning, and elite power brokers but also in a cultural milieu with embedded views of etiquette, taste, aesthetics, and definitions of social justice, to say nothing about other facets of political and economic history. What people select to eat can be as much a taste preference or social status preference as a practical ecological decision, and that decision can make the difference between rice paddy intensification or broadcasting wheat and barley. Marcus opens up a highly significant area of inquiry demanding immediate attention for a truly complete and timely review of land, water, and labor management.

Erickson's chapter is perhaps the most provocative in this book. He specifically challenges conventional views of the political economy and intensification

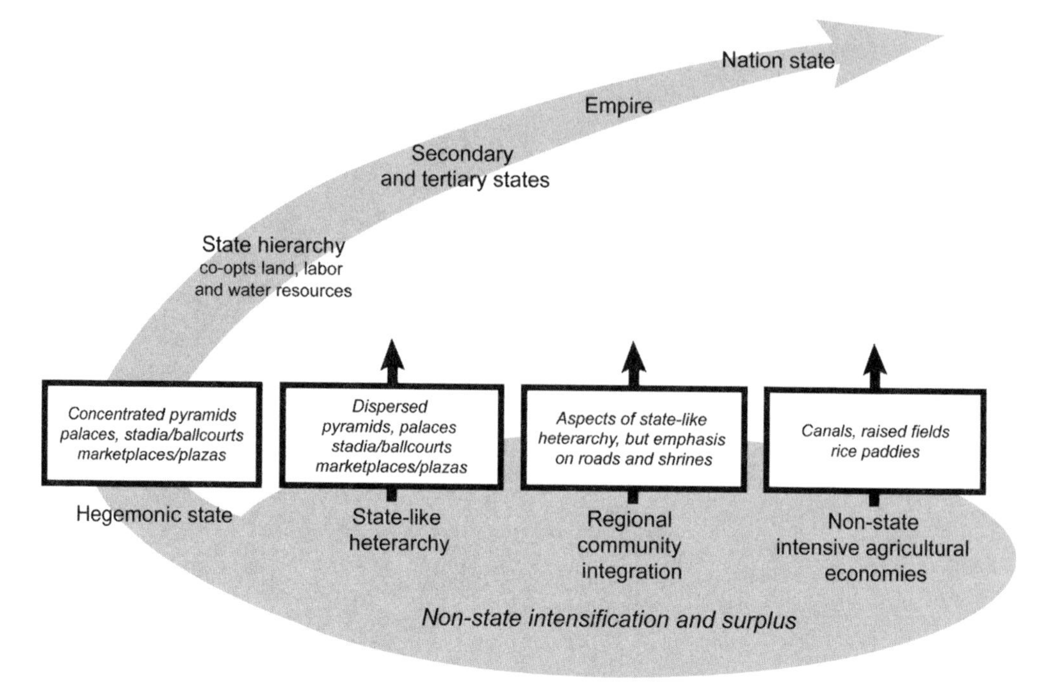

Figure 15.1. Kinds of engineered landscapes and associated socioeconomic and sociopolitical organizational schemes appropriated by hegemonic states with time.

as presented most eloquently by Stanish and through implication by Williams for their regions in Peru. Erickson, working nearby in Bolivia, brings to the table a notion of the political economy that emphasizes the broad backs of a support population. Erickson's view is that with or without the state, peasant communities are quite capable of intensifying the harvesting or exploitation of their landscapes. His focus on the built environment away from sizable urban aggregations or towering monumental architecture significantly alters our understanding of the political economy and intensification. For some time Erickson (1993) has supported a bottom-up view of social complexity, suggesting that surpluses derived from agricultural intensification are generated nearly as readily by small-scale communities composed of smallholders as by a state's policy of coercion or forms of tactical cajoling. What is underdeveloped in Erickson's presentation is what these small communities do with surpluses if they do not reinvest them in labor-intensive civic construction projects, private estates, or elite trappings of status and control. In societies that have the wherewithal to produce surplus but do not direct it toward statelike projects, what comes of it? Given the wide interpretive distances represented by the Andean contingent in this book, it might be useful to dip into the ethnographic literature for a complementary view.

Thomas Abercrombie (1998) has made much of "pathways of memory and power" as practiced by the present-day Aymara of highland Bolivia and their ideological investments to a landscape—especially roads and shrines. What if Abercrombie's assessment of these bottom-up self-organizing communities (communities that in the past intensified through raised field construction and maintenance) reflect an ancient ideological orientation to a highly integrated complex society but without the hierarchical order of a state? Here we return to my Maya example and the human interdependencies built into certain landscapes. Lofty pyramids or palaces may be one outcome of surplus generation, but surpluses may not materialize (or be invested in monumentality) in the same manner when a community is organized around labortasking.

It is fitting that I end with Erickson's view of agriculture, polity, and society, in which the notion of a polity without clear boundaries or rigid definition underscores process—the same processes that draw archaeologists to the early state and its concentrated landscape manifestations of pyramids, palaces, stadia/ball courts, or marketplaces/plazas (Figure 15.1). Nevertheless, other landscape signatures are as intellectually revealing though often not as visibly pronounced or aesthetically pleasing (roads and shrines to raised field systems, rice paddies, and canals). Without pyramids and palaces ideology becomes a slippery topic in archaeology—but this need not be the case. Perhaps by broadening our analytical views of landscapes to move beyond residential sites and monumental civic centers, we can best comprehend the vitality of social complexity.

REFERENCES

Abercrombie, T. A.

1998 *Pathways of Memory and Power: Ethnography and History among an Andean People.* University of Wisconsin Press, Madison.

Adams, R. McC.

1955 Developmental stages in ancient Mesopotamia. In *Irrigation Civilization*, edited by J. H. Steward, 6–18. Pan American Union, Washington, DC.

1966 *The Evolution of Urban Society: Early Mesopotamia and Prehistoric Mexico.* Aldine, Chicago.

1978 Strategies of maximization, stability, and resilience in Mesopotamian societies, settlement, and agriculture. *Proceedings of the American Philosophical Society* 122:329–335.

Ashmore, W., and A. B. Knapp (editors)

1999 *Archaeologies of Landscapes: Contemporary Perspectives.* Blackwell, Oxford.

Bender, B. (editor)
1993 *Landscape: Politics and Perspectives*. Berg, Oxford.

Boserup, E.
1965 *The Conditions of Agricultural Growth*. Aldine, Chicago.

Bradley, R.
2000 *An Archaeology of Natural Places*. Routledge, New York.

Brookfield, H. C.
1984 Intensification revisited. *Pacific Viewpoint* 25:15–44.

Chayanov, A. V.
1966 On the theory of non-capitalist economic systems. In *On the Theory of Peasant Economy*, edited by D. Thorner, 1–13. Richard D. Irvin, Homewood, Illinois.

Denevan, W. M.
1992 The pristine myth: The landscapes of the Americas in 1492. *Annals of the Association of American Geographers* 82:369–385.
2001 *Cultivated Landscapes of the Native Amazonia and the Andes*. Oxford University Press, Oxford.

Dunning, N., S. Luzzadder-Beach, T. Beach, J. Jones, V. Scarborough, and T. P. Culbert
2002 Arising from the Bajos: Anthropogenic change of wetlands and the rise of Maya civilization. *Annals of the Association of American Geographers* 92:267–283.

Dunning, N., V. L. Scarborough, F. Valdez Jr., S. Luzzadder-Beach, T. Beach, and J. G. Jones
1999 Temple mountains, sacred lakes, and fertile fields: Ancient Maya landscapes in northwestern Belize. *Antiquity* 73:650–660.

Erickson, C. L.
1993 The social organization of prehispanic raised field agriculture in the Lake Titicaca Basin. In *Economic Aspects of Water Management in the Prehispanic New World*, edited by V. L. Scarborough and B. L. Isaac, 369–426. JAI Press, Greenwich, Connecticut.

Fernea, R. A.
1970 *Shaykh and Effendi: Changing Patterns of Authority among the El Shabana of Southern Iraq*. Harvard University Press, Cambridge, Massachusetts.

Gilman, A.
1981 The development of stratification in Bronze Age Europe. *Current Anthropology* 22:1–23.

Glick, T.
1970 *Irrigation and Society in Medieval Valencia*. Harvard University Press, Cambridge, Massachusetts.

Halperin, R. H.
1994 *Cultural Economies: Past and Present*. University of Texas Press, Austin.

Harding, G.
1968 Tragedy of the commons. *Science* 162:1243–1248.

Hirsch, E., and M. O'Hanlon (editors)
1995 *The Anthropology of Landscapes: Perspectives on Place and Space*. Clarendon, Oxford.

Kirch, P. V.
1994 *The Wet and the Dry: Irrigation and Agricultural Intensification in Polynesia*. University of Chicago Press, Chicago.

Kowalewski, S. A.
2003 Scale and the explanation of demographic change: 3,500 years in the Valley of Oaxaca. *American Anthropologist* 105:313–325.

Maass, A., and R. Anderson
1978 *And the Desert Shall Rejoice: Conflict, Growth, and Justice in Arid Environments*. MIT Press, Cambridge, Massachusetts.

Marcus, J., and K. V. Flannery
1994 Ancient Zapotec ritual and religion: An application of the direct historical approach. In *The Ancient Mind*, edited by C. Renfrew and E. B. W. Zubrow, 55–74. Cambridge University Press, Cambridge, UK.

Morrison, K. D.
1994 The intensification of production: Archaeological approaches. *Journal of Archaeological Method and Theory* 1:111–159.
2005 *Oceans of Dharma: A Political Ecology of Place*. University of Chicago Press, Chicago.

Netting, R.
1993 *Smallholders, Householders: Farm Families and the Ecology of Intensive Sustainable Agriculture*. Stanford University Press, Stanford, California.

Park, T. K.
1992 Early trends towards class stratification: Chaos, common property, and flood recession agriculture. *American Anthropologist* 94:90–117.

Park, T. K. (editor)
1993 *Risk and Tenure in Arid Lands: The Political Ecology of Development in the Senegal River Basin*. University of Arizona Press, Tucson.

Polanyi, K.
1957 The economy as instituted process. In *Trade and Market in the Early Empires*, edited by K. Polanyi, C. Arensberg, and H. W. Pearson, 64–94. Free Press, New York.

Possehl, G. L.
1998 Sociocultural complexity without the state: The Indus civilization. In *Archaic States*, edited by G. M. Feinman and J. Marcus, 261–291. School of American Research Press, Santa Fe, New Mexico.

Sahlins, M. D.
1955 *Social Stratification in Polynesia*. University of Washington Press, Seattle.
1961 The segmentary lineage: An organization of predatory expansion. *American Anthropologist* 63: 322–345.

Sanders, W. T., J. R. Parsons, and R. S. Santley
1979 *The Basin of Mexico: Ecological Processes in the Evolution of a Civilization.* Academic Press, New York.

Scarborough, V. L.
1993 Water management in the southern Maya lowlands: An accretional model for the engineered landscape. In *Economic Aspects of Water Management in the Prehispanic New World*, edited by V. L. Scarborough and B. L. Isaac, 17–69. JAI Press, Greenwich, Connecticut.
1998 Ecology and ritual: Water management and the Maya. *Latin American Antiquity* 9:135–159.
2003a How to interpret an ancient landscape. *Proceedings of the National Academy of Sciences* 100:4366–4368.
2003b *The Flow of Power: Ancient Water Systems and Landscapes*. School of American Research Press, Santa Fe, New Mexico.
2005 An overview of Mesoamerican water systems. In *Precolumbian Water Management: Ideology, Ritual, and Power*, edited by L. J. Lucero and B. W. Fash. University of Arizona Press, Tucson, in press.
In prep. Colonizing a landscape: Water and wetlands in ancient Mesoamerica. In *The Early Mesoamerican State: Essays in Honor of Barry L. Isaac*, edited by V. L. Scarborough and J. E. Clark. University of New Mexico Press, Albuquerque.

Scarborough, V. L., and F. Valdez Jr.
2003 The engineered environment and political economy of the Three Rivers region. In *Heterarchy, Political Economy, and the Ancient Maya: The Three Rivers Region of the East-Central Yucatan Peninsula*, edited by V. L. Scarborough, F. Valdez Jr., and N. Dunning, 3–13. University of Arizona Press, Tucson.

Scarre, C. (editor)
2002 *Monuments and Landscape in Atlantic Europe: Perception and Society during the Neolithic and Early Bronze Age*. Routledge, New York.

Sherratt, A.
1980 Water, soil, and seasonality in early cereal cultivation. *World Archaeology* 11:313–330.

Stanish, C.
1994 The hydraulic hypothesis revisited: Lake Titicaca Basin raised fields in theoretical perspective. *Latin American Antiquity* 5:312–332.

Trigger, B. G.
1990 Monumental architecture: A thermodynamic explanation of symbolic behavior. *World Archaeology* 22:119–132.

Wittfogel, K. A.
1957 *Oriental Despotism: A Comparative Study of Total Power*. Yale University Press, New Haven, Connecticut.

INDEX